Neopoetics

Neopoetics

THE EVOLUTION OF THE LITERATE IMAGINATION

Christopher Collins

Columbia University Press New York

Columbia University Press
Publishers Since 1893
New York Chichester, West Sussex
cup.columbia.edu

Library of Congress Cataloging-in-Publication Data
Names: Collins, Christopher, author.
Title: Neopoetics : the evolution of the literate imagination / Christopher Collins.
Description: New York : Columbia University Press, [2016] | Includes bibliographical references and index.
Identifiers: LCCN 2016013378 (print) | LCCN 2016024578 (ebook) | ISBN 9780231176866 (cloth : alk. paper) | ISBN 9780231542883 (e-book)
Subjects: LCSH: Semiotics. | Visual pathways. | Language and languages—Origin. | Poetry—Psychological aspects. | Poetics—History—To 1500. | Evolutionary psychology. | Brain—Evolution. | Neurolinguistics.
Classification: LCC P99 .C5698 2016 (print) | LCC P99 (ebook) | DDC 302.2—dc23
LC record available at https://lccn.loc.gov/2016013378

∞

Columbia University Press books are printed on permanent and durable acid-free paper.
Printed in the United States of America

c 10 9 8 7 6 5 4 3 2 1

Cover design: Martin Hinze
Cover image: © Ian M. Butterfield (Rome) / Alamy Stock Photo

Hoc genio opus dedico
Savangunci fluminis
ob permulta munera
inque honorem numinis:
confluamus suaviter

Contents

Preface

I think I may safely presume to speak for you, too, when I say that we all take reading and writing for granted. These skills were at first difficult for us to learn, of course. Managing to make sense of and reproduce those arcs and angles was not bequeathed to us by biological evolution as were walking, reaching out, and grasping. Yet by the age of five or six, most of us had mastered the basics of literacy and were ready to improve and expand upon them. By now, reading and writing are so ordinary to us that we forget what an extraordinary achievement of cultural evolution they actually are. We have gotten to a point of such easy familiarity with this medium that, like the proverbial fish that can never know what water is because that is simply the condition of its existence, we are no longer aware to what extent literacy, the transparent medium we swim in, has modified our awareness of the world. For instance, when in my first sentence I wrote that "I think I can safely presume to *speak* for you, too, when I *say* . . . ," you probably didn't think there was something odd about a silent page of print talking to you in the voice of a person who was not really present.

The Evolutionary Perspective

A child's capacity to acquire language is a genetically imprinted trait that took millions of years to set in place. To make the distinct sounds we use to construct meaningful utterances our ancestors needed to evolve finely sequenced neural circuitry linking particular regions of the brain to particular muscles in the diaphragm, chest, larynx, tongue, and jaw. Though speech was not their inevitable outcome, these adaptations made possible the evolution of the "language-ready brain," as Michael Arbib (2012) has called it. But at that point it was also necessary for those men and women living in Africa some 300 to 200 thousand years ago to realize there was an advantage in sharing their knowledge and intentions using symbolic signs in the form of arbitrary, conventional mouth sounds.

In my last book, *Paleopoetics* (2013), one major question was, "How did language transform our primate brain?" I approached this from different angles, e.g., human evolution, anthropology, linguistics, neuroscience, cognitive psychology, phenomenology, and semiotics. In the course of my research, I came to believe that the transformative process attributable to language had been in progress a long—a *very* long—time before the first sentence was ever uttered. The hierarchical complexity of our brain that made language possible had been in place because over many millions of years a set of neural systems had evolved, some general, some specific, that could be reconfigured to perform new tasks, such as imitative learning, tool making, gestural communication, and vocal speech.

That word "paleopoetics" connotes the evolved skills associated with the making of things, especially the making of such cultural products as stories, songs, and rituals, which, in a later literate context, we have come to know as novels, poems, and dramas. (The *-poetics* in that word comes from the Greek verb *poiein,* meaning to "build" or "create"). Wishing to shed new light on these made things, I needed first to establish a link between biological and cultural evolution, and I chose to do so by examining how tools extend the reach, strength, and therefore the survivability of their users. In the process of biological evolution, our ancestors first developed the ability to locate objects to use *as* tools, then later learned to modify found objects with which to *make* tools.

Here, then, with the manufacture of tools we have the beginnings of culture as a conscious extension of the human body. Distinct from a found tool, e.g., a stick lying under a tree that is used to knock down nuts from a high branch or a nearby rock used to break their shells, deliberately shaped artifacts are instruments designed to extend the powers of the user to perform specialized tasks and for that purpose are saved both as tools and as models to be imitated and improved upon. The better the toolkit, the further and more effective that extension becomes. Since the nature of human culture is social, multiple individuals using tools of various sorts create an interlinked field of operations, a *co*-operative community. If we recognize that language is a means of extending individuals' thoughts, feelings, and intentions outward into a shared space, we may regard language as a tool, as well. It would then follow that our human ancestors, once they placed language in their cultural toolkit, were able to use it to fashion from it new, more specialized tools, i.e., verbal artifacts.

I should hasten to add that by the phrase "verbal artifact" I do not mean what the New Critics at the mid–twentieth century or the neoformalists at its close might have meant, viz., a self-contained, self-sufficient, self-referential objet d'art. As I use the phrase, I define it in terms of physical artifacts unearthed as evidence of cultural evolution, e.g., stone hand axes, arrowheads, bone flutes, or ceramic bowls—products skillfully crafted to serve human needs. A hammer, for example, is a self-contained artifact, to be sure, but not an entirely self-referential one. One can hang an antique hammer on the wall and admire its hand-forged design and its gracefully tapered haft, but, when one does so, this tool loses its essential nature as an extension of the human hand that is used to pound nails or other materials. To regard a verbal artifact as an autonomous, self-referential object, one must first remove it from its instrumental function, which, as a piece of language, is to signify meanings beyond itself, and then focus only on its nonsignifying elements, such as its phonemes, rhyme patterns, and rhythmic structures.

I realized that the claim I made in 2013 that these verbal artifacts are not only analogous but equivalent to tools might raise some eyebrows. Isn't one of the properties of aesthetic objects said to be their uselessness, their end-in-themselves-ness, and isn't that what distinguishes the artistic from the practical? Well, maybe not, I thought. And if an artwork, verbal or otherwise, serves as a tool, it must do so as an extension of its user. So, just as we feel our arm stretch outward, into, and through the stick we use to reach an apple high in a tree, we must be able to feel ourselves extending into and through a work of art. Art, then, is the ability to make things that we value and preserve not simply for themselves, but because by using them we are able to discover and expand our innate human powers—sensory, kinetic, emotional, conceptual—powers that we might otherwise never know we had. When not in use, an artifact is an inert object, but, when we use it, it transforms itself from an object into an instrument. The painting in the darkened hall, the violin in its box, the musical score or the novel on the bookshelf—all these are mere objects until they are seen or played or read. They come to life over and over again but only when *our* life flows through them.

When art objects such as these are taken up and used, they become instruments of imaginative play. Whenever I have inserted that word "imagination" in a book's subtitle, I have meant the capacity to form mental simulations of reality. Children do this when they role-play. Scientists do this when they create hypothetical models. Insofar as words have the power to stand for every perceivable object in every sensory modality, words are the preeminent medium of the imagination, and to name a thing or concept by assigning it an arbitrary mouth sound or a series of written squiggles is an instance of human play behavior. So, when we think of metaphor, metonymy, or irony as "wordplay," we should acknowledge them as play within play.

The Emergence of Writing

As the evolution of the language-ready—and the imagination-ready—brain was a long process, so, too, was the evolution of the "writing-ready" brain. Though a comparatively recent innovation, no older than five thousand years, writing builds on a skill vastly older than language, i.e., fine-tuned manual control. Since the higher apes, from which our line diverged about six million years ago, can grasp branches for clubs and stones for hammers by wrapping their fingers around them in the power grip and also manipulate smaller objects with their fingertips in the precision grip, we assume our common ancestor could do so also. Moreover, since manually dexterous hominid apes lack fine vocal control, it is quite likely that early humans would have first used their hands to communicate with one another. We still, needless to mention, rely on gesture with, and sometimes in place of, spoken language.

Gesture, manually produced and visually received, may have first taken the form of shapes that resemble what they mean. If we want to communicate with people whose language we do not share, we have to fall back on gesturing and may, for example, move our hand to represent a bird in flight or a person walking up a slope. When we do so, we use *iconic* signs. If we share a common gestural code, we may show a raised palm to signify a friendly greeting or form a circle by touching the tips of index and thumb to convey a sense of approval. These gestures, conventionally linked to their meanings, are termed *symbolic* signs. Be it icon or symbol, a hand gesture is a visual message that is as fleeting a thing as any auditory message conveyed through speech.

Writing requires fluid hand and arm movements as does gesturing. Here we have a physiological link between these two actions: a written mark is a momentary manual gesture, but one that also leaves a lasting trace. For example, all it takes to create the meaning of a quadruped and preserve it as a frozen gesture is to drag one's fingertip in moist ground to represent in lines the spine, four legs, and a circle for the head. Then others who come along may look at it and see what was meant. Perhaps one could add a few more strokes to distinguish what kind of quadruped it is—antlers or horns, perhaps, or the long tail of a lion.

Writing undoubtedly began as iconic picture making such as we find in petroglyphs and cave paintings. Assuming a human community had the ability to exchange auditory symbolic signs, i.e., spoken words and sentences, its members could view the depicter's frozen gesture and then comment on it. If these pictures were sufficiently numerous, interpreters could convert them into a narrative—a hunt, say, or a migration. With that level of elaboration, pictographic writing was born.

It is not my intention in this book to trace the origins of writing, its gradual transition from images of referents to stylized hieroglyphs and ideograms that signified through those pictures particular speech sounds—first syllables then, later, separate phonemes. My project instead is to understand how writing altered the means by which verbal artifacts are preserved, passed on, and reused by others and how it thereby transformed the nature of these instruments.

Like other instruments, verbal artifacts sometimes break down, need repair, get lost, and must be recrafted. In an oral culture they are subject to variation, sampling, and improvisation, for, unlike material artifacts, such as axes or figurines, artifacts made of words exist only in the minds and mouths of storytellers and singers, who inevitably reshape them from generation to generation. Performers and audiences may suppose that a traditional story or song is transmitted faithfully, but this is almost never the case. As anthropologists and ethnomusicologists have attested, oral artifacts mutate, and, like evolving biological species, only those variants that best serve the needs of their audience survive and are reperformed.

Whenever and wherever writing is introduced, this process of oral transmission begins to change: certain traditional oral compositions, once transcribed, become standard texts, and all other variants gradually vanish. This does not mean, however, that literacy simply replaces orality. We still talk together, amuse ourselves with jokes, and even pay professionals to do so for our entertainment. As the popularity of theater and film demonstrates, our oral-audio-visual brain is as active and insatiable as ever.

Though writing is a superior means of stabilizing and preserving verbal artifacts, it has never been foolproof. This was especially the case before the invention of printing. Until about five hundred years ago, book publishing had to rely on scribes, hard working and undercompensated, who, when their minds drifted or their vision blurred, might insert textual errors or, when their minds were overactive, might contribute their own emendations or glosses that became difficult thenceforth to disentangle from the original document.

The Cultural Perspective

The specific cultures I use to illustrate the impact of literacy on the making of verbal artifacts are those of ancient Greece and Rome. If I were sufficiently knowledgeable, I would have widened my scope to include non-Western literacies, such as Arabic, Indian, and Chinese literatures. I trust, however, that the early effects of writing on Western culture have similarities to those experienced in other societies. Moreover, by grounding my study in the rapidly advancing science of the brain, I hope to compensate for what otherwise might seem a limited Eurocentric perspective: cultures may differ, but the nature of the evolved human brain is the common patrimony of all.

When we narrow our cultural perspective to Greece in the late seventh century BCE, we discover that the new literacy had generated a new name for the singer of songs—this title was *poet*. People who claimed this new designation might not necessarily be proficient (hand)writers, but, if they were successful composers of verbal artifacts, they could hire scribes to take down their dictation. Nor were they necessarily expert performers, but they could distribute copies of their works to singers or to troupes of actors for public performance. What distinguished an early literate author from a traditionally attributed source was the fact that he or she was a "maker" whose name, place, and time might now be inserted within the work itself. By that measure, Homer, whose epics bear no self-referential traces, may have been a skillful arranger of traditional narrative episodes, a master of poetic diction, and a renowned singer in an oral lineage but was not likely the original author of the stories that now bear his name. Hesiod, on the other hand, did refer to himself by name, place of birth, and livelihood; seemed anxious to be identified with his works; and may have composed them in writing. The very word "author," deriving from the Latin noun *auctor* and the verb *augēre* (to increase, as in "augment"), implied one who added to a people's cultural wealth by inventing something new. As a composer of written verbal artifacts, an author was the origin of a palpable sort of increase, the proliferation of multiple material copies of a given text.

The other role that literate culture created was that of the *reader*. But if classical Greece tells us anything about the effects of the introduction of literacy on an oral society, reading is slower to evolve than authorship. Though an estimated 70 to 90 percent of Athenians in the fifth and fourth centuries BCE could neither read nor write, most were avid theatergoers and could sing from memory the popular songs of the day as well as portions of epic verse. (Plato, who had little patience with popular culture, decried the influence of the "theatocracy.") Perhaps the most valuable function of early writing for verbal artifacts was to provide public performers, rather than private readers, with authoritative scripts.

Even at this stage, however, as it served the needs of the oral, performing arts, writing was changing the verbal artifact. Authors still wanted their songs and dramatic speeches to be memorable, but now they no longer needed to rely on the old mnemonic structures of preliterate culture, e.g., formulaic phrases, coordinate clauses, and repetition. As soon as performers had written copies to refresh their memory, a new level of verbal complexity could be introduced. Authors now could startle audiences with elaborate compound words, difficult allusions, and embedded subordinate clauses.

The origins of writing, as of every other culturally evolved innovation, still lie hidden in the biologically evolved brain that, despite its plasticity, retains clear anatomical traces of its own evolution. Written artifacts and the imagination required to animate them have at their heart the sensorimotor networks woven together over 200 million years of mammalian evolution. Because these foundations still lie within us here and now, my concern with prehistory and early cultures is not prompted so much by a fascination with the past as by a desire to comprehend the ancient depth of the living present that carefully chosen words can suddenly illuminate in the mind of a reader.

Poetics and Poemics

A project such as this does, however, present certain challenges. If we hope to understand a literary text created in another time, another tradition, or both, we must do so in terms of cultural contexts that include such information as contemporary and historical knowledge, cosmological and religious assumptions, ethical judgments on class and gender, and conceptual metaphors unique to its language—all of which differ from our own cultural inheritance. In addition, we must understand this text as framed by specific literary values, e.g., its genre as related to other then-recognized genres, its intertextual relation to works within its selected genre, its stylistic links to oral discourse, and the degree to which it reflects certain critical norms. In short, as much as we might like to use our own insights and methods to penetrate its meanings, we must also read the work within the cultural matrix from which it emerged.

If, for example, I choose to discuss a Greek poem written in the early second century BCE in terms of the eye movements and working memory used in reading, the auditory areas of the brain used in simulating vocal sounds, and the motor neurons simulating vocal articulation, I am applying knowledge unavailable to second-century readers. I am taking an *etic,* i.e., relatively objective, outsider stance toward the text. If, on the other hand, I approach the text identifying as exclusively as possible with an educated reader of the Hellenistic era, this *emic* approach would foreground other elements. (These terms, "etic" and "emic," derive from the phon*etic*/phon*emic* distinction and were first applied by the linguist Kenneth Pike to the contrasting anthropological viewpoints of outside observer vs. inside participant.)

"Poetics," as Aristotle used the word, was indeed an etic research project that might also have been entitled a "natural history of poetry."

My application of current neuroscience to classical poetry in this book is also meant to stress the "etics" of "poetics." But while a neurocognitive etic approach is valid and, arguably, less disruptive to the meanings of ancient texts than other modern but ideologically anachronistic points of view, this etic perspective is not sufficient unless it is complemented by an emic stance. Fortunately, the special strength of a science-grounded po*etics* is that it is quite compatible with a culture-grounded po*emics*.

Unlike *Paleopoetics*, which had no texts to present as evidence, *Neopoetics* has rich textual resources available to it. But, I submit, applying the methods of cognitive poetics to historically contextualized texts requires something analogous to binocular vision. With one eye we must view a text in terms of the culture we are studying. With the other eye we must view it in terms of the most advanced scientific insights we can find, assuming that the reader-related processes of the brain, e.g., imagination (in the various sense modalities), memory (in its various systems), and the basic emotions have remained, over the last sixty thousand years, relatively constant across cultures. To elaborate this optical metaphor slightly, imagine we are looking through a stereoscope at two photos of a landscape, each slightly spatially displaced. The image to the left represents an etic view, based on cognitive neuroscience; the image on the right represents the emic view, with all cultural specifics of a given time and place. When we first look with both eyes through this optical instrument, we may see one and then the other image, but if we wait until our eyes become fully balanced, we suddenly glimpse that landscape three-dimensionally projected. This is the vision that cognitive or, as now we might call it, *neuro*cognitive poetics gains when merged with cultural-historical poemics. This is, at least, my hoped-for effect. It remains to be seen if I have succeeded.

Acknowledgments

I am grateful to those scholars who have read and commented on the manuscript that has now become this book: Michael Corballis and Michelle Scalise Sugiyama. Thanks, too, to Brett Cooke and Richard A. Richards, who reviewed my initial proposal. I hope they see reflected in the final product the thoughtful insights they each contributed. I wish also to express my gratitude to those associated with Columbia University Press: to Patrick Fitzgerald, publisher for the sciences, for his continued encouragement of projects, like mine, that endeavor to demonstrate how science and the humanities may illuminate each other; to Robert Fellman, copy editor, for his careful eye for style and syntax; to Martin Hinze art director, for her cover design; to Milenda Lee for the book's interior design; to Ryan Groendyk, editorial assistant, for his ever-helpful e-mails; to Marisa Pagano for the catalogue description of my book; and to Michael Haskell, supervising production editor, for ensuring that all these separate operations were perfectly sequenced, scheduled, and executed.

Neopoetics

One

Innovating Ourselves

Since the evolution of speech, writing has arguably been our species' most consequential innovation. By "innovation" I mean any successful, alternative way of doing something. Some of these changes were the result of gene mutation, the prime factor in natural selection, as over time alternative procedures aided their users to survive and reproduce and were thereby passed along to their descendants. Other innovations, learned by imitation, were transmitted through cultural evolution. Nonhuman animals show little or no capacity for innovation. Our nearest evolutionary cousins, the chimpanzees, for example, have the innate ability to hurl things, yet they cannot learn how to shape objects and accurately throw them. Human children are not only born throwers but have the innate capacity to learn how to design, construct, aim, and accurately throw any number of projectiles.

While we acknowledge how writing has shaped the way we now conceptualize our world, our experienced past, and our projected future, we should also recognize that no innovation, writing included, has wholly changed any piece of our biologically evolved equipment. Writing, as a skill that takes several years to master, is built upon a child's biologically hardwired capacity to learn its caregivers' language, for every human innovation is, after all, an innovation of *something* and is deemed successful only when it allows us to perform that something in a better way. If, for example, the object of innovation is the power of locomotion or vision, its enhancement is important only because that pre-evolved power is itself important. Thus the domestication of the horse and, later, the invention of the wheel and axle were significant innovations only because swift and agile locomotion was and continued to be vital to human survival. Similarly, optical lenses, photography, and television were significant only because visual perception has always been our major link to the world we live in.

Before we consider the invention of writing we might therefore ask ourselves what long-established means our early ancestors had of knowing and communicating their knowledge. In this first chapter I will examine a set of the cognitive skills, some prelinguistic, some language based, that they relied upon to interpret their world. The interpretation of signs, I will argue, is the preexistent power that linguistic systems and, later, writing systems were designed to innovate.

How We Got This Way . . .

. . . is an incomplete sentence, a thought we humans have always been trying to finish. Being a curious, storytelling species, we have always wondered where we came from. We may either accept a traditional origins narrative, one, for example, that tells how an extraterrestrial being suddenly made the first man and woman, or we may accept a terrestrial explanation and believe we got this way through some slower process of change. The most accepted terrestrial-origins story has it that we descend from prehuman hominid forest dwellers that first appeared in Africa some fifteen million years ago. Some seven to six million years ago our evolutionary line diverged from that of the great apes, specifically from genus *Pan*, now represented by chimpanzees and bonobos. These apes have so far survived within their various biological niches despite their vulnerability to predators, among them human hunters, but seem to have shown little change since our ancestors split from them. Our ancestral line, on the other hand, has led to toolmaking, speech, writing, cities, art, and science.

Most of us were first introduced to those primate cousins of ours when, as children, we were taken to a zoo and shown them in the place they had been collected. One of the first things we noticed was that they could, when they wanted to, stand up on two legs, look us straight in the eyes, and grimace. We thought they were grinning because they knew they were naked and were acting like misbehaving children. Except for the big ones, the gorillas, these apes were about our size, the size of children, big children, like the people in the building up the street who didn't speak or, when they spoke, we couldn't understand. They, too, behaved badly sometimes and made us laugh. Our parents told us they were born that way and could not grow up normally.

When we were a little older, we may have heard someone say that the human race was somehow related to these apes. But it was hard for us to acknowledge them as kin. Though we felt kindlier toward them whenever we visited them, they still embarrassed us a little. We gradually came to understand why they were locked up, why, for their own good and for ours, society had to confine them to what seemed to us to be institutions for the zoologically insane.

When we were old enough to learn about evolution, we were shown drawings of creatures that seemed half-ape and half-human. These were disturbing pictures. These "people" were still as hairy and naked as apes. The males held sticks. The females clutched infants. They walked stooped over, but, when we could see their faces, their eyes looked startled and confused. We came to understand that these early humans were as yet unable to speak. They were literally *dumb*. Unlike us, who can share opinions, argue over what we believe, and tell jokes and stories, they seemed resigned to a grim, humorless existence, and we were profoundly glad we weren't born that long ago.

Evolutionary time, we were informed, had indeed been very long. If we took a string representing 2,000 years, stretching back from today to the time of Christ, then added to it 125 more strings of similar length, we would arrive back at the time, about 250,000 years ago, when the neural wiring of the brain and the vocal tract had evolved to the point that some groups of humans might have begun communicating through symbolic sounds. This era had been preceded by an even longer time: if we now added to that string a skein of years ten times as long, we might arrive back at the moment, two and a half million years ago, when someone in East Africa first struck two stones together to make a cutting blade.

Over the recent centuries, our model of terrestrial time, like our model of cosmic space, has had to undergo major revisions. As a way of appreciating how radical these revisions have been, let me offer a thumbnail account of how our model has changed.

The earliest Western historiographers visualized world time as a series of ages. The two most influential of these timelines arranged these ages on a downward slope. The Greeks imagined a linear succession of Golden, Silver, Bronze, and Iron Ages, with a chaotic Age of Heroes inserted between the last two; the biblical narrative began in Eden and proceeded to recount the history of a single people, first, through tales of patriarchs and prophets, then, of kings, disunity, exile, and finally utter subjection. In both traditions these downward trajectories sometimes ended with a promised reversal of fortunes. Vergil in his *Fourth Eclogue*, drawing comfort from a cyclic view of time, declared that a new series of ages was at hand, beginning with a new Golden Age. Postexilic Jews and, later, Christians also envisioned an upswing, inaugurating a long age of theocratic peace and righteousness.

The later concept of the "Renaissance" owed much to the cyclic narrative. This "rebirth" was visualized as a revolution of the wheel of time *back* not to Eden but to the glories of Greece and Rome. When the linear model of time reasserted itself over the cyclic model, the return to a glorious past lost its appeal, and the direction of time became future oriented, anticipating the gradual, irresistibly rising path of progress. Toward the end of the seventeenth century, the "Quarrel of the Ancients and the Moderns" pitted those who revered classical authority against those who dared to believe in the superiority of contemporary science, morality, and letters. This led to a new division of "universal history" into three ages of "civilization," the ancient, the medieval, and the modern, preceded by the "state of savagery."[1] The current civilization of Europe was deemed one of unprecedented enlightenment. In the late eighteenth century, when the founders of the American Republic inscribed "*Novus ordo seclorum*" (a new series of ages) on their national seal, they borrowed some of the words of Vergil's "messianic eclogue" but added that significant word "*novus*."[2] The founders admired the achievements of Greece and Rome, but, in the spirit of the Enlightenment, this moment of theirs was to be the start of a wholly *new* series of ages, not a replay of any past ones.

When the idea of biological evolution emerged in the early nineteenth century, it appeared in the familiar context of this three-age timeline of civilized history proceeding out of a savage state of nature. What made evolutionary thought such a radical departure was that biological science was beginning to focus its attention upon that savage state, realizing, to its amazement, how very deep into prehistory it extended and that, in some respects, it had laid the foundation for what moderns regarded as "civilization."

The unearthing of carefully crafted bifacial blades among the bones of now-extinct animals indicated that primitive humans had been far more advanced than club-hefting, stone-heaving "ape-men." These first toolmakers, paleoanthropologists have since agreed, split off some four million years earlier from certain apes that had exchanged the relative security of the forest canopy for the chance to scavenge carcasses and sometimes run down small animals on the open grasslands. A number of species evolved from these apes, most notably the *Australopithecines.* These bipedal hominids had knapped rocks to form crude tools (Harmand et al. 2015), but their later descendants, *Homo habilis* and *Homo ergaster,* learned to modify such implements, customizing them for particular functions.

Toolmaking is now considered so significant an innovation that we use it to mark the beginning of genus *Homo* and its dozen or so species, all of which have gone extinct except *Homo sapiens,* now represented by only one subspecies, *Homo sapiens sapiens.* We celebrate this innovation, but, as I suggested earlier, whenever we inquire into the nature of any innovation, we need first to consider what action is being targeted for innovation, not merely its resultant modification. A new tool, for example, is usually designed to serve a preexistent activity and is replicated only if its use saves time and energy and optimizes results.

Most innovations also involve the conversion of some general-purpose skill into a specialization, which in turn implies a narrowed application. *Homo habilis,* accordingly, did not initiate tool use. The ancestors of modern great apes could no doubt improvise some tools, found objects such as branches, stalks, and rocks, and the sharp-eyed, sure handed *Australopithecines,* as we now know, made stone implements. There is, at least, no evidence that they modified objects for special tasks, e.g., cutting, peeling, puncturing, chopping, or fashioning other tools. When lithic manufacture produced the Oldowan chopper (ca. 2.5 mya), this marked an innovative advance in a long continuum of hominid tool use, significant, yes, but hardly unprecedented. Toolmaking was a culturally evolved innovation that built upon an already biologically evolved capacity to use found tools.

Stone technology did, however, make *Homo* a more adaptive genus, one able to compete more successfully with other hominid genera, defend itself from predators, enlarge its foraging territory, and increase its numbers. Then, as changing weather patterns affected vegetation and the migratory patterns of the grazing herds, early humans, equipped as they now were with specialized, portable hunting and butchering tools, could now range farther afield in pursuit of game. By 1.5 million years ago, bands of one hardy species, *Homo erectus*, armed with bifacial Acheulian hand axes, were able to migrate in waves out of Africa northeastward over what we now call the Middle East, some spreading southward along the coast of India to the Far East, others venturing northward into Eurasia and Europe. Those who eventually evolved into archaic *Homo sapiens* began their migration a million years later. The most adventurous of them all, our own subspecies, began spreading northward out of East Africa a mere seventy thousand years ago.

In addition to stone tools, our Paleolithic ancestors fashioned devices, simple and compound, out of other materials, e.g., wood, vegetable fibers, bone, antler, horn, and hide. We have suggestive fragments, but little evidence of their design or precise use (Camps i Calbet and Chauhan 2009). Yet if stone artifacts are any indication, their makers would not have been inspired by the pure joy of innovation. These were pragmatic folk: if a particular kind of tool seemed to work, they reproduced it millennia after millennia. Paleolithic technology existed for one purpose only—to support a population of hunters and gatherers large enough to defend itself from predatory competitors such as hyenas and big cats. The size of prehuman hominid groups has been estimated at fifty members, comparable to the average chimpanzee troop, but by the time of *Homo erectus* (1.5 mya), it may have grown to an average of 111 (Aiello and Dunbar 1993). For such numbers to survive they needed ready sources of food, secure shelters, effective weaponry, and, equally important, the social cohesion necessary for coordinated action.

The "modern synthesis" of Darwinian evolution began with a reassessment of the gene theory of Gregor Mendel and continues today through research in molecular genetics and genomics. One of the principal implications of this synthesis is that what drives biological evolution is not the simple version of natural selection, in which the "fittest" individuals live long enough to reproduce and do so plentifully enough to leave the next generation with more copies of themselves. In a far more complex version, random gene mutations occur, some potentially helpful, others detrimental, most merely neutral. Over time in given populations these mutated genes are transmitted to offspring, and the traits they give rise to help determine the future of both the genes and the individuals that carry them. If these traits are appropriate to—i.e., fit—the ecological niche of a given animal, it may survive. It may also spread and, if isolated long enough, become a whole new species.

Resistance to the notion of our evolution as a seamless continuum from ape to human might be motivated by unscientific biases, but there are still legitimate reasons for dividing this continuum. An observation of nonhuman primates reveals many innate traits but few, if any, that individual monkeys and apes can acquire or teach to other conspecifics. Humans, on the other hand, have an apparently unlimited capacity to acquire new traits in the form of knowledge and technique and to share these with others. They have, it seems, an innate ability to teach, learn, and preserve learned traits transgenerationally. There seems to be a difference between biological and cultural evolution—but is it real or only apparent? And, if real, on what is it based, and what is its significance?

Unlike biological evolution, which may be visualized as a vertical transmission of traits, cultural evolution is a horizontal process. This means that within a short time one person can convey information to several people who can then go off and share that information with others. Cultural evolution, "modified by learned innovations not just random mutation," is a much more rapid process than genetic transmission "because we can actively choose superior variants either by direct experimentation with variants or by rules of thumb like imitating the successful or prestigious" (Richerson and Boyd 2010, 295). This power to send and receive cultural information is, nevertheless, an innate, genetic trait passed on from parents to children. Though some cultural material may spread in the form of memes by a kind of viral contagion, the most useful sort is gained through an effortful process of observation, learning, and practice. But, effortful as it often is, its rewards more than compensate for that effort.

How did we get this way? That is, how did we produce the cascade of innovations that has made us the dominant species on the planet? The "dual-inheritance" theory, as its name implies, claims that every one of us inherits both a genetic and a cultural legacy. Our genetic code provides a rich down payment of biological traits, including a knack for understanding and imitating others of our kind. This enhanced social instinct has made it possible for us to amass and share that second inheritance, culture, with all its innovative benefits (Henrich and McElreath

2003, 2007). Broadly speaking, our genetic makeup has become our *innovandum*, culture our *innovation*, and it is culture that has bestowed on us a peculiarly human power to innovate ourselves.

Our first inheritance has come down to us through that winnowing process known as natural selection, that long series of trials and errors, but our second inheritance we receive relatively risk free. All we have to do is pay attention, learn, and practice. Even if we do make mistakes, their consequences are rarely life threatening. Since the success of human societies depends on their capacity to store and dispense usable information, cultural evolution presupposes the invention of mutually accepted sign systems. In short, before we could learn what to do, we first needed to learn what we mean.

Methods of Mindsharing

The structure of early Paleolithic society was built on a foundation of long-established primate social skills, such as alliance building and deference toward dominant leaders, but, once on the open savannas, humans could no longer rely on any alpha male for protection. A vertical social hierarchy would now be maladaptive. Foraging, hunting, and defending the group from larger, stronger, faster predators, they could survive only by collective efforts and, to coordinate such efforts, they were obliged to develop new and more effective means of communicating.

Whatever their form, communicative codes would have had to be built upon earlier-evolved, hardwired cognitive skills. One of these, termed "theory of mind," is an individual's general inference that just as he or she has thoughts, intentions, and desires, so do others. Since chimps exhibit this genetic skill, it is safe to assume that our Paleolithic ancestors possessed both it and its more explicit follow-up, "mind reading." The latter skill is one's ability to infer *particular* thoughts, intentions, etc., from an observation of *particular* behavioral clues, e.g., facial expression, "body language," or tone of voice, all of which can be conveyed wordlessly.[3]

In recent decades interest in mind reading has been spurred by research into so-called mirror neurons, the brain cells, situated mainly in the premotor cortex, that fire in a coordinated response when a primate perceives another conspecific moving its hand to reach out or grasp an object. Mirror-neuron activity is thus a special instance of eye-hand coordination, for as another's outward hand movement is visually perceived, the perceiver's own motor cells responsible for hand movement also fire. This mirroring reaction, it has been proposed, is the neurological basis of primate (and human) social intelligence, that suite of skills that helps us assess one another's needs, desires, and intentions and choose our own response accordingly. Long before spoken language, our ancestors could read one another's involuntary affect in its causative context and determine what action was appropriate—the more precise the reading, the more timely and efficient the response.

Social intelligence is related to an even older, broader cognitive skill: all animals must be able to read their environments, i.e., interpret natural *indexical* signs. So, depending on their biological niches, they may, when they smell smoke, infer fire; when they feel a damp wind, sense an approaching rainstorm; when they hear a specific sound, prepare to respond to friend or foe, predator or prey. To respond effectively to an emergent situation an animal must activate a complex set of motor programs, some instinctual, others learned. These routines, termed "motor schemas," may include running forward, swerving, backing up, leaping, grasping, kicking, or biting. For primates, humans especially, they include arm and hand movements, e.g., reaching, grasping, thrusting, swinging, throwing, etc. As with all motor skills, these schemas are organized and stored in procedural memory (Donald 1991, 150).

Motor schemas can be combined in complex ways to enact intentional indexical signs that communicate thoughts and intentions to others. Among such indices is aggressive display behavior meant to indicate a threat, even when this is ritualized behavior that does not necessarily proceed to an attack. Territorial scent marking and mating behaviors may also be classified as intentional indices, as are primate orofacial gestures (e.g., stares or bared teeth), sounds that indicate pleasure, postures that indicate submission or comfort, and touches that indicate alliance or affection.

Some indexical signs, e.g., tracks left by an animal, were not intended to be read by others. Pointing is an intentional index denoting the wish that others look or move in a particular direction. Pointing is "pointless" unless it has a stable vector, an object within the here and now, and a socially relevant situational context. When a person points to a set of animal tracks, this is a socially situated action requiring a sign sender and one or more sign receivers, and it is, in this case, intentionally used to draw others' attention to *un*intentional, natural indices.

But not all useful entities exist in the perceivable present. Some belong to the past, to the present elsewhere, or to some imaginable future. Limited to indexical signs, humans would have been unable to deal with situations temporally and spatially absent. They could not have shared knowledge of a honey hive that one of them found *last year* or of a leopard sighted on the *other side* of the hill, nor could they make a promise to distribute meat from *next month's* kill. To exchange such information, prelinguistic humans would need to evoke the mental images of these absent objects via *iconic* signs, much as we do when we play charades, except no one then could respond by calling out "honey!" or "leopard!"—much less the concept "food sharing!" To form an effective communicative code, iconic signs would have had to be relatively transparent and convey immediately recognizable meanings.

When we think of how an iconic sign functions, the first example we think of is probably a picture. If we recognize what a picture represents, we can say that that is the picture's referent and assume that the person who fashioned that image had that same referent in mind. But this would hardly be the way early humans would have shared information—there are simpler, more direct ways to accomplish this. One might be to make an imitative sound—a roar for a leopard, a hoot for an ape, a whistle for the wind, and so forth. Since iconic vocalizations like these could only refer to sound-making entities, one might have to use pantomime to imitate animal postures and gaits, the flight of birds, or the swimming movement of fish. But hand movements might prove the readiest available, the most energy-efficient, indeed, the handiest means of communicating what one has in mind. As we know, an ape has little voluntary control of its vocal organs, but its hand is a precise and flexible effector. It is most likely, therefore, that early tool-using humans would have first used iconic hand gestures, not voice, to refer to objects that were not present or perceivable and so could not be pointed out by an indexical gesture.

Indexical signs would sometimes cue mental imaging, e.g., interpreting an animal's tracks may evoke some image of the specific beast, but iconic signs manually transmitted would always exercise the powers of imagination. What the communicator "had in mind" would be designated by the gestural icon, which then would evoke a correspondent mental image in the mind of the receiver. To achieve this communicative effect, both sender and receiver, moreover, would need to possess a similar archive of mental images, a common prelinguistic lexicon—a common *iconicon*.

Philosophers have long felt the need to account for the way we recognize things and imagine them when we think of their names. Plato called them "ideas," or the abstract archetypes of things. Kant called them "schemata." George Lakoff and Mark Johnson (1980) argued that, rather than being eternal or transcendental entities, these images derive from our embodied experience of our three-dimensional environment. Mental images, they claimed, are constructed out of simpler elements, "image schemas." These are not in themselves innate, yet all animals that need to navigate their habitats, find food, avoid danger, and reproduce their kind must have the innate capacity to build an archive of these images. It is important to keep in mind that image schemas are preconceptual spatial features and that concepts are higher-order elaborations of elemental features such as *inside, outside, over, across, up*. As such, these schemas are prelinguistic: words evoke them now only because words, as symbolic signs, designate concepts that are constructed out of selected image schemas. By analogy, the name of an entrée printed in a menu designates a particular dish that has been prepared from a precisely selected set of ingredients. We might not be able to recognize all the elements that went into the lobster bisque, but, when it arrives at our table, we know what we ordered and can distinguish it from our companion's grilled salmon.[4]

The theory that iconic signs, gesturally transmitted, could serve as an efficient communicative code, plausible as it seems, does, however, present some problems. One standard objection is that tool users would have to stop their work and drop their implements when they communicated. Another objection is that communication would be possible only when receivers had an unobstructed view of the sender. This means that they would need to stand close enough to see and interpret correctly the sender's manual "air pictures." Likewise, if they were walking behind a leader, the effectiveness of that person's gestures, manual and orofacial, would be diminished. A third objection is that communication would be impossible in darkness, which, before humans mastered fire, was a time of heightened danger. Assuming that tool use, travel, and nighttime security, all crucial to the survival of early humans, would be adversely constrained by a reliance on iconic gesturing, critics of the gestural theory make valid points.

I would add one other objection: iconic gesture, compared to vocal communication, would be a perilously *slow* means of sending messages. The reason for its slowness is that it entails at least three separate steps. First, the sender must arouse the attention of the receivers. Second, they must approach close enough to "read" manually shaped icons. Third, they must all construe those icons in the same way before they could jointly react appropriately. Consider, for example, this scenario: the sender, while foraging food, has observed her young son in danger. The message she wants to send is this: "A leopard is stalking my son in the pine stands on the other side of the hill. Help save him." Would a gesturally conveyed message be effective in such an emergency? Would the child survive?

Before I try to address the objection that this scenario illustrates, I will reconsider some of the semiotic issues I have already introduced.

In his well-regarded first book, *The Symbolic Species* (1997), the anthropologist Terrence Deacon endeavored to account for the fact that we, the only surviving species of genus *Homo*, are unique among animals in our capacity to communicate through arbitrary signs, i.e., symbols. Why do we deploy these signs when we speak, read, and do math while our genetically close cousins, the apes, have only a few emotive cries and no more than twenty to thirty gesticulations to convey their states of mind? How did we get that way?

Agreeing with linguists such as Noam Chomsky, Derek Bickerton, and Steven Pinker, Deacon premised his study with the proposition that language, as a symbolic system, could not have been preceded by any code intermediate between animal signals and fully grammatical speech. "It is tempting to try to conceive of language as an interpolated extreme of something that other species produce, such as calls, grunts, gestures, or social grooming" (1997, 34), but Deacon deems such speculation fanciful. This was not a new position, of course. It had long been assumed that speech was the decisive difference between humans and (other) animals. Though evolutionary thought, as Darwin advanced it, implied that modern humans were the result of a gradual process of change over several million years, a number of eminent scholars found it inconceivable that symbols could be derived from primate vocalizations or visual signals. "Language," the philologist Friedrich Max Müller had concluded, "forms an unpassable barrier between man and beast" (1889). He had already (1868) dismissed all claims to the contrary, derisively labeling them the "bow-wow" theory (that words began as imitative sounds), the "pooh-pooh" theory (that the earliest words were emotive interjections), and the "yo-he-ho" theory (that language derives from sounds used to coordinate strenuous group labor).

It does seem incontrovertible that all modern humans, members of the subspecies *Homo sapiens sapiens,* have the innate capacity to acquire a symbolic code, while nonhuman hominids lack this capacity. Even when raised in human households and trained to communicate with humans through visual symbols, their communicative abilities eventually reach an unpassable barrier, a fact consistent with the theory that humans owe their language to a process of genetic mutation (Chomsky 1968, Bickerton 1992). The question remains, however: was this a single, all-or-nothing genetic event, a toggle switch somewhere in our DNA that, once flipped on, remained thenceforth in the open position?[5] Even if at some point a decisive genetic mutation did occur, might it not have been preceded by a set of necessary preadaptations? At any rate, a symbolic system, such as language, could emerge only in a "language-ready brain" (Arbib 2009). Moreover, the prelinguistic traits of gesture and intonation continue to accompany speech and are indeed essential to oral communication (McNeill 1992; Armstrong, Stokoe, and Wilcox 1995).

With that in mind, I found myself in *Paleopoetics* (2013, 123–126) taking issue with Deacon's semiotic approach. In *The Symbolic Species* he had disqualified iconic gesture as a communicative vehicle by choosing to define an icon restrictively as a sign that operates when a given object is perceived to be a given entity only because it *looks like* that known entity. As an example, he presents us with a bird that cannot spot a tasty moth on a tree trunk because the colored markings on the insect's wings so well blend into the bark that all the bird can recognize is "bark, bark, bark" (1997, 76). Because appearances can be deceiving, iconic cognition is inherently defective, whereas indexical cognition, entailing, as it does, prior experience and stored knowledge, is much more trustworthy. Thus, any animal that can interpret indexical signs is less likely to make snap judgments. It will instead consider the probability that if X is perceived, Y might be hidden close by or that, if X occurs, its cause might be Y and its predictable effect might be Z. Here again Deacon treats index, like icon, as a natural sign, a phenomenon that an observer interprets, but not as a sign *intentionally* performed to communicate a meaning.[6]

The Coevolution of Signs and Technology

Table 1.1 is intended to summarize what I have considered major prelinguistic innovations. I have arranged them as adaptations of those two fundamental biological functions, sensation and motion, which can be conceptually reduced further into input/output and stimulus/response. The components that I have paired as opposites in this table I regard as both distinct and complementary.

The first pair listed, *perception* and *action*, are simply the outward engagement of the conscious sensorimotor brain with the object world through the activation of the sensory and motor cortices plus the afferent and efferent neural systems. Perception occurs when a set of *sensa* is selected and recognized as a token (X) corresponding to the type (Y) stored in semantic memory, the system that stores our knowledge of the world. Action occurs when a perceived set of identified tokens (X) constitute a situation that elicits a certain set of motor responses (Y). Though perception usually precedes action, action supplements perception, as when a

TABLE 1.1
Stages in the coevolution of signs and technology

	Sensation (input)	Motion (output)
1. Conscious function	**Perception.** (Token X recognized as type Y)	**Action.** (Situation X provokes action Y)
2. Use	**Using found indices.** (Objects used to discover other objects or events: X understood to imply Y)	**Using found tools.** (Objects used to transfer force to other objects: X used in order to affect Y)
3. Mimesis	**Deixis.** (Showing *what*: pointing in order to direct gaze toward objects or other indices)	**Teaching.** (Showing *how*: sharing tool use technique)
4. Production	**Sign making.** (Types converted first into reusable gestural icons, later into conventional visual symbols, finally into conventional vocal/auditory symbols, i.e., language)	**Tool making.** (Types converted into reusable tools & other artifacts).

perceived object needs to be approached and manipulated for closer inspection.

The relation of perception to action appears to have been built into the vertebrate nervous system, making it possible for animals to evolve new behaviors. As Michael Anderson (2014) has argued, the plasticity of the human brain permits new tasks to generate networks that link to older, long-established centers and pathways. The neural structures responsible for the so-called higher cognitive processes would therefore be found more widely distributed in the brain. This explains why speech production and reception are centered but not narrowly confined, the one to Broca's area, the other to Wernicke's area: not only are these two more recently evolved functions linked to each other, but each is also linked to specific motor, tactile, auditory, and visual areas.

The second level of engagement with the object world, *use*, presupposes an animal's ability to interpret an object as contextually connected to other objects or conditions not currently perceived. The percept X thus functions as an indexical sign, inducing the perceiver to infer Y, as when a shape, color, etc. of a particular leaf indicates the presence of an edible root hidden below it or when a scent in the air indicates the vicinity of a predator. The reader of a natural index infers, "if this, then also that," and may then proceed to specify the "that." On the motion side, a situation emerges that requires that some available object (X) be found to deal with it, a stick perhaps to dig for a clam (Y) or a branch as a weapon to repel a predator (Y).

Most animals are capable of some indexical cognition, and some, including the higher primates, can manipulate some natural objects as found tools. But when we move on to the third level, *mimesis*, we enter an innovative stage, the stage where humans excel. Here information is shared by intentional transmission and intentional reception. Someone shows others where to look or what to do—or, more precisely, what to perceive or how to do something. In typical *deixis* a person manually directs others' attention, and they comply by optically shifting their gaze, i.e., choosing to imitate with their eyes the movement and direction of the pointer's hand. In *teaching*—hands-on teaching, that is—a person performs a series of actions that others watch and then reperform. Since both kinds of mimetic communication typically involve both visual and manual operations, they are instances of eye-hand coordination. But what makes this a specifically human innovation is that this coordination is cooperant: the hands and eyes that work together now belong to *different* persons.

Mimetic behavior, the prerequisite for culture, can include a variety of forms. Visual attention shifts, for example, may be cued by deictic gaze shifts and head bobs, not only by finger and hand pointing. Auditory attention shifts may be cued in others by cocking the head in a certain direction, cupping a hand behind one ear, or closing the eyes. The complementary character of perception and action in mimesis is perhaps best demonstrated in gathering and hunting activities. When older, more experienced people lead the way and the others follow, the leaders do two kinds of pointing: one kind induces motor response by showing others where they are headed, or the leaders may show this simply by walking in that direction; the other induces perceptual responses by showing some wayside detail, e.g., animal tracks or a berry bush. This showing teaches the novices how to look and then how best to deal with what they find.

Built upon the preceding three, the fourth level of innovation introduces a new sign function, the icon. By icon I mean an intentional act or object recognized jointly by sender and receiver as representing something else to which it is similar. My definition of icon thus differs from Terrence Deacon's in that mine assumes that the viewer fully understands the difference between the signifier and the signified and that an icon functions as a sign only when it stands *for* the reality it represents, not when it stands *as* that reality. Semiotic understanding presupposes that we and others store accessible mental images composed, as we now understand, of elementary image schemas and that these schema-constructed mental images are not to be confounded with the perceivable phenomena we encounter in the world. To mistake one for the other is to misperceive, and to do so continually is to become delusional.

The social use of icons, as I described earlier, involves the intentional externalization of mental images as visible signifiers. Thus, a hand gesture in which the thumb is out and the fingers are extended and rapidly flexed and arched while the hand swings smoothly across the body, could iconically represent a bird in flight but would never be mistaken for a real bird. Like mental images, gestural icons are schematic representations distinct from their meanings. In that respect they function as iconic signs displayed in social play: when a mother and child pretend to use bananas as telephones, to borrow Alan Leslie's (1987) example, each knows quite well the difference between the two objects.[7]

Gestural icons can, of course, be improvised on the fly. We still perform this charadelike tactic as a desperate last resort when we need to communicate with someone who does not understand our language. But if we have tried to do so, we know how imperfect that improvised thought transmission can be. Imagine how chaotic a sporting event would become if the referees continually improvised their signs. If prelinguistic humans used gestural signs, we may be sure these gestures were soon conventionalized. If so, they would also have been abbreviated, and, if abbreviated, their iconicity, over time, would have become effaced until they eventually became *symbols,* i.e., arbitrary gestures, capable of transmitting meaning in the shortest time possible (Collins 2013, 129).

This sequential progression was observed in Nicaraguan schools for the deaf in the 1980s and 1990s. It was found that while finger spelling and lip reading were largely unsuccessful, the children on their own had produced a rudimentary system of hand gestures. In the 1990s a younger cohort of students took this pidginlike language and invented a grammar for it equivalent to that of a creole language. In gestural terms, this was a shift from a predominantly iconic code to one that used abstract symbols to structure its lexicon (Senghas 2010). This link among iconic abbreviation, symbolic signs, and syntax suggests a spontaneous, self-forming process as lexicon generates syntax (De Vos 2014).

Natural selection would indeed have favored populations that could quickly communicate individuals' needs and intentions. The only way that slow communicators could survive on the open grasslands would be to avoid circumstances that require the immediate mobilization of the group. They would, for example, avoid competing directly with rival predators and instead warily scavenge for scraps. Should they challenge the big cats and the hyena packs, they would likely become prey themselves. In the scenario I sketched earlier, the foraging band might have saved that distraught woman's son, but only if she had been able to inform them clearly and quickly. If so, he might have lived and, perhaps, passed his genes on to us.

When we think of communicative codes, we normally associate conventional, arbitrary signs, i.e., symbols, with vocal utterance and associate indices and icons with manual gestures because speech is a symbolic code for which gesture serves now as a descriptive or emphatic accompaniment. But there is no reason to assume this was always the case—that gestures could never convey symbols. Once we dissociate sign function from communicative medium, we can consider the possibility that humans were a "symbolic species" long before they became a speaking species. This progression of signs—from index to icon to symbol—is a communicative continuum connected to but not determined by the physiology of transmission. In this innovative series, one sign type after the other becomes salient, but, when a new sign function does become foregrounded, it never eliminates what it has superseded. Instead, it converts the older sign function into new and other uses. In this respect, cultural evolution models itself upon biological evolution and, as Michael Tomasello has pointed out, employs

> a kind of cultural ratchet, as each version of the practice [of any skillful behavior] stays solidly in the group's repertoire until someone comes up with something newer and more improved. This means that just as individual humans biologically inherit genes that have been acquired in the past, they also culturally inherit artifacts and behavioral practices that represent something like the collective wisdom of their forebears. To date, no animal species other than humans has been observed to have cultural behaviors that accumulate modifications and so ratchet up in complexity over time.
> (2009, xi)

By characterizing this fourth semiotic level as "production," I am linking *sign making*, beginning with the creation of an iconic, protosymbolic code, with *toolmaking*, the crafting of task-specific implements. In both sorts of innovation something is standardized, saved, and reused. Sign making and toolmaking each presuppose a specific mimetic aptitude not only to imitate the behavior of others but also to replicate that action in accord with internally stored motor schemas. Both behaviors, we should not forget, also draw on earlier-evolved cognitive processes. Both match token with type: the gestural icon, when performed, is a token representing a type, i.e., the mental image it signifies. Similarly, the mental image of the finished tool is the type that guides the worker as he or she shapes the token (Adornetti 2014b).

Human communicative codes, I suggest, developed in tandem with technological culture. In each line of innovative progression, the change was from general to specialized function. Stone tools began with general-purpose hand axes and were later customized for specific tasks, e.g., slicing, digging, throwing, puncturing, etc. The modification of natural objects that produced specialized stone, wood, and bone tools and, through their use, other helpful artifacts was in every case a serial procedure that followed a prescribed sequence. To acquire any of these early technologies, one had to learn to (1) narrow one's attention to one thing at a time, (2) memorize the proper sequence, and (3) confine one's focus to one perceptual channel, usually vision. Our broad multimodal openness to the perceivable world was never lost, but it could be held in abeyance whenever a narrowing of focus seemed necessary. Transmitting this knowhow to others entailed a similarly sequential series of signs: the teacher must have the means to signify that first one must find A, then find B, then apply B to A to produce C, then D, and so forth. This series of signs, like the steps in a recipe, had to be remembered in order. If gestural signing had been used, the visual sequence would have to be committed to memory; if vocally taught, the meaningful words would need to be rehearsed in their proper sequence. With practice, the learner could eventually perform these steps by accessing procedural memory and, though the separate steps no longer needed to be called to mind through visual or auditory imagination, the proper linear sequence would be correctly replicated.

The cognitive skill that most differentiates humans from other primates might well be our mastery of serial attention in the transmission of both object information and social information. Like the step-by-step teaching of a technical skill, the conveyance of commands, requests, advice, and gossip took a form that highlighted the before-after, cause-effect relations in human behavior. If you do A, B will do C, which will lead to D and E, etc. Despite the fact that this narrow serialization often leads us to identify before and after with cause and effect, the post hoc ergo propter hoc fallacy, it has proved a useful means of passing along proverbial wisdom.

Insofar as language is an innate human attribute common to speakers of all separate tongues, we would expect it to possess some universal features. One obvious feature is its medium, phonological effects produced by serial actions of the human vocal tract (Hockett [1960] 1982). Another general feature pertains to the sequential syntactical organization of those sounds (Greenberg 1990). What separate elements are thus organized is a much-debated topic, but it is interesting to note that for Joseph Greenberg, the linguist who devoted much of his research to enumerating universals, the majority of his list of forty-five involve sequential order.[8] I think it is reasonable to posit that, broadly considered, the two principles common to all spoken language are serial phonology and sequential grammar.

What has come to be termed the Sapir-Whorf hypothesis claims that the ways by which different natural languages categorize environmental features, name colors, and define spatial and temporal dimensions exerts influence on, and may even determine how, speakers perceive and conceptualize the world (Sapir 1921, Whorf 1956). Proponents of linguistic universals, beginning in the 1960s, have strongly critiqued this linguistic relativism and, stressing the common elements in all languages, elements that they locate at deeper structural levels, have accused the relativists of jumping to conclusions based on superficial analyses (Chomsky 1965, Pinker 1994). Taking a more nuanced view of these competing claims, most current theorists have come to accept that native speakers of separate languages (Saussure's *langues*), insofar as these have different lexical and syntactical features, do tend to perceive and process the world differently. These speakers are not, however, constrained by their languages to do so—they can, with a little effort, view phenomena in other ways. In short, one's native language influences but does not determine how one thinks. On the other hand, as a universal hardwired human trait, the language faculty (Saussure's *langage*) does have a number of universal elements (Regier and Kay 2009, Wolff and Holmes 2011, Zlatev and Blomberg 2015).

This debate raises a number of provocative questions. How is language related to the process of thinking? Is thought equivalent to inner (subvocal) speech? Do we need language to solve problems and construct concepts? If so, to what extent does our native language influence our ability to do so? Are there some thought processes that work efficiently without linguistic mediation? Is visual imagination one of these processes? To what extent do writing and reading writing affect the relation of language to thought?

I will raise these questions again in later chapters, but for now I will propose that since language at every level of analysis is an efficient medium of both communication and thought, we can speak of human consciousness in large measure as *linguistic consciousness.* We live immersed in language. We are so comfortable with our word-defined world that we tend to forget that no other animal shares this world, which, after all, is merely a universe of discourse. Our dogs and cats look up at us quizzically as we make a myriad of mouth sounds but forgive us our babbling so long as we can still open food tins and set down bowls of water. As we move about daily upon our inexplicable trajectories, other animals eye us with justifiable suspicion. We are to them as space visitors would be to us: powerful, obscurely motivated aliens who use a secret code to make things happen.

From Pronouns to Speech Acts

Martin Buber in his classic meditation *I and Thou* ([1923] 1958) tells what his experience has been communicating with his cat. The cat, whose mood has been a blend of disinterestedness and anxiety, becomes intrigued by the man's attention and, "lighting up under the touch of my glance, indisputably questioned me: 'Is it possible that you think of me? Don't you really want me just so you can have some fun? Do I concern you? What is it that comes from you? What is it that surrounds me? What is it that comes to me? What is it?" This moment of intimate contact soon fades, however. As Buber interprets this encounter, "for the space of a glance the world of *Thou* had shone out from the depths, to be at once extinguished and put back into the world of *It*" (97–98, translation slightly modified).

Once our species became fully capable of theory of mind, we could sustain the mutuality of the personal second-person pronoun and could regularly enter into Buber's I–Thou relationship—even before we had words for "I" and "Thou." The primary semiotic rules of human engagement require the pronoun paradigm, specifically the first two personal pronouns that, as Benveniste ([1966] 1973) maintained, form the minimal elements of discourse, which in this case is prelinguistic discourse. If we can trace the main features of this ur-grammar, we may identify those *innovanda* that fully grammatical language improved upon, those earlier-evolved primitives that remain its foundational features.

As I have already suggested, indices in the form of directed gaze and finger pointing were likely our first intentional signs and, as referential gestures, would have marked the onset of what Merlin Donald (1991, 2001) has called the mimetic stage of human evolution. Pointing, which implies the command "Imitate my hand with your eyes," closely coordinated first- and second-persons in perceiving environmental details. The teaching of stone knapping and other culturally transmitted skills would have entailed the additional tacit command "Watch what I do, then do what I did." Later, with the introduction of iconic gestures, humans added a third-person reference and were finally able to communicate information about nonpresent entities.

Since this pronominal ur-grammar describes a system of perceptual and motor activities, it may be usefully explored from the perspective of Austin's and Searle's speech-act theory. In figure 1.1, I have fitted John Searle's taxonomy of speech acts onto the familiar grammatical structure of the three personal pronouns.

In what follows, I will not be considering pronouns as "standing for" particular nouns but rather as specific roles in a speech situation: *I* represents the speaker; *You*, the person(s) spoken to; and *They* (he, she, it), the person(s) or thing(s) spoken about. In applying speech-act theory to this pronoun paradigm, I will focus first on the producer of speech acts, the *I*, and this first person's "illocutionary force" and, later on, its "perlocutionary effects," including the second person's (addressee's) responses.[9] Accordingly, the *I*, when making self-reflexive utterances, produces *expressives*.[10] When the *I* commands, begs from, prays to, etc. some specific *You*, singular or plural, the speaker produces *directives*. When the *I*, speaking to *You*, refers to *They*, the *I* produces *assertives*. I will first place this three-part model in an evolutionary context, then comment on those other two speech-act types, *declaratives* and *commissives*.

1. Expressives. "I feel X." Self-reflexive utterances.

   Declaratives. "What I say effects X." Utterances that enact their meaning.

2. Directives. "I command (request, suggest, etc.) that you do X." Socially manipulative utterances addressed to second persons.

   Commissives. "I promise you I will do what I say." Utterances that oblige the speaker.

3. Assertives. "I say that X is true." Utterances of belief about third persons.

Figure 1.1 Searle's five speech-act types set within the pronominal paradigm.

Expressives, as Searle (1983, 166) defines them, are self-reflexive utterances of "feelings and attitudes" that do not need an audience. When feelings are expressed to others, no specific action on their part is explicitly indicated, though some action may be implied. For example, "I'm cold" may be interpreted, depending on the circumstances, as an indirect directive, a polite request for a warm wrap, for the closing of a window, or for another log to be put on the fire. In presymbolic ur-grammar, expressives might be communicated by some emotive vocalization or gesture denoting some affective state that others could read as an indexical sign.

When a first-person singular (*I*) addresses a specific second person (*You*), singular or plural, with the purpose of inducing that party to respond with a specific action, this utterance is termed a *directive*, a speech act that may take the form of a command, request, a prayer (as to a deity), etc. In primate societies, nonhuman and human, directives presuppose social information; i.e., the communicator must know the other's status, what kind of directive is permissible, and what kind of response may be predicted. Directive speech thus underlies the art of rhetoric. In the earliest form of ur-grammar it could take the form of any manipulative act, e.g., a pushing away, a pulling toward, a threatening posture, a directed gaze, or, among humans, a pointing finger or hand (Kennedy 1992). All these prelinguistic acts are intentional indexical signs that direct others' actions, just as we language-endowed primates do when we say "Pay attention," "Do what I'm telling you to do," or "Look over there and see for yourself."

Alison Wray (1998) in her formulaic theory of human protolanguage speculated that the earliest spoken language was a set of arbitrary vocal sounds, one-word "sentences" used to convey intentions within a close-knit group. Her invented vocabulary illustrates how this communicative code might have worked: *mabu* might mean "keep away"; *madu*, "take the stick"; *mebita*, "give her the food"; and *ikatube*, "give me the food." Each utterance was a holophrastic statement, a complete thought that could neither be divided into meaningful segments (morphemes) nor syntactically combined with other formulaic utterances (51–52). As these examples show, the vocabulary in this protolanguage, composed wholly of directives, could refer only to immediate pragmatic social circumstances. References could not be displaced in space and time: no one could tell stories or report on what existed beyond the immediate purview of the *I* and the *You*. "There would be no reference or description for its own sake (*this is a tree; the tree is tall*)" (51). There would be no way to say "*the world looks lovely tonight* or *I wonder if it will rain tomorrow* or *it wasn't like that in my day*"(52). We would find this kind of discourse quite limiting, yet we should remind ourselves of how much our own familial discourse, especially from parent to child, is also composed of holophrastic directives, e.g., "wipeyourfeet," "brushyourteeth," "cleanyourroom," "takeoutthetrash," "comeinandeat," "timeforbed."

According to this scenario, some time after 500,000 years ago, among archaic *Homo sapiens* species, including *H. heidelbergensis* and *H. neanderthalensis,* vocal sounds, already used in formulaic directives, were becoming more distinctive and controllable to the point that humans found they could further increase the speed and energy efficiency of symbolic information by designating separate sounds for separate referents. While retaining the formulaic style for use among intimates, some populations now began communicating with others using separate words and a compositional grammar (Wray and Grace 2007). The sequence of communicative innovations that I am proposing is shown in figure 1.2.

As each subsequent stage in this sequence was introduced, it incorporated the essential elements, the *innovanda,* of the stage that preceded it. Assertives could have been first introduced as icons (stages 2 and 3), but they became fully operative only with the invention of compositional speech. Having the advantage of symbolic signs, speech acts of this type could now efficiently name and archive the contents of the world together with their actions and interactions. Reliable informants of useful object information became esteemed and rewarded by those who sought them

1) Indexical gestures
2) Iconic gestures
3) Protosymbolic (abbreviated) iconic gestures

} ur-grammar (stages 1–3)

4) Formulaic-symbolic speech
5) Compositional-symbolic speech (lexicon + syntax)

Figure 1.2 Human communication: a hypothetical sequence of stages.

out. Though assertives could represent such information, this knowledge of the world would always have been deeply implicated in the economy of social information. For instance, the ability to name an absent person and then tell when and to whom that person did a particular action opens up that conduit of social information we know as gossip. Whatever we think of this behavior, we no doubt indulge in it as much as did our ancestors—and for the same reasons: it helps us form alliances and reinforces our communal sense of right and wrong (Dunbar 1996). Being able to communicate about a third person without drawing his or her attention (without, for example, having to point out that person) enhances the power of this practice as a means of social regulation. It also places a burden on talebearers. Since assertives, unlike expressives and directives, have truth value, transmitters of gossip must prove trustworthy sources or become themselves objects of gossip.

In figure 1.1, I have inserted the remaining two speech acts a bit to the right, *declaratives* between expressives and directives and *commissives* between directives and assertives. Each, I suggest, absorbed features of its two adjacent speech-act types.

Declaratives, as Searle describes them, are utterances that create the outcomes they name, as in "I sentence you to prison," "I hereby pronounce you man and wife," or "I apologize." These can sometimes assume an abracadabra-like magical power, as in "I curse (or bless) you" or, more often, invoke societal authority, as in "By the power vested in me, I hereby . . . " A key adverb for declaratives is "hereby" because it situates the causative force of its words in the utterance itself. This speech act does not confess a feeling or a wish, as does an expressive ("I hate you and would like to see you go away"); it does not command compliance, as does a directive ("leave the country and don't come back"); nor does it state a fact or belief, as does an assertive ("I banished him this morning"). I see the declarative speech act, as in "I hereby banish you," as deriving partly from expressives, in that the speaker is the source of the power being exerted, and partly from directives, in that this power is intended to create a predetermined effect.

Within a prelinguistic ur-grammar, a declarative might take the form of any dominance display. A chest thumping, stamping, ground slapping, or charging about would be distinct from a facial expression of affect and from a directive signal—it would instead be an index intended to convey the "power vested" in that individual. In a speech-mediated society, this authority becomes a role conferred on a person by a god, some tribal elite, or popular suffrage. The power of the declarative word is "vested" in certain classes of persons who often do indeed wear ritual vestments and wield ritual accoutrements, e.g., ankhs, scepters, crosiers, or gavels, when performing their offices.

Commissives, utterances that promise, vow, or guarantee some action, also appear to be hybrid speech acts. As agreements to comply with certain commands, they derive from directives, but, as specifying particular third persons, objects, and undertakings projected into the future, they also derive from assertives. Commissives are not simply acts of compliance to a directive but rather are voluntary self-commitments and, while they do indeed bind the speaker to perform, or desist from, a future action, they may also bind others. The speaker promises "I will keep my word," but others, too, must keep theirs. Moreover, commissives not only oblige the speaker to pay back a good quid for a good quo. In oaths of vengeance, they serve a retributive function in righting wrongs. A commissive is, moreover, a personal pledge of trust that replicates the paramount social contract implicit in the symbolic code of language: we all promise to share information about the world by means of this communally agreed-upon vocabulary and syntax, and we furthermore assert that, when we do so, what we say is true.

In his account of the evolution of language, Terrence Deacon (1997) proposed that "in order for cooperative provisioning [and sexual access] to be a stable strategy," public rules must be established that encourage self-commitments to future actions (396–397). For this, no indexical signs would do; only a symbolic code could formulate a promise. True, an object or action that one promises to someone is not likely to be present and perceivable during the act of promising, but, as I have argued, it does not follow that only symbolic linguistic signs can signify what is promised. Since it is clear that an iconic gesture can also signify an absent item, promises could also have been conveyed iconically. Nevertheless, humans did eventually become a vocally "symbolic species." Full grammatical language innovated on earlier modes of communication. Insofar as hominid societies had been cohesive units that survived through cooperative efforts of foraging and defense, commissive speech acts would have reinforced the mutual trust necessary to coordinate these efforts.

It is not enough, therefore, to analyze communicative signs simply from the perspective of the sender: they must be understood as social functions that also require the response of sign receivers. The dialogic nature of discourse is especially apparent in interrogatives, which are therefore difficult to assign to a single speech-act category. Searle (1983) termed them directives because they are requests for information, but he quickly found he needed to rethink that assignment. For one thing, perlocutionary effects of questions take on the character of any of the five speech-act types. "How do you feel?"–"I feel apprehensive" (an expressive answer). "Have you reached a verdict?"–"We find the defendant innocent" (a declarative answer). "Shall I run for office next year?"–"No, don't" (a directive answer). "Do you promise to come with me?"–"Yes, I promise" (a commissive answer). "Where did you grow up?"–"I grew up in a small town in Appalachia" (an assertive answer.) If one agrees with Mikhail Bakhtin (1986, 145, 168) that speech is essentially dialogic, one must conclude that every utterance is meaningful only as an answer to an implicit question and that this answer itself potentially generates further questions.

In conversational discourse, responses tend to be normative reactions. As the sender's utterances must be relevant to real-world contexts, the receiver's reactions (actions, utterances) need to be relevant to the sender's message (Sperber and Wilson 1986, Grice 1989). This would equally be the case whether the signs employed were indices, icons, or symbols. Thus *expressives* usually elicit sympathetic responses because the sender is likely to reveal feelings or attitudes in the presence of persons expected to share similar sentiments; *directives* would be addressed to persons likely to comply with particular commands, instructions, or requests; and *assertives* would be addressed to persons interested in the topic and inclined to believe the sender. Of course, their reactions would not always be predictable, but sign senders who choose their addressees usually pick those that share their own perspectives and goals and expect them to react cooperatively to the message.

This is also the case when *declaratives* and *commissives* are paired. For example, couples choosing a person to officiate at their marriage tend to select one that shares their beliefs and speaks institutionally efficacious words. The officiator, too, requires that the vow takers submit to the transformative power of his or her declaratives. Their promise of compliance to the marriage contract takes the form of commissives that are made in the presence of the community and of the god(s) that sanction(s) their vows, which are primarily promises made to each other. Their commissives thus prompt the officiator's declaratives.

Declaratives and commissives are utterances that normally involve more persons than simply the speaker and the hearer, for besides the *I* and the *You* there is the *We*, physically or virtually present. The sociospatial extension that these two speech acts provide is matched by their temporal extension. Neither act is limited to the present moment or to the present participants. A declarative, be it a naming, a declaration of war, a prison sentence, or a consecration, marks the beginning of a line of future events. A commissive, be it a religious vow, a New Year's resolution, a solemn oath, or a swearing of vengeance, is similarly projected into the future. Both speech acts can also be past acts that may be looked back upon as defining the meaning of subsequent events. For these reasons declaratives and commissives are essential connective features of narrative. Consider the function of these two speech acts in the Homeric epics and the Hebrew Bible: the commissives of Achilles to absent himself from battle and of Odysseus to return to his kingdom and his queen, the biblical declaratives that begin with Creation and occur whenever a covenant is made between God and humans, and the biblical commissives when God promises punishments and rewards and when Israel pledges itself to upholding the laws of Moses. Narratives that demonstrate the efficacy over time of these two speech acts link human events as meaningful consequences of these verbal events.

Like commissives, declaratives also presuppose a social structure of traditions, hierarchical authority, property, customary laws, and public contracts, but both commissives and declaratives have elements that may also derive from much earlier primate behaviors. Because declaratives are said to conform to the positive template "*This utterance does what it means*," declaratives apparently cannot take the negative, though every other speech act can (Haverkate 1990).[11] "I hereby do NOT proclaim you king" would, I admit, be an anomalous speech act, not to mention a rather awkward moment for all involved, but there is one context in which declaratives may take that construction. That context is social play, and the signifier that inserts the negative into the declarative formula is what Gregory Bateson (1972) termed a *metacommunication*.

In January 1952, Bateson visited Fleishhacker Zoo in San Francisco. Having examined the rough-and-tumble play in which two monkeys performed aggressive actions, apparently with no intent to cause the effects normally denoted by this behavior, Bateson concluded that both participants must have somehow first exchanged a metacommunication that, in effect, inscribed a "play frame" around their interactive episode. As I analyzed this (Collins 2013), the indices of aggression had been framed by a shared understanding that these signs were not indices at all but mere *icons* of aggression, i.e., a mimesis of the combat these actions normally represent. If put into words, this metacommunication would be something like this: "I do NOT declare you my mortal enemy," that is, "Everything I do next is NOT what it seems." The verbal spell remains in force only as long as the metacommunication is mutually agreed to. This ur-grammatical function of declaratives, I would argue, is the semiotic tactic that lies at the root of human play and fictionality, without which role testing and performative imagination would be impossible.[12]

Speaking to the Mind's Eye

Besides their use in communicating thoughts, gesturing and speaking have another thing in common: both are motor behaviors. As Ulric Neisser phrased it, "To speak is to make finely controlled movements in parts of your body, with the result that information about these movements is broadcast to the environment. For this reason the movements of speech are sometimes called *articulatory gestures.* A person who perceives speech, then, is picking up information about a certain class of real, physical, tangible . . . events" (1976, 156; my italics). While this is an important insight into speech production, it is also important to recognize that what the person perceiving speech is *not* perceiving are precisely those articulatory gestures. Those minute muscular movements responsible for producing phonemic differences are for the most part hidden from sight in the speaker's throat and mouth—they are heard but not seen. In gauging the impact of language on human communication, we need to consider the ways our visual brain has adapted to this new nonvisual channel of information.

As a system of acoustically delivered symbolic signs, language has subsumed the functions of both indices and icons, hitherto largely conveyed through visible gesture, and, by doing so, it has radically innovated those older modes of communicating. If two of us are facing the same array of objects, I can simply tell you to look at some detail without having to point to it. Using assertives, we can now also engage in gossip in the visual presence of the "gossipee" by resorting to "conspiratorial whispering" (Knight 1998). We can even transcend the present. Indexical gestures, e.g., joint gaze, pointing, and other ways of entraining mimetic routines, were and still are confined to the perceptual here and now. Articulatory gestures, however, being mouth sounds only arbitrarily related to things and actions, demonstrated new and special powers. By taking over the job of indexical signs, assertive speech acts represented, for the mind to imagine, mental displays within which particular details could be pinpointed and shown to be either spatially parallel to other details or temporally sequential in a before/after or cause/effect relation. When strung together by a rule-governed syntax, they gave humans the option to take leave of the perceived present and time travel into the remembered past and the projected future or to space travel into distant, hypothetical, or even counterfactual realms.

The grammatical features that we use to share our experiences conform to most of the procedures of visual perception. That is, the ways by which we perceive the world are partly replicated in the ways we imagine the world, and the ways we imagine it are partly replicated in the structures of language. Nouns, for example, represent figures, i.e., bounded objects separable from the ground that surrounds them, and because words can be transmitted much faster than iconic gestures, our working memory can retain a series of nouns long enough to form a ground in which to place each new noun as a potentially figural image (Hurford 2007, 95). Nouns also correspond to fixations, those momentary actions when the two eyes focus on the single object before a saccadic eye movement shifts their attention to another point in the visual field. If, for example, I should tell you about a walk I took one winter day into a wood of tall straight trees, I would first evoke in you a generalized series of repeated forms that would give your mind's eye no particular place on which to fixate. But if I then asserted that I saw a small bird that kept flying and perching just ahead of me on my way, I would give you something to focus on, albeit an erratically shifting fixation point, a horizontally moving figure crisscrossing an already established ground of motionless vertical trunks. If I then said this bird flew behind a woodpile and *then* proceeded to describe this perfectly crafted four-foot-by-eight-foot cord of maple logs, I would finally provide you with a stable figure to visualize, in this instance a human artifact of horizontal logs standing out from a surrounding ground of upright trees and flitting bird.[13]

Consider, too, how linguistic features correspond to other eye movements. In the sentence "the bird is behind the woodpile," the mind's eye is asked first to form the image of a bird, then shift to an image of a woodpile, and make that shift as instructed by the preposition "behind." Up until the bird vanishes from view, it retreats from the observer, who must simulate the "vergent" movement that the two eyes perform when adjusting to changes in depth. The behaviorist psychologist B. R. Bugelski (1971), in critiquing attempts to correlate language with visual perception, claimed that prepositions have no image content—one cannot form a mental image of an "across" or an "under." Yet this fact actually confirms the vision-language correlation, for a saccade is an action that presents no visual content thanks to a mechanism termed *saccadic suppression*. When locative prepositions prompt us to simulate saccadic movement, it makes perfect sense that we cannot visualize anything during that shift from one verbal image to another. On the other hand, to visualize the flight of the bird from *in front of* to *behind* the woodpile we need to simulate an additional eye movement, "smooth pursuit," which keeps a visual object in view while presenting its background as a blur (Collins 1991a, 103–115).

Intransitive action verbs, such as *fly, jump, run, limp, crawl, sidle, slide,* and *swing,* in English termed "manner verbs," do seem designed to evoke vergence and pursuit eye movements. Accordingly, if the context governing the verb indicates that we are observing the action of someone or something from a removed vantage point, we simulate visual images of this action. But suppose the context invites us to empathize with the agent. Now we may also simulate these actions through locomotor, as well as oculomotor, responses and feel ourselves physically engaged in the events described. In the first case, we would draw upon image schemas to create suitable mental pictures of the action; in the second, we would draw upon motor schemas, to the extent perhaps of tensing the same muscles we ourselves would use to execute these actions. Then again, if the verbal cues were compelling enough, we might, alternately or in parallel, view this other from the outside *and* from the inside of the moving person or object.

Brian MacWhinney (2005) has referred to our visual simulation of a sentence as the "descriptive mode" and our motor simulation as the "enactive mode." In the descriptive mode we locate imagined objects in an *allocentric* frame of reference, one in which we visualize them disposed before us in a particular spatial configuration, each having size constancy and measurable distance from other objects. In the enactive mode, we project an *egocentric* frame of reference, one in which we visualize objects in spatial relation to ourselves, objects that seem to grow larger as we approach them, diminish as we retreat, block our passage, move toward us or away, etc. (Collins 2013).

In MacWhinney's view, this imaginative shifting, or "perspective taking," has significant implications for an evolutionary view of language. He makes the strong claim that perspective shifting was not a byproduct of grammar but that it was longstanding selective pressures to shift perspective that determined the emergence of grammar. This is not a trivial chicken-and-egg conundrum. If language was "a new machine built out of old parts" (Bates and MacWhinney 1988, 147; MacWhinney 2009, 417–418), then humans recognized this need long before communication became compositional. When, for example, I point in some direction, I want you to engage in perspective taking, i.e., to note my perspective and then make yours as similar to mine as possible. Theory of mind and pointing would therefore be two prelinguistic *innovanda* that provided the old parts for the new perspective-sharing machine of grammatical language.[14]

With the advent of symbolic communication and third-person referencing, it now became possible to ascribe theory of mind to persons displaced in time and space. One could now say to someone about others, "He thought she was telling the truth, but he was wrong," or "She watched as the sun rose over the mountain," thus animating verbal human representations with the same consciousness demonstrated by physically present speech participants. When, a moment ago, I told you about my walk in the woods (here we'll assume that I orally narrated this anecdote), my words did the work of iconic signs without depending on visually conveyed iconic signing. My articulation of linguistic symbols was intended to prompt you to take my perspective and produce a corresponding string of mental images, icons that more or less resembled the images I was giving voice to. Of course, I could, and probably would, have accompanied my spoken words with iconic gestures, my forearms moving up and down to represent the trees of the forest, my hand demonstrating the flight of the bird, and both hands shaping the dimensions of the woodpile (McNeill 1992). If I did, it would have been to emphasize the mental images I was trying to convey and perhaps impress you with my earnestness. Yet watching me and these visible gestures of mine would tend to distract you from the mental images I wanted you to form by yourself—from your imagining that *you* were seeing what I was imagining *I* was seeing.

As the essential imaginative function of grammar, perspective taking profoundly transformed human communication and still remains at the heart of oral discourse. Its potential, however, was not fully realized until the advent of written texts, for oral discourse continued to constrain it by the visual presence of speakers. When language is spoken, we have *auditory* input running parallel with *visual* input (gestures, facial expressions, body language, etc.). While these two perceptual operations occupy the foreground, verbally evoked mental images must fend for themselves. During oral performances, hearers could learn to associate themselves imaginatively with the visual perspective of the storyteller: to see through his eyes, or, rather, through his verbalized mind's eye. Hearers could likewise venture into the mind of imagined speakers and visualize what these characters were seeing. Yet while they exercised their imagination, hearers also remained connected to the visually perceived source of the words they heard and the place in which they were uttered. In the chapters that follow I will trace the stages by which the visual imagination gradually untethered itself from the physical presence of the spoken word.

Summary

In this first chapter I began with a discussion of innovation, a concept broad enough to apply to successful changes in the course of both biological and cultural evolution. Whenever these innovations occur, they alter the way the human subject operates in the object world. One may regard technological improvements, from better mousetraps to better computers, as innovations, but, as the title of this chapter indicates, my focus here has been on "innovating ourselves." By that phrase I mean the ways our remote ancestors' brains gradually changed in response to selective pressures—grew larger, formed new neural networks and larger, more complex social networks, developed new means of communicating, and launched the collective enterprise of cultural evolution.

In exploring this transformative process, I have pointed out that when the human brain forms new pathways and networks and thereby becomes capable of new functions, it builds upon structures already in place. It is important to remember that biological evolution is a very long-term process and that while a species can go extinct and structures within an extant species may become modified over time, these older neural structures rarely if ever become obsolete. As newer structures and functions become more prominent, the older architectures, heretofore essential to survival, become deployed for other, subsidiary tasks. We therefore still find within us the hard-won legacy of many million years of hominid evolution.

The course of human evolution, as I have outlined it, is marked first by increasingly accurate ways of interpreting natural signs and then by the creation of sign systems by which to communicate with one another. Semiotics, with its basic distinctions of indexical, iconic, and symbolic signs, provides us with a helpful framework, indeed a logical sequence of communicative modes. Accordingly, as gestural icons became simplified, they would have eventually approached the status of symbols, i.e., arbitrary signifiers only distantly resembling their signifieds, a process of conventionalization that prepared humans to use nonvisual symbols, namely, syllabic sounds, to communicate thoughts and intentions.

Within this sequence, we can also recognize the distinct properties of the two alternative sensory channels over which semiotic information may be transmitted: a gestural channel perceived by vision and a vocal channel perceived by hearing. Both channels may be open and operative at the same time, but one will be dominant while the other is recessive.

The needs that seem to drive the evolution of sign sending appear to be prosocial—the need to share resources, bond with others, obtain their protection, cooperate with them, and regulate the behavior of *un*cooperative individuals. The basic discourse model of first-person sender and second-person receiver, ever shifting as the role of sender passes from one to another conversant, suggests that communication has always been about social negotiation. In my discussion of Austin's and Searle's theory, I pointed out that speech acts, too, could be arranged in an evolutionary sequence that reached completion with the appearance of third-person reference and assertive speech acts. When humans reached this point, they finally possessed the means to share their thoughts of entities not immediately perceivable— objects and events projected into a past, a future, an elsewhere in the present moment, or into concepts that can exist only in the mind. With a vocabulary of third-person reference and a grammar that could organize those words, they had now arrived at the stage at which they could send signs that prompted visual images to appear in one another's mind—images of places and persons who were not physically present. With that the play of imagination had commenced.

Two

Narrative Memory

The ability to transmit symbolic signs through gestures and, later, vocal sounds eventually evolved into fully grammatical language. But as the language-ready brain was millions of years in the making, so it took perhaps as long as 200,000 years for oral cultures in separate locations, e.g., Mesopotamia, the Nile and Indus valleys, northern China, and Mesoamerica, to have the institutions in place to support the production and storage of written records. It then took millennia more before writing expanded from administrative recordkeeping to historiography, philosophy, natural science, and what we now generally mean by "literature." All the while, writing remained a resource of privileged elites. Not until recent centuries and in industrialized societies was general literacy a goal governmentally promoted.

In this chapter, as I continue to explore the forms of sign reading that preceded writing, I will focus on units of language larger than conversation and simple speech acts. I will begin by examining informal storytelling as the conversion into speech of personally experienced events in the form of sequences of visual images. These are the images we visualize whenever we tell ourselves or others *what, where,* and *when* significant incidents have touched, moved, and changed us. The structures of formal narrative, namely, its linear sequence, centralized personal perspective, emotional power, fictivity, and treatment of time, closely reflect, as we shall see, the peculiarities of the *episodic* memory system.

There is nothing automatic, however, about narrative recall. Humans have always had to rely on mnemonic techniques that exploit the properties of language—its capacity to associate the names of places and persons and then use these to organize and retrieve episodic sequences. We still depend on proper nouns to tell stories from our own past. When we stop in midnarration, it often is to ask ourselves, "Where was I that spring?" or "What was that family's name?" Our listeners might be satisfied with any place name or personal name we choose, but it is we who need the mnemonic support that correct names provide. In a preliterate oral culture of storytellers, memory for names would be an absolute necessity. In such cultures this was also a matter of personal concern, for, unless one "made a name for oneself," one could not expect to be remembered by later generations.

Episodic Memory

Culture is, in large measure, a language-encoded long-term storehouse of information. This human archive is built, however, upon older systems, those structures of the brain that support long-term memory systems, namely, procedural, semantic, and episodic memory. Each has its own set of functions: *procedural* memory, to learn and execute motor skills; *semantic* memory, to learn and access general knowledge, beliefs, facts, and conventional signs such as language itself; and *episodic* memory, to store and retrieve events personally experienced in specific times and places. Of these three memory systems only the last appears to be unique to humans and must therefore be a relatively recent adaptation of the primate brain (Tulving 1983).

Each system has played a crucial role in the evolution of language. Procedural memory, essential to gestural communication, had to be further fine-tuned to make phonemic speech possible. Semantic memory, which mammals and birds employ in identifying food and sources of danger and in navigating their environment, was recruited by humans when they developed symbolic, i.e., conventional and arbitrary, systems of signs. The final memory system to evolve, episodic memory, with its temporal and spatial identifiers, prepared the scaffolding upon which sentences might be organized into coherent, storable, and retrievable discourse.

If, as I proposed in the preceding chapter, a protolanguage of symbols, gestural and spoken, preceded syntax, and if semantic memory preceded episodic memory, an evolutionary link between these two pairs suggests itself: lexicon with semantic memory, syntax with episodic memory. If we regard semantic memory as now our "mental thesaurus" (Tulving 1972, 386), it would seem that its archive of image schemas and types formed the basis for symbolic signs, be they gestural emblems or spoken words. It seems plausible then to regard the episodic-memory system as the structure underlying syntax. At any rate, this is what the linguist Talmy Givón has proposed:

> The rise of lexical-phonology gave an enormous boost to the mental organization of *homo sapiens'* semantic memory. Likewise, there is no doubt that the rise of grammar gave an enormous boost to the episodic representation of states and events, and especially their coherent concatenations, in the human brain. But the basic neural-cognitive capacities were already in place before our ancestors climbed out of the trees.
>
> (2009, 307)

If grammatical language recruited the neural capacities of episodic memory to form "representations of states and events," then narrative would be a likely format within which to find structures and functions patterned upon those of episodic memory. To test this hypothesis, I will now consider ten features shared by episodic memory and narrative: (1) linearity, (2) perspective, (3) emotion, (4) confabulation, (5) time, (6) place, (7) person, (8) performance, (9) iconism, and (10) writing.

Linearity

While we are living them, episodes can be quite complex affairs. Occurring in particular moments and in particular settings, they may entail multiple modes of perceptual input as well as speech elements, vocal tone, gestures, joint gaze, and pointing. Such episodes may also involve multiple persons engaged in different, overlapping activities, some of which we may process as parallel, some as serial, information, at empathetic moments shifting our perspective from ourselves to that of others and then back again as circumstances dictate.

A recollected episode, on the other hand, is a sequence of images that represents an incident as a discrete unit with a beginning, a middle, and an end (Ezzyat and Davachi 2011). The original incident may have had multiple parallel details and a rich, three-dimensional complexity that normal episodic retrieval is incapable of representing, but a remembered event, dim and narrow as it now seems, demands the full attention of the retriever to isolate it as a kind of figure against a ground.[1] What kind of a figure is a remembered episode? Insofar as its seriality is its salient feature, it is a line that struggles to define itself as it proceeds forward through a finite series of points until it finally fades and vanishes. But what, then, is its ground? What is the ground of a one-dimensional figure, that series of points that constitutes a line? Since a line cannot be surrounded, its ground must be its *before* and its *after,* those fade-away extensions at either end. Were it not grounded in this fringe of forgetfulness, no episodic unit could be remembered.[2]

That oral narrative shares linearity with episodic memory is readily apparent, for, as every phoneme, syllable, word, or sentence is replaced by other sounds, the narrow band of meaning extends forward until it reaches its end. If it is preserved as what I call a "verbal artifact" (Collins 2013), it is because it is retold by others and perhaps passed down through subsequent generations. If so, it will then be understood not in the context of only one rememberer's lived experiences, his or her store of episodic units, but in the context of other collectively preserved narratives and in the even larger context of that people's culture—its institutions, ethics, folkways, etc.

Perspective

A recollected episode, as a pared-down replica of what we observed while that event took place, is a first-person replay that we project on the screen of our mental cinema. Yet some researchers have taken exception to this assumption, for, while it is true that most recent experiences are retrieved as though occurring within one's external visual field, more remote episodes sometimes include an image of oneself within that scene (Nigro and Neisser 1983; Schacter 1996; Holmes, Waters, and Rajaram 1998; Rice and Rubin 2009). These researchers conclude that episodic-memory retrieval may take a first-person or a third-person perspective or a mix of both.[3]

The perspective taking that we practice when we remember our own past now changes when we convert those linear visualizations into grammatical language and organize them into a narrative. First persons do not readily turn into third persons in traditional narratives, nor are the actual storyteller or audience members persons in the story. When these participants take the perspective of others, as when the narrator in voice and gesture acts out the role of a character or when hearers identify strongly with a certain character, this is a free choice and a temporary perspective taking. Whether heard or read, a narrative allows us to move alternately into and away from characters: we take on their third-person perspective in all its visual, motor, and emotional details and make it our own first-person perspective, then distance ourselves and view them again wholly as third-person figures.

Emotion

By the time Endel Tulving completed his groundbreaking study, *The Elements of Episodic Memory* (1983), he had devoted a mere two paragraphs to the connection between emotions and episodic memory. Though his avoidance of the topic was typical of what we now look back upon as the *dis*embodied stage of cognitive psychology, he did then predict that this connection would prove fertile for future research. Future research, his own included, did indeed establish that episodes with heightened emotional content tend to be retained longer because, it is believed, experiences that arouse emotions are encoded in more detail and, if consequential, are linked to other long-term memories.

The main function of emotion is to mobilize the various systems of the body, i.e., the respiratory, circulatory, reproductive, and skeletomuscular, to confront suddenly emergent circumstances. Anger, for example, prepares one to fight; fear prepares one to flee, but these emotions are not quite the same as the reflexes they set into motion (Ben-Ze'ev 2000). Emotions make one aware that one is preparing to do something and give one a brief time to reconsider one's options (Shearer 2001). Since "negative" emotions such as anxiety, fear, and anger are often occasioned by perceptions of danger, they become associated with long-lasting episodic memories to a much greater degree than "positive" emotions such as pleasure and contentment. This is because remembering how dangerous circumstances have been faced and overcome has a very real survival value (Allen 2008).

For these reasons, the retrieval of episodic memories is typically accompanied by affective states that may sometimes be emotionally overwhelming, though they are usually milder experiences—subtle mixes of feelings and of even subtler moods. Hearing or reading narratives or viewing dramas also evoke a range of emotions. Aristotle recognized this as an essential factor in fictive performance when he wrote that "tragedy is an imitation of an action. . . . [that] through pity and fear effects the purgation of these emotions" (*Poetics*, 1449b27–28, Butcher's translation). What he meant by the *katharsis* of these *pathêmata* (negative emotions) has been debated for thousands of years. Did he interpret the purging of these emotions in terms of medical science, ethics, religion, or some other fourth-century system of belief? We can, though, safely assume that Aristotle attributed these emotions to the audience and that he thought their purging, whatever that meant, had a beneficial effect. Later in the treatise he does give us some clues concerning those two emotions that tragedy evokes. Neither pity nor fear could be aroused for a main character who was an out-and-out villain, "for pity is aroused by unmerited misfortune, fear by the misfortune of a man like ourselves" (1453a5–6).[4]

Having cited this passage, Harold Skulsky (1958, 156–157) proposed that these emotions were "hypothetical." Learning to assume this "as-if" stance seems difficult enough. "What is more difficult [is that] this pity, in contemplating the misfortune *in another* and this fear, in experiencing misfortune *as another*, though mutually exclusive, must be felt at the *same time*. Thus the emotions are not only actuated histrionically (*dia mimêsiôs*), but experienced by an act of the imagination which can only be termed histrionic." In other words, the emotional state of the role the actor is performing is accompanied by the audience members' performance of pretending to view and at the same time *to be* that character.

This notion that enacted narrative can create a double perspective, that persons of the *theatron*, the viewing place, can remain in their seats while simultaneously being drawn onto the stage as actors, is an insight with broad implications for our understanding of theory of mind, empathy, mind reading, and poetics. As a gloss to Skulsky's insight, I will quote a passage by Henry David Thoreau, from "Solitude," chapter 5 of *Walden*:

> With thinking we may be beside ourselves in a sane sense. By a conscious effort of the mind we can stand aloof from actions and their consequences; and all things, good and bad, go by us like a torrent. We are not wholly involved in Nature. I may be either the driftwood in the stream, or Indra in the sky looking down on it. I may be affected by a theatrical exhibition; on the other hand, I may not be affected by an actual event which appears to concern me much more. I only know myself as a human entity; the scene, so to speak, of thoughts and affections; and am sensible of a certain doubleness by which I can stand as remote from myself as from another. However intense my experience, I am conscious of the presence and criticism of a part of me, which, as it were, is not a part of me, but spectator, sharing no experience, but taking note of it, and that is no more I than it is you. When the play, it may be the tragedy, of life is over, the spectator goes his way. It was a kind of fiction, a work of the imagination only, so far as he was concerned. This doubleness may easily make us poor neighbors and friends sometimes.
>
> (Thoreau [1854] 2004, 134–135)

Thoreau accepts the concept that theatrical experience involves the doubleness of first- to third-person perspective shifting, but he takes this mental artifice out of the theater to show how it functions in the everyday life of the mind. As we have seen, narrative, whether told or acted out, does indeed exploit the everyday life of the mind by recruiting the structures of episodic memory to re-present social interactions and perspectives.

Confabulation

The more emotionally arousing the experience, the more likely its details will be stored in long-term, episodic memory, but this relatively vivid archive is nonetheless fragile and subject to slow but inevitable decay. While it is happening, an ongoing episode has its own identity and is still clearly contextualized by the circumstances that preceded it, but it is rarely recalled in its simultaneous details and full emotional force. If we wait a few days and then try to access that episode, we find not only that its before and after have faded but that the central event now shows gaps. This is why Tulving (1983) stressed the effortful character of episodic-memory retrieval. Like museum conservators faced with the task of restoring some ancient, newly excavated mural, we must pore over the data and mentally plaster and repaint the event using semantic memory, our knowledge of the world, to represent to us how events *must have happened*.

Once semantic memory fills in these gaps, we have completed the initial stage of confabulation, the construction of what is inevitably to some extent a false memory. But there is more to come. In retrospection, as I have pointed out, we also transform the lived episode into a linear sequence of actions and then, if we choose to tell others this story by transposing this image sequence into the medium of language, we process it even further. What in our private recollection was only a tendency to sequentialize is now a rule firmly enforced. Wish as we may to convey the multiple simultaneous actions of multiple agents, our success is limited both by our medium and by the capacity of our hearers to convert word-by-word input into an action scenario. We may say, "While X was doing A, Y was doing B," but, if we want to add another level of complexity and tell how halfway through Y's action, Z began doing C, we and our hearers will find ourselves juggling too many balls in the air at once.

Time

Episodic recall requires some temporal marker. Linearly connected as they are, the contents of episodic memory require before-and-after markers, without which one's autobiographical self collapses into chaos. Temporal markers are also obviously essential to every narrative. As Paul Ricoeur put it: "Indeed, I take temporality to be that structure of existence that reaches language in narrativity and narrativity to be the language structure that has temporality as its ultimate referent. Their relationship is therefore reciprocal" (1980, 162). Because of its importance, I will discuss the mnemonic of time at greater length in the next section of this chapter.

Place

Everything that has ever happened to us happened to us somewhere. Place, together with time, is therefore an essential feature of every episodic memory (Tulving 1983). Just as temporal sequence provides the medium for action and change, three-dimensional space provides the medium for multiple, simultaneous actors and events. Place, as setting, is also an indispensable feature of every narrative and, as such, requires further discussion in this chapter.

Person

In episodic memory, the dominant perspective is that of the self of the rememberer. In narrative, however, this centrality is often shifted to a third-person character. In mnemonic terms, this character provides the vital link between moments of time and, as a traveler through space, between separate places. Person is a topic I expand upon in the final section of this chapter.

Performance

In episodic memory this corresponds to the retrieval of a stored event. Every time we rehearse a past experience, we strengthen its trace, though we may also alter it through confabulation. In narrative, we reinforce our hearers' and our own recollection of a story by retelling it, thereby keeping it in circulation. In an oral society, successful performance of a person's actions in time and place requires a complex set of skills that we may categorize as "performance style." This mnemonic is analogous to sexual reproduction: unless a collectively stored narrative is regularly reperformed, it cannot survive. (I will examine this mnemonic in chapter 3.)

Iconism

This is the visual display of places and persons such as we find in sacred architecture and the depictions of divinities and ancestors. These displays serve to evoke narratives that explain significant events of the past that are thus represented.

Writing

This too is visual display, but not of iconic signs. Even though a script may have originated in stylized visual representations, over time its use as the representations of speech sounds induces it to become a code of arbitrary signs, i.e., symbols. In short, writing becomes a system of visual symbols representing a system of auditory symbols. (I will examine both iconism and writing as visual instruments of memory in chapter 4.)

At this point I have concluded my brief introductory discussion of the ten features of narrative memory. The first four, common to episodic memory and to narrative, I will reexamine in various contexts in the succeeding chapters. The second six, as *mnemonic* features, I will now, as promised, begin to present in more detail.

The Mnemonics of Time

From time to time, we mental time travelers need to realize that only one time really exists. Not only is there no time like the present—there is no time *but* the present. This present is not, however, experienced as a dimensionless instant. As William James estimated, in duration it can last from a few seconds to almost a minute, within which there is both retrospection and prospection. As Edmund Husserl (1991) put it, the mind is a tensile thing that in the present moment stretches backward in retention and forward in protention. That very word "moment" derives from the Latin word *momentum*, which means "movement," in this case the perpetual movement of consciousness from the partly past to the almost future. It is only when we contemplate the completely past and the completely future that we seem to take leave of the present and its contingent circumstances. But this too is illusory, for when we remember or anticipate some wholly nonpresent action, the only action our mind performs is the present action of remembering or anticipating. When, for example, I tell you what once happened to me when I was twelve, the events told are long past, but the telling is always a present act of the imagination.

As the fifth trait that narrative shares with episodic memory, temporality comprises a number of interesting phenomena, first and foremost the variable duration of remembered units of time. In simple terms, a particular episodic memory is a time- and place-specific unit of remembered experience, e.g., as characterized in the sentence "We had a good time last Saturday in the park, didn't we?" An online episodic event is one in which we are currently engaged in some action that requires us to keep in mind a continuum stretching from the recent past to some point of completion in the near future, e.g., "This is proving to be an interesting conversation." The grammatical tenses help out, too: a remembered experience usually takes the simple past tense, and an ongoing experience takes the present progressive.

Henri Bergson ([1910] 1960) regarded the latter tense as "pure duration,"

> the form which the succession of our conscious states assumes when our ego lets itself live, when it refrains from separating its present state from its former states. . . . [It] forms both the past and the present states into an organic whole, as happens when we recall the notes of a tune, melting, so to speak, into one another. Might it not be said that, even if those notes succeed one another, yet we perceive them in one another, and that their totality may be compared with a living being whose parts, although distinct, permeate one another just because they are so closely connected?
>
> (100)

The musical analogy reminds one of William James's ([1890] 1950) "stream of thought." As an extended episodic present, it also corresponds closely to what Merlin Donald (2007) has called the "intermediate-term memory," a lengthened mode of working memory (Ericsson and Kintsch 1995, Baddeley 2000).

These two temporal phenomena, grammatically expressed in the past tense and the present progressive, have led psychologists to posit two distinct kinds of duration judgments, retrospective judgment (how long a remembered episode seems when we retrieve it) and prospective judgment (how swift a current episode seems as we are experiencing it in real time). Of course, the two are related in origin: every remembered episode has to have once been a current episode, its status in long-term episodic-memory storage largely determined by the amount of attention and emotion it aroused when it actually happened. But when what was once a prospective experience is converted into an episode retrievable by retrospection, its estimated duration undergoes some radical transformations.

To cite just one example, a visit to an unfamiliar city will be experienced as fast moving, full of detail, and therefore of long duration while we are prospectively engaged in it. Yet, when we return home, it often seems to have shrunk in remembered length and, for quick retrieval, is tagged simply as "my trip to San Francisco" or "my visit to Copenhagen" unless, of course, we supplement episodic memory with souvenirs and photos.

In order to measure durational distortions, psychologists usually compare two "times," one characterized as "objective," the other as "subjective." Objective time is regarded as a constant, the tick-tick-tick of external chronometric time and the body's internal, somewhat less accurate, timing mechanisms, the various rhythms of its biological clock.[5] Subjective time, in both prospection and retrospection, is variable and is made so by those other variables—*attention*, the greater or smaller cognitive load occasioned by an online event (Brown 1985); *complexity*, the number of significant details involved in an event (Ornstein 1969, Flaherty 1999); and *storage size*, the amount of "space" required to archive a retrievable event in long-term memory (Ornstein 1969).

"Objective" and "subjective" are admittedly broad, generalizing terms, but they do capture the overall difference between those two "times." We do not have to specify some absolute units proceeding at an absolute rate to agree that "objective" time is time as no single consciousness measures it. It is unaffected by the interests and emotions of any subject, whereas "subjective" time is shaped by the inner states that a subject experiences. These contrasting times have much in common with those two ways we store our knowledge of the world, namely, in semantic and episodic memory. The latter distinction has led to the recognition that "remembering" and "knowing" both enter into our estimates of retrospective duration. Of remembering, William James wrote: "Retrospect depends obviously on the multitudinousness of the memories which the time affords. Many objects, events, changes, many subdivisions, immediately widen the view as we look back. Emptiness, monotony, familiarity, make it shrivel up" (James [1890] 1950, 624). "Knowing" tells us that the time we stood waiting for our seat in the restaurant was more like ten minutes than the half-hour it felt like and that the time it took to ski down our first slope was more like one minute than a quarter-hour. In short, we learn to manage both kinds of time tolerably well.

We live our lives looking forward, but, since the future never reveals itself until it becomes the present, this prospective experience is one of ever striving toward an unseen goal. On the other hand, when we look backward at our life and choose to retrieve some portion of it, we locate it on a linear timeline after and before other filed-away events. The episodic memories we store on this graduated sequence are organized in what we regard as objective time. We *know* this past of ours in the same detached way we scan our CV, but, when we *remember,* when we "go back" and "relive" a given episode, our experience suddenly loses its cool, retrospective distance. It now resumes the character of prospective action with all the willful, risky swirls and eddies of living experience.

Before we enter into an act of recollecting any past duration, we must shift into what Tulving (1983) called "retrieval mode." This prepares us to engage in the effortful act of composing a narrative for ourselves, a string of images of persons, places, and actions, organized by speech or by speechlike linkages of *before, then, after, while,* etc. If we observe ourselves in this process of self-narration, we become aware that some intervals of time are more information packed than others and seem to swell in retrospect, whereas other intervals, scant of detail, serve as summary introductions, conclusions, or logical bridges between one major episode and another. When we encounter gaps within our narrative, we shift from remembering to knowing and do so by consulting our semantic-memory store, our general knowledge of the world. In this process we may even make things up—confabulate. For example: When I went from that scene in the hotel lobby to that scene at my friend's house, I must have taken some form of transportation—was it subway, bus, taxi? I can't recall, so I'll suppose it was a taxi and insert some brief imagery of streets seen through a car window (Rajaram 1993).

When we proceed to tell someone what we remember, we reconstruct it even further so that it has optimal coherence and interest for our audience, adding context when necessary and shaping our story so that it makes a point. The result is not identical to the story we retrieved from episodic memory, but its temporal components will be similarly shaped. When logically necessary, we fill gaps, i.e., forgotten intervals, with confabulated transitions and edit its remembered intervals by expanding some and shortening or deleting others. Yet the way we shape time when we narrate to others will still conform to the ways the human brain shapes episodic content in the process of retrieving it.

Narrative Time

Henri Bergson once remarked that, whatever time is, its most useful function is to prevent all things from happening at once.[6] But it was not before grammatical language finally came along to organize symbolic signs that our ancestors could have understood the length and breadth of this dimension and mentally inhabit "other times." With the emergence of full language, statements of fact and belief, assertives, could now incorporate the earlier four speech acts, plus other assertives, and in so doing begin to convey some of the complex personal interactions that had long characterized our intensely social species. A storyteller, for example, might utter the following sentence as an assertive speech act composed of other embedded speech acts: "The chieftain now tells his people [*assertive*] that he feels sorry [*expressive*] that he banished his friend [*declarative*] and, promising to restore him to his former place of honor [*commissive*], orders his men to search for him tomorrow [*directive*] and beg him to return [a second embedded *directive*]." One could also form metacognitive sentences, e.g., "the chieftain knew that his friend felt wronged by him," which allow the speaker and hearers to shift perspective to one mind and from that mind to another's.

Grammar also assigns time slots (tenses) for events. The chieftain now feels sorry about something he did in the past and promises to undo that act in the near future, a future that involves a four-part sequence—first the search, then the implied finding, then the begging, and finally the returning. Thanks to the grammatical capacity of assertion (on the part of a storyteller) to range freely over past, present, and future tenses, long sequences of action may be communicated as interwoven threads of thought, emotion, and action.

Over the past decades, as cognitive psychologists have been studying how and why autobiographical time intervals seem to swell and shrink in retrospect, literary theorists were asking some of the same questions about fictive time. Since the structure of narrative is based on the structure of episodic memory, it should come as no surprise that narratologists and psychologists came to a number of similar conclusions about the consciousness of time. But before we see how convergent their research has been, I will present a diagram that represents how narratologists such as Seymour Chatman (1978), Gérard Genette (1980), and Mieke Bal (1985) have classified various subjective durations in terms of "pacing," or "tempo." Like the psychologists, these narratologists proposed two kinds of time, which they called "story time" and "discourse time." Story time is the objective series of events, the unidirectional chronology. We understand that this time is ticking relentlessly in the background, but we do not find it doing so in the narrative. The poem or novel that we hear or read comes to us as discourse with its temporal form preshaped as discourse time.[7]

Like the remembered contents of episodic memory, events in discourse time have variable speeds and durations. Narrative theory has settled on five distinct paces: (1) *ellipsis*, (2) *summary*, (3) *scene*, (4) *slowdown*, and (5) *pause*. *Ellipsis* is a portion of time we are aware of only as a gap, a period of story time that is not narrated in discourse time, as when the narrative leaps from autumn to spring with no mention of what occurred in winter. *Summary* is an interval of hours, days, or years that is quickly narrated in a sentence or two ("the winter passed uneventfully"). The pace of discourse time conveyed through summary is represented in a verbal passage that is much briefer and quicker than the corresponding interval could have been in chronometric story time. In *scene*, however, the ratio is closer to one to one, as it is in dialogue, for, when direct speech is embedded in the narrative, its duration is neither compressed in time nor expanded. *Slowdown*, sometimes also called *stretch*, occurs when the discourse lengthens as it describes action in slow motion, as it were, an effect that often suggests an intensified focus on significant details. Finally, with *pause*, the flow of story time stops entirely, and a spatial description, a train of associated thoughts, or some other digression is supplied. This word-filled pausal insertion has no necessary reference to the passing of story time and is filled wholly with discourse time, making it the mirror opposite of ellipsis, which has no discourse time but is all blanked-out story time. See table 2.1.

To make these pacing distinctions a bit more concrete, I will connect them to an actual narrative. For this purpose I have adapted book 19 of the *Odyssey* and present it now in outline form:

A. Disguised as a beggar, Odysseus lay in the great hall pondering how best to slaughter the Suitors who had so long besieged his wife. (Summary)
B. A time gap in the narrative. (Ellipsis)
C. Later in the evening, Penelope pitied the stranger and called upon the old servant, Eurykleia, to bathe his feet. (Summary)

TABLE 2.1
Narrative pacing

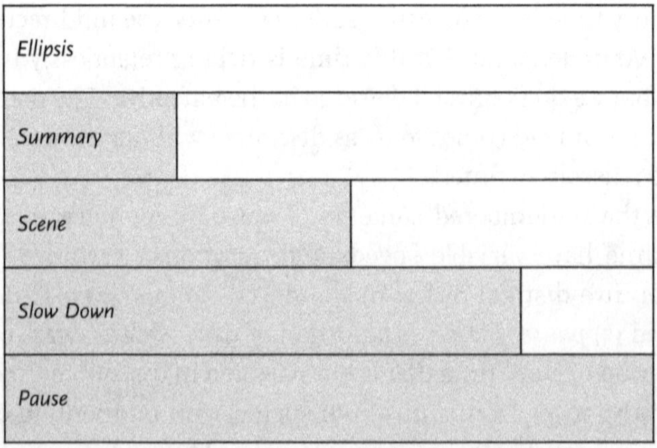

D. "The gods hated my master, the lost king, as they must have hated you," Eurykleia said. "I am glad to wash your feet because you so remind me of Odysseus." "Yes," he answered, "many have noted how closely I resemble that poor lost wanderer." (Scene)

E. Then, as she lifted his leg over the basin to wash it, she recognized a scar on his thigh, the same scar that Odysseus had ever since, as a boy, he was gored by a wild boar's tusk. This happened when he was visiting his grandfather, the crafty old Autolykos, and joined him and his sons on a hunt on Mount Parnassus. The boar charged the boy, who held his ground and speared the beast, but not before it struck him that deep gash. (Pause)

F. Eurykleia let go his leg, and his foot fell down in the basin. The bronze clattered. It tipped back to one side, and its water spilled onto the ground. (Slowdown)

G. Odysseus made her swear to keep his true identity a secret as he prepared to trap the Suitors of his wife and reclaim his kingdom. (Summary)

Figure 2.1 represents the durational structure of this narrative as a left-to-right progression of events. As it indicates, story time and discourse time normally share the same starting and ending points. While the pause (interval E) may appear as a flashback, this expanded moment, whether generated in the mind of Eurykleia, Odysseus, or "Homer," is directly cued

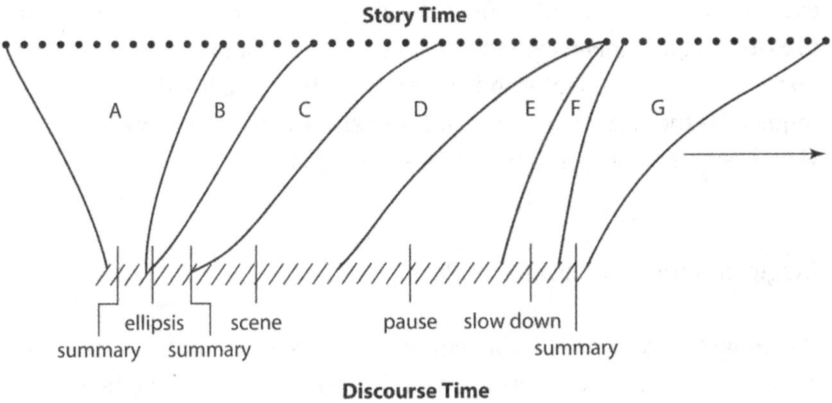

Figure 2.1 Story time and discourse time in *Odyssey* 19: contrasting durations.

by the preceding action and has consequences for the action that follows. Note also that the story time is here marked along the bottom by a dotted top line representing the objective "ticking" of chronometric time but that discourse time, marked by forward slashes, also keeps time—the linear time of phonemes, syllables, and words. It is this latter time that precisely determines the variable subjective durations of narrated intervals.

Composers of successful narratives seem to have learned how to simulate the structures of episodic memory by building durational variation into their tales. When we retell, rehear, or reread such a work, we therefore enter into a kind of retrieval mode, which entails, among other conventions, a willing suspension of disbelief and a readiness to take the perspective of the narrator as principal rememberer. We have not actually participated in or witnessed the events narrated, but we can give them life by treating them in the manner in which we would be retrieving them from our own long-term episodic memory and would relive them. Paradoxically, however, when we actively recollect our own past or when we hear or read a narrative, we are not so much looking back as looking forward. We simulate them as prospection as they emerge in the present progressive flow of time that Bergson called "pure duration."

Though narrative structure is modeled upon the structure of episodic memory, its content is quite different. Remembering the major events in one's own life is one thing, but remembering the events in one's cultural past in narrative form takes special procedures. To access externally stored, communally shared memory, one may have recourse to external *aides-mémoire*. Literate cultures use visual symbols in written texts to store this content, but there are other, much older visual techniques. In the next two sections, I will survey some of these alternative visual resources, starting with indexical signs.

Magic and the Mnemonics of Place

We know that for verbal information to survive in an oral culture it must be retold. This means it has to remain in the memory of the hearer long enough to be communicated to others and seem interesting enough to them to warrant a further retelling. To be stored by others, the content of an oral composition has to be extraordinary enough to compete successfully with other oral texts for a niche in the communal memory. We would suppose that the stranger, more mind-boggling details there are in an oral narration, the more memorable it becomes. But this is not necessarily the case. If it were, our dreams would be eminently memorable. In an oral culture, verbal artifacts cannot survive simply by packing themselves with bizarre and miraculous incidents. They must include counterintuitive details, yes, but they must also reflect the consensual reality of their audience (Boyer and Ramble 2001, Norenzayan et al. 2006). In a collectively confabulating oral culture, narratives that feature gods, ghosts, talking animals, and superhuman heroes have a selective advantage over flatly realistic tales (Guthrie 1980). As paper, ink, and typeface have been elements necessary to the survival of information in a print culture, a judicious seasoning of magical and monstrous goings-on are necessary for a tale to survive in an oral culture.

This raises another question. If the oral message is so dependent on the oral medium, can there ever be a message distinct from this medium, a factual content with truth value that is extricable from the necessary format in which it is conveyed? If the Greek word for truth (*alêtheia*) has any bearing on this discussion, the question of separating truth from its confabulated husk of fable is meaningless in an oral culture, for, in the absence of the critical habits that literacy promotes, the "true" is simply the unforgettable (*a* = not + *lêth-* = forgetting). What we generally refer to as "mythology" is thus the genre of discourse that memory-dependent oral transmission naturally engenders.

When an episodic memory that has been first told and then retold by others achieves some relatively stable form as a traditional tale, it takes on further traits. Oral narrative relies on vivid episodes that reinforce unquestioned, preestablished beliefs, often ethnocentric and xenophobic in their moral judgments. Characters are simple and recognizable, their motives interpreted through action and through the stated judgments of other characters. Narrative thinking tends to equate priority with causality (if event A precedes event B, A is assumed to have caused B) and outcomes with intentionality (if C is the ultimate result of A's acting upon B, the entire sequence had to have been designed by the initiator of A). Finally, narrative thinking has a great tolerance for internal contradiction: being a linear discourse that ties together a sequence of separate episodes unfolding over time, it may present a character doing or saying X, then later on doing or saying something quite opposite (Y). Behavioral inconsistency is usually ignored because of the distance between passages, but, when a contradiction is too glaring to ignore, some rationalizing episode may be inserted to bridge the gap between X and Y, e.g., the character was under a spell, drank a potion, or was misled by a deceiver. Factual, as distinct from behavioral, inconsistencies are of least importance in narrative logic, since facts are treated as interchangeable modules of information, elements of form rather than of substance.

Before we can approach the topic of indexical mnemonics, we need to consider why in the first place have preliterate worldviews incorporated beliefs in magic, talking animals, ghosts, gods, and the like.

One of the great breakthroughs in primate evolution was theory of mind (see chapter 1). Realizing that one shared a world with conscious, intentional others, some helpful, some hostile, one would need to identify them in the environment, especially those that posed mortal danger. Evolutionary psychologists have posited what they call an "agency-detection device," a putative brain module dedicated to mobilizing the body to defend itself against danger on the basis of fragmentary indexical clues, e.g., a rustle in the grass or a moving shadow. More often than not the sound would only be the wind and the shadow that of a tree bough. Still, those endowed with a sensitive agency-detection device were more likely to survive and reproduce than those less alert to these clues (Barrett 2004, Atran and Norenzayan 2004, Meschiari 2009b).

Some cognitive scientists of religion have suggested that this device over time became "hyperactive" and began to read intentional agents in every phenomenon—in clouds, in the sun and moon, in trees and mountains—an anthropomorphic impulse that led directly to the concept of spirits and gods (Barrett and Lanman 2008, Geertz 2011). It was, I might add, also possible that religion evolved as a way to modulate this otherwise paralyzing startle reflex by conceptualizing agents even more powerful than saber-toothed tigers, more lethal than vipers, invisible beings with whom one could form alliances, much as primates had always done with more powerful conspecifics.[8] As the Psalmist says, "the fear of the Lord is the beginning of wisdom," and god-fearing populations generally proved more successful in their pursuits than their edgy, all-fearing neighbors.

Having invented powerful, invisible allies, humans needed to understand these beings' intentions. One could interpret the intentions of predators by such indexical signs as growling, heavy breathing, posture, and gait, but how could one read the mind of unperceivable superhuman beings?

One way to do so was to keep track of unusual weather events (lightning strikes, whirlwinds, floods, and earthquakes) and celestial anomalies (eclipses, comets, and fireball meteors). These events were usually interpreted as indices of the gods' displeasure. There were yet other indexical signs observed in unusual phenomena, odd patterns of bird flight, streaks of color in the entrails of sacrificed animals, and the casting of lots. These *divinatory indices* were classified as "acts of god(s)" that pointed to the plans they had for humans in the immediate present or the near future. These extraordinary events, if interpreted as speech acts, would be *expressives* (displays of anger or resentment), *directives* (commands to perform or refrain from particular acts), *declaratives* (pronouncements of covenantal union or the abandonment of the people to their fate), and *commissives* (promises of benefit or, more likely, threats of punishment).

The one kind of speech act that these sudden, unpredictable divinatory indices could not be identified with is the *assertive,* and for this we must turn to the durable features of the landscape. Boulder-strewn mountain slopes, cliff-walled gorges, and springs of fresh water—were they not also the work of conscious beings of prodigious strength? They must have placed their marks on the earth at some earlier point in time, some originary age, for no one can see such changes to the earth's surface happening now. Long ago these same beings had also placed high in the sky the sun, moon, and constellations, the movements of which they continue to regulate. What these lights now are and what they do must somehow indicate why the gods had long ago put them there. We may term these features of earth and sky *etiological indices* (Meschiari 2009b).[9]

Our search for the early mnemonic structures of narrative has now narrowed to objects and locations that function as indexical signs pointing back to the past events that long ago produced them. Here again, we might usefully reflect on the way we access our personal past. When we try to recall an episode from our own past, we always need to situate it in a particular place. This is an effortful reconstruction. But if we physically revisit the place where that particular episode happened, the objects we see there are indexically mnemonic, cueing our mental images and often prompting a series of other vivid recollections. We have probably all experienced that rush of returning memories occasioned by a return to our hometown or to places we knew in childhood. We are surprised we remember so many details as we walk that landscape, and, if we have a companion, we feel an irresistible urge to point things out and tell stories about what once happened there.

Sometimes regarded as separate modes of conveying information, showing and telling are effective complements, their content doubly encoded, both visually and verbally (Paivio 1971, 1990).[10] Before writing, show-and-tell was the most efficient mode of transmitting object information, with every knowledgeable person acting as a guide to others, especially to younger kin. If, for example, the activity was food gathering, the leader would name particular roots, tubers, sprouts, berries, nuts, etc. and show where they might be found. The learning process would entail recognizing indexical signs, e.g., how a leaf or flower indicates the place where a root or tuber is hidden or how animal trails often lead to a stand of fruit trees (Sugiyama 2001).

Most of the information this guide imparted he would have learned from his or her own elders, for this is how cultural knowledge is distributed. The knowledge, for example, of amanita mushrooms and hemlock roots, is best not acquired through trial and error. There is other knowledge, too, that is best learned secondhand—knowledge of floods, volcanic eruptions, earthquakes; murders, suicides, deadly battles; and encounters with ghosts, monstrous beasts, and shape-shifting gods. Cultural narratives cluster about and are evoked by indexical landscape features, such as mountains, springs, caverns, forests, and cliffs. Wherever humans have settled, there are those who can show strangers these indices and tell their meanings in the form of narratives. As Yeats wrote, "all races had their first unity from a mythology that marries them to rock and hill" ([1938] 1999, 107).

Like rock and hill, natural features have been woven into our earliest shared narratives, but they are not the only landmarks. Human ruins litter the surface of the earth, leading people for thousands of years to wonder, "who made these walls and who tore them down—and why?" When correct answers were difficult to come by, it was much easier to fabulate them in the form of etiological myths.

Who built them? When conquering hordes destroy a long-established city, they often lack the technology to rebuild it on the same scale, so legends grow that gods or giants had built those structures. Hence the cyclopean architecture of Mycenae and Tiryns, the walls of Ilion built by Poseidon and Apollo, Mont St. Michel heaped upon the sea by the ogre Cormoran, and the menhirs and dolmens of northern Europe erected by some race of giants. Even geological formations, e.g., the Giant's Causeway in Northern Ireland, could be cited as proof that indeed "there were giants in the earth in those days."[11]

The "agent-detection" impulse is especially aroused by the remains of an ancient city. We now have a number of natural explanations for these sites, e.g., overuse of scarce water resources, disease, and climate change, but mythic explanations always favored intentional acts: some more powerful agents had to have destroyed the city—but why? Out of a scattering of massive stones and broken statuary the preliterate imagination composes a narrative over centuries, gathering motives and incidents from any traditional source that seems to fit.

Homer, believed to have lived four centuries after the fall of Troy, presumably had a wealth of preexistent legends to work with. Where did all those stories come from? If any was based on eyewitness accounts passed down from generation to generation, we can be sure it was revised and augmented by countless other traditions, mainly from Mycenaean Greece, Ionia, and perhaps Anatolia. If the setting of Homer's *Iliad* was a city first excavated by Heinrich Schliemann, this city had been destroyed by fire around 1200 BCE and partly rebuilt three times before it was abandoned around 950. Thereafter it remained a ghost town for two centuries (Vermeule 1964). During that period, now referred to as the Greek Dark Age, its ruins would have provoked travelers' questions and locals' interpretations (Minchin 2016).

It is not inconceivable that Homer, if he actually existed, was one of those travelers (Hertel 2003). "How could such a massive city have been built?" he might have asked. "Certainly it was made by gods and demigods." "Yes," some local guide might say, "it was founded by King Laomedon to whom Zeus had sent Apollo and Poseidon to cut and lay the squared stonework of its walls. There is a tale of divine rage when the king refused to pay them and other tales of a sea monster sent to devour a princess, of Heracles sent to save her, of magical horses, of the king's sons murdered, all but the young prince, Priam. There stood the royal *megaron*, here the Temple of Athena, and yonder the Scaean Gates, that broad stone structure where war chariots once burst forth. How was such a city destroyed? Ah, gods and heroes laid it low long, long ago, and this is how it all happened . . ."[12]

Before I continue to trace how landmarks served as mnemonic indices for narratives, I need to step back for a moment and acknowledge that, while the ancient connection of landmarks to narratives was quite important, the earliest purpose for this connection was probably not to memorialize a people's past. Mentally mapping one's environment and learning to navigate it in speed and safety had to have been more crucial to our ancestors' survival than narrative skill. In the early relationship between stories to places, stories would have played a significant but ancillary role: a traditional story may have been preserved not because it was entertaining or particularly meaningful but because it served as a mnemonic for the route. Travelers' ability to store a long detailed list of places in a narrative would have helped them find their way from one landmark to another and in the process avoid environmental dangers or trespassing on lands of hostile neighbors. Outcroppings of rock, small streams, marshes, headlands—these were given names so that itineraries could be orally *told*, i.e., sequentially enumerated, or *re-counted*. The notion that narrative could serve as a mnemonic aid to landmark recognition is a core insight of "narrative cartography," a recently emerging subfield of anthropology. What follows is a very brief excursus on cartographic mnemonics. My point in inserting this is to highlight the functional relationship of journeying to storytelling by teasing apart some of the tangled threads of visuomotor and verbal memory.[13]

The maps we now carry with us on trips are light, foldable paper sheets marked with readable symbols. They are two-dimensional planar representations of space, but the earliest roadmaps were one-dimensional linear charts, itineraries, also called "strip maps" (MacEachren 1986). For travelers, be they traders, generals, or pilgrims, a chart representing a sequence of movements apparently made more sense than the spread-out projection of a landscape. The earliest maps, however, would not have been physical documents at all but mental maps. Australian Aborigines, for example, think of their land as mapped out by their ancestors when they first walked across it during the time of creation, the Dreaming.

> Each of these Dreaming tracks [also known as "songlines"] has a wealth of associated mythological stories which describe the particular geography of that particular stretch of land and the natural resources that may be found in it. These narrative records of the creation of the land are thus, in one sense, oral maps of the country, and they are sometimes recounted or sung when Australian Aborigines move through their country.
>
> (Klapproth 2004, 69)

Recent studies have shed further light on the relation of place to narrative in oral cultures. For example, place names, or toponyms, among American Indians often contain references to specific events that once occurred there (Basso 1996, 50–57). This is also true of Hawaiian place names (Louis 2004), Mayan (Solari 2013), Andean Quechua (Howard 2002), Amazonian (Santos-Granero 1998), Anglo-Saxon (Howe 2008, Trubshaw 2012), and Middle Irish (Tarzia 2012). As allusive lists, they sometimes acquire independent aesthetic value. Keith Basso (1996, 45–46) reports that Apache cowboys have been known to murmur sequences of place names to themselves ("talking names") for the pure pleasure of it. "I ride that way in my mind," one explained.[14]

The Mnemonics of Person

Named sites always seem to have served as hooks upon which to hang a single episode, e.g., the ford of Jabbok and the naming of Israel, Plymouth Rock and the landing of the *Mayflower* Pilgrims, but they are less helpful in organizing extended episodic sequences (Rubin 1995). Episodes may be associated with single places, but a series of places may themselves be linked through the journeying of one particular person or people, e.g., the kingdoms Heracles visited to perform his Twelve Labors or the places along the way from Aztlan to Tenochtitlan in the Aztec migration narrative. Combining personal names with place names is important to preliterate societies because such information helps maintain traditional boundary lines and family lineages. As for narrative, the addition of person to a series of places, i.e., a single protagonist to a familiar itinerary, doubles the mnemonic power of that narrative, thus increasing its chances of long-term survival.

As a mnemonic device for travelers, narrative cartography continued to be practiced, but at some point a form of "cartographic narrative" must have emerged that we might call "journey narrative," i.e., any story whose plot structure is based on a person's journey through a series of different settings. As a genre, the journey narrative includes quest narratives and adventure stories, such as the *Epic of Gilgamesh*, the *Odyssey*, the *Ramayana*, the *Aeneid*, the books of Exodus and Numbers, the Synoptic Gospels, the *Divine Comedy*, *Don Quixote*, *Pilgrim's Progress*, and *Gulliver's Travels*. Its basic plot involves a departure from home followed by a series of adventures, i.e., arrivals (in Latin, *adventus*) at new locations in each of which the traveler must undergo a test, an ordeal similar to an initiation or a rite of passage.[15] The linear map, or itinerary, continues to function as a mnemonic for the linear narrative, but it is the character of the traveler and the meaning of his or her actions that now become foregrounded.

For any social species like ours, other persons obviously represent a major focus of attention. Kin, allies, friends, foes—these loom large in our thoughts and memories. But what prompts us to remember and retell the experiences of wholly *other* persons, particularly persons we have never met? True, we might profit from learning how they confronted certain situations, but it seems to me we need some more immediate payoff, some experience of pleasure. To understand the origins of narrative pleasure we might for a moment consider another pleasurable narrative process, namely, gossip.

I have already noted that gossip occurs when several persons in the *I–You* relationship meet in private to discuss a third person, a *He* or a *She*, who is absent or out of earshot. In typical gossip, the sharers of secrets automatically qualify as an in-group and enjoy whatever privileges such membership entails, such as the opportunity for social bonding and alliance building. We share this information because we, you and I, feel we need to defend ourselves against this "other," a person rarely powerless or socially inferior, who may either legitimately enjoy certain advantages but has more likely acquired them through cheating, free riding, bullying, or outright theft. Gossipers are not inclined to shift perspectives, to enter into the point of view of this gossipee, who remains firmly consigned to third-person status. The way this person becomes (in) famous is an expanding process: first A, the gossip initiator, reveals information about X, the targeted other, to B, C, and D. Then B tells E, F, and G while C tells H, I, and J, and so forth. As this piece of gossip proliferates, transmitters further down the line become less responsible for the truthfulness of the story, a fact that relieves them of the risk of bearing false witness (Giardini 2012). What gossipers, talebearers, and recipients of such information most enjoy is violating the privacy of gossipees, and they gossip with less hesitancy when this pleasure is unmixed with anxiety.

Vergil's allegorical set piece on Fama in book 4 of the *Aeneid* (173–195) analyzes how rumors spread, in this case rumors about the sexual relationship of Aeneas and Dido, queen of Carthage. Fame, here *ill*-fame, begins small out of fear (176) because the first tellers of a scandalous story may be held doubly liable: they will lose credibility among the recipients of their story should it be proven false and risk reprisals should the gossipee learn their identity. But soon this little imp swells into a monstrous birdlike creature, an embodiment of all the peering eyes, chattering mouths, and avid ears of the populace (181–183). She is everywhere, perching on roofs and towers, vigilantly watching, grasping as fiercely onto fabricated and distorted messages as onto truths, feeding the people with ambiguous talk, enjoying herself (*gaudens*) as she sings out facts and nonfacts in equal measure (184–190). Fama has no plan. She is no villainous Iago. She is a swollen, brainless thing, impelled by the joy of being huge, powerful, and invulnerable—features that today we find all too familiar in electronic social media.

There is, I would suggest, a link between the characters targeted by gossip mongers and the main characters of oral and literary narratives. Most narratives usher hearers or readers into virtual *theatra*, seeing places, where they are granted privileged views of ordinarily hidden actions. They are, like Fama perched atop the turrets or, as the saying has it, like flies on the wall, witnessing the most secret, intimate moments of persons who normally guard their privacy jealously. In the Homeric epics, for example, we get to witness the strategizing of battlefield commanders, their dreams at night, their visions of gods, their emotional breakdowns, their laments. Here we are present as a grief-stricken king pleads for the corpse of his slain son and as a husband and wife become reunited after twenty years of hardship and separation. In other narratives we find ourselves present during moments of terror alongside heroes as they encounter giants, lions, dragons, ghosts, and gods. Invisible voyeurs, we are unscathed by mayhem and unsullied by unspeakable events—a man who marries his mother, a queen who has sex with a bull, an unwitting king who is fed the roasted flesh of his only son. Part of the pleasure of these narratives is that they let us ordinary folk imagine ourselves witnessing the private lives of exalted personages. We not only feel ourselves their social equals but also, when we pass judgment on their behavior, their moral superiors.

We are fascinated by the stock characters we encounter in traditional oral narratives, e.g., the wise soothsayer, the proud king, the elegant queen, and the young, impetuous hero. These types, however, have little of what we would recognize as complex characters.[16] In Homeric epic they are represented by symbolic names and epithets. Odysseus, whose name probably means "troublemaker," was also known by his epithet *polutropos,* meaning much-traveled, versatile, or wily. In Vergil's epic, Aeneas is *pius,* ritually correct or dutiful. In the biblical Pentateuch, Abraham is the friend of Yahweh, Moses his obedient spokesman, and Jacob a trickster who, after a strange supernatural encounter, renames himself Israel, the "Wrestler-with-God." Heracles is best known for his twelve labors, but none of them changes the hero or reveals him in a new light: "Heracles" is simply the name that links them all, a mnemonic rather like a label used to tag a particular number of items, in this instance, adventure episodes.

Storied figures from the remote preliterary past tend to display personalities that by modern standards are flat and featureless. In the process of oral transmission whatever seemed inessential appears to have fallen away, implying that personal peculiarities have no lasting mnemonic value. We remain, however, curious about the unique traits that must have characterized figures such as Noah, Gilgamesh, Rama, Agamemnon, Helen, Orpheus, King Arthur, and Robin Hood—assuming these names designate real flesh-and-blood persons. Some scholars have believed it possible to reconstruct plausible character profiles for these figures from the small scraps of evidence that remain. Others assumed that the bards who sang their exploits were concerned with ideal types and what we regard as psychological complexity they rejected as distracting inconsistency. Yet others have concluded that the preliterate heroes and heroines of legend simply lacked the consciousness of self that we possess, and so the way they are presented to us was exactly what they were.

Bruno Snell put forth this latter hypothesis in his influential study *The Discovery of the Mind* (1953), claiming that Homer's epics represent a stage in cultural evolution before the Greeks discovered mind or intellect (*Geist*). Basing his conclusion on the words Homer used to refer to cognitive processes, Snell maintained that the men and women the poet depicted either were, or *thought* they were, composed of independent parts, their bodies of separate limbs loosely allied, and their conscious functions distributed in organs such as the liver, heart, and lungs. In times of crisis, if no god intervened, whatever "self" existed needed to dialogue with one or several of those autonomous inner organs. A person was therefore not yet an integrated individual but rather a "dividual."

> The intellect [*Geist*] was not "invented," as a man would invent a tool to improve the operation of his physical functions, or a method to master a certain type of problem. As a rule, inventions are arbitrarily determined; they are adapted to the purpose from which they take their cue. No objective, no aims were involved in the discovery of the intellect. In a certain sense it actually did exist before it was discovered, only not in the same form, not *qua* intellect.
>
> (vi)

Mind, or intellect, with all its self-reflexive, agential implications, would not be discovered, Snell argued, until some two centuries after Homer, when the Greek lyric poets presented themselves as fully formed, emotionally complex individuals. Later explored by the tragedians through the physical presence of stage actors, the hierarchical structures of personhood were finally defined by Plato and Aristotle in their theories of mind.

Yet personhood must have existed long before it was "discovered" in post-Homeric Greece. We are able to tell one another apart visually and auditorily because our brain dedicates considerable neural resources to recognizing faces and vocal timbre so long as these persons are relevant to us. Although people pass as unidentifiable humans if we are not acquainted with or need to recognize them, we know family, friends, neighbors, and coworkers as many-sided conscious individuals like ourselves. It is unbelievable that preliterate humans were ever unable to perceive the people they knew as individuated selves.

When a name is pronounced, a specific individual is implied. But when in a given population there are more persons than available names, an additional name may be added to specify the person referred to, such as "A from B" or "A, son of C." This ambiguity of names does not, however, apply to persons about whom narratives cluster. In Greek oral tradition, for example, "Heracles," "Achilles," "Jason," and "Odysseus," to name just a handful, may have patronyms and epithets linked to them, but they do not need them: each figure is one of a kind, and a single name is all the identification required. Except for the two Ajaxes, every principal has a unique name. In the *Odyssey*, by my count, there are only seven instances of names shared by two or more persons, and these are very minor characters. In the *Iliad* there are fully one hundred and ninety-two such names that are shared by two or more persons. Over half of them are shared by fighters on both the Trojans and the Greek side, and nearly all these multiply named men are mentioned only once, either when they are killed in battle or when they are referred to as the father or grandfather of a fallen warrior. So, adding insult to injury, persons who appear in only one episode, usually *in articulo mortis*, are more likely to have to share their name with someone else. Conversely, the more episodes attributed to a single person, the more likely that person will have a single unique name. This suggests that, in the process of oral transmission, if multiple tales have become linked to a single person, that unique name has served as a mnemonic ligature for these tales.

The person for whom mnemonics most matters is, of course, the teller of the tale. When that tale is an epic, retrieving the words of the narrative for the delight of an audience was quite a remarkable feat. Oral performance also affected the way persons were portrayed. This is a point that Joseph Russo has stressed in his critiques of Snell's theory of Homeric psychology (Russo and Simon 1968, Russo 2012):

> The depiction of the thinking mind as more open to external influence, and to dialogue with subordinate components of the conscious

> self, than is the case in real life situations, can be explained as a set of preferences developed within the oral performing tradition for *externalizing what is internal and making hidden processes visible, lively, and entertaining*—all in the service of seizing the attention of an audience and bringing it, as it were, *inside* the narrative rather than remaining distanced spectators of the unfolding story.
>
> (Russo 2012, 15)

Insofar as a person in an oral narrative has an identity, this is constituted by his or her actions, which are the actions of an absent third person disclosed by a now present first person, the narrator. It is the narrator's job to conjure up that absent person by invoking his name and mentally leading the present audience into those places, also invoked by name, where this person performs the actions that make the story culturally significant. The fact that oral narrative is *about* a third person, the hero, but is told *by* a first-person narrator, a doubleness that I have touched upon earlier, holds some interesting implications for the travel story as the ur-genre of narrative. The hearers of such a narrative must be simultaneously aware of a nonpresent person, belonging to the past, who transits from one arrival point to another, *and* of the present person who now is leading them through each of these changing settings.

The Greek word for narration, *diêgêsis,* is instructive in this regard because this verbal noun does in fact mean a "leading through," being derived from *dia* (through) and the verb *hêgeomai* (lead). *Diêgêsis* can also mean description, a guiding of the mind's eye among the details of an atemporal space, but the content of a *diêgêsis* is usually a series of actions (Collins 2013). This suggests that when this word acquired the meaning of "narrative," Greek speakers regarded the narrator as something of a tour guide, a person who took others with him into an unfamiliar region and then showed and told them what had happened there.

Semantic memory is the brain's assembly of general knowledge, facts, and beliefs, most of which we learn from others. These individual pieces of information are readily retrieved when prompted by sensory stimuli but are difficult to marshal into larger, more complex wholes. Narratives organize this cultural knowledge by converting the mnemonic structures of episodic memory, such as time, place, and person, into spaces in which to display the contents of semantic memory. Performers lead their audiences through the halls and galleries of this vast storehouse, but it is only through reperformance, the fourth mnemonic, that this knowledge becomes the shared memory of a people, their *alêtheia,* their unforgotten truth.

Summary

With grammar came what we recognize as "full language," the communicative code that, among other innovations, made time thinkable not only as the flowing present but also as the vast archive of the past and as the veiled vistas of the future. The ever-changing present inspired the conceptual metaphor *time is travel,* which implied that one could look back down one's path to an already experienced past or peer ahead into an undisclosed future. Though humans undoubtedly could reflect on past experiences using episodic memory, their ability to share such recollections must have been quite limited before the emergence of grammatical language.

In this chapter I first examined the way the episodic-memory system organizes our autobiographical past as separate events, each assigned its particular time and place, and then proceeded to locate the features of this system in narrative. As a verbal artifact patterned upon episodic memory, a narrative is a *linear* sequence of components that can represent the *perspectives* and *emotions* of third-person agents. Like episodic retrieval, which uses semantic memory to fill in the gaps through *confabulation,* narrative freely fictionalizes. These four features are common to both episodic memory and narrative. Six additional features are *time, place, person, performance, iconism,* and *writing.* The first three of these are common to episodic memory and narrative, and the last three are unique to narrative, but all six are mnemonic features used to transmit and preserve narratives.

Time, an indispensable marker for personal autobiographic memories, keeps events from blending into one another. But remembered time is not objective, chronometric time. It is recalled instead in a multitude of durations, speeds, and degrees of specificity. These subjective variations are reflected in the various "times" of narrative, i.e., *ellipsis, summary, scene, slowdown,* and *pause.* Place, a necessary marker for the retrieval of episodic memories, is no less necessary for narratives. Memory of place, a crucially important human skill for safely traveling within a given landscape, may be stored in story form, a practice termed "narrative cartography." This mnemonic introduces the figure of the traveler as a means of mentally mapping particular routes and landmarks. If we reverse the relation between route and person and now make the traveler our focus and the itinerary the mnemonic device, we have a new genre of discourse, the "cartographic narrative," i.e., the basic adventure tale or travel narrative. Here the actions of a particular person moving through a particular series of places can be a way to represent communal values and models of good and bad, successful and unsuccessful choices and outcomes. By supporting narrative, episodic memory supports the construction of the self as a responsible agent subject to the moral judgment of others, a precondition of membership within a social unit. It is also a mnemonic means of organizing, storing, and retrieving the general contents of semantic memory. Starting with the next chapter, we will be exploring the ways in which the eighth mnemonic, performance, refreshes and reinforces this remembered knowledge. Then, in chapter 4, I will begin examining how visuality has been used to store narrative information first in iconic representations, then in writing.

Three

The Dancing, Singing Daughters of Memory

> Muses of Helicon—they are the first we should honor in singing,
> They whose haunt is the great and sacred Helicon Mountain,
> They who circle the violet spring of Hippocrene, softly
> Dancing, and likewise circle the altar of Zeus, the almighty.
>
> —Hesiod, *Theogony* 1–4

Storytelling also circles—it comes back and starts over. This is as true in literate cultures that cherish traditional religious or secular narratives as it is in preliterate cultures. Stories, unless they are wholly improvised, are not merely performed: they are *re*performed and for that reason must be *re*collected. In an oral culture, storytelling may not require verbatim recall, but the skillful reperformance of a well-known narrative does require a very good memory for times, places, and persons. This is especially the case when that performance includes rhythm, melody, and movement. One can appreciate why, in some cultures, performers, before they begin, say a prayer to whatever power they believe will help them remember their words, tones, and timing. For the ancient Greeks, the divine patronesses of this performing art were the Muses, the nine daughters of Zeus and of Mnemosyne, goddess of memory.

While it is true that Memory and her daughters personify the aptitude needed to retrieve and retell narrative sequences, they also represent other, subtler neurocognitive processes implicated in performance. In this chapter I introduce several statements made by the early Greek poet Hesiod (fl. 700 BCE) concerning the Muses' function and their art. Then, for a means of analyzing these statements, I turn to contemporary neuroscience, where two musicological issues have received much recent attention and debate: the relation of music to language and to human emotions.

Memory and the Mnemonics of Performance

Aristotle's short treatise, its title traditionally translated into Latin as *De memoria et reminiscentia*, draws a useful distinction between *mnêmê*, our capacity to store memories, and *anamnêsis*, our ability to access them. Many nonhuman animals exhibit *mnêmê* to the extent that they can learn to link a current perception with a former event, what we term conditioned-reflex learning. Having some form of *semantic* memory, mammals and birds can be mindful of the world around them, but most researchers agree they lack *episodic* memory, i.e., they cannot voluntarily retrieve and mentally relive specific past events. This ability, as Aristotle also maintained, is uniquely human.

Episodic memory is not, however, exclusively retrospective. When one person tells another to remember to do something not *now* but in the future, that speech act is a directive. If the other agrees, that promise to remember and to perform an action is a commissive that seals a social contract. In our evolutionary past the earliest kind of contract may have been marriage, a commitment that could only have been made through a fully grammatical language that included a future tense (Deacon 1997). This perhaps sheds light on that curious Greek word for suitor, *mnêstêr.* Though the men who wooed Penelope did not appear to be socially responsible, the noun *mnêstêr,* if it is at all connected with that host of other *mnê-* words, must have meant something like "rememberer," a person who would pledge to remember his commitment to his future wife and to her family. (Cf. the name Hypermnestra, the only daughter of Danaus who remembered her marriage vows and did not, as Danaus commanded, kill her husband.)

In our daily conversational speech we often remind one another of things to do. The English verb "remember," when used in a command, usually functions as a future imperative, e.g., "remember to lock the door when you leave the house." The addressee is not commanded to do anything *now,* just to remember the speaker's words at some specified future point and then, and only then, perform the commanded action. (In psychological literature this is termed "prospective memory.") There are thus three elements to this idiom: (1) "remember," the command to bear in mind some future action; (2) that action expressed in an infinitive phrase; and (3) the specific circumstance or point in time meant to cue that action. The interval between the moment the command is spoken and its future fulfillment may be minutes, hours, or days, but rarely months or years. As a future imperative, "remember" has a more restricted time horizon, for, to make sense, this command must apply to a period during which one's words can be still easily retrievable.

The Romans had, as an inflected grammatical form, the "future imperative," and could express it directly in their verbs, but if they wished to stress the mental aspect of this imperative, they could say "*memento* [+ infinitive]"—"keep in mind" (some future action or event) or "*Esto memor*"—"be mindful (of someone or something) from this point on." Of course, the Greeks too had a word for "mindful"—*mnêmôn*. When this adjective was used as a noun, it meant an official trained to remember (with or without written memoranda) municipal statutes, contracts, case histories, and the like—hence our word "mnemonics." For us, concerned as we are with the performing arts as the Greeks understood them, there is that other cognate to consider—*mnêmosunê*. Hesiod in the *Theogony* gives a prominent place of honor to the goddess Mnemosyne as the mother of those goddesses that preside over the performing arts, the Muses.[1] *Mnêmosunê* signified a number of mental activities, e.g., the act of thinking about something, the state of being generally aware of something, recollecting it, and reminding others of it. Though it usually referred back to some past event, it did so as a turning of the mind in the *present*, and, as I have just noted, it could also refer to *future* events.

The earliest mention of the word *mnêmosunê*, in fact its only mention in the Homeric epics, occurs in the *Iliad* (8.181), where it serves the function of a future imperative. Hector, assured by an omen of Zeus's favor, decides to launch a surprise attack on the ships that the Achaean Greeks had pulled up on the beach. Before riding off into battle, he shouts to his men: "when I come to the hollow ships, bethink you then of blazing fire [literally: may some mindfulness of blazing fire arise], so that I may set the ships afire and slay those Argives near the ships, who will be bewildered by the smoke." In other words, "when you see me at the ships, remember then to come with torches."

By the word *mnêmosunê* the early Greeks obviously referred to something akin to what we mean by "memory" and the activity of recollection (Havelock 1986, 100). But we cannot simply assume that their understanding of memory was the same as ours. When prescientific thinking selects certain apparent aspects of a particular cognitive process, it interprets them in ways congruent to a larger cultural belief system. Early Greek theories of memory, while based on observable and introspected phenomena, interpreted the process as including input from outer agents (gods and other unseen beings) and from semiautonomous inner resources, e.g., the *thumos* and *phrenes*, soullike organs residing in the chest. Nevertheless, despite the gulf separating early Greek folk psychology from modern cognitive models, we assume that the same architecture of the brain that had evolved over 2.5 million years to encode and decode experience, e.g., the hippocampus, amygdala, parahippocampal cortex, frontal cortex, and the visual and auditory systems, operated a mere three thousand years ago just as it does today. We may therefore view archaic and classical interpretations of *mnêmosunê* in the light of contemporary neuroscience. But before I venture to do this, I will comment further on the specifically Greek concept of *mnêmosunê*.

The wise Titaness Mnemosyne, the personification of memory, is never depicted as a performer. She neither sings nor dances. Yet she represents the inner powers that the Muses, her singing, dancing daughters, externalize. As she is the power, the *dunamis*, they are the embodied act, the *energeia*. As immortal offspring of Mnemosyne, they inherit the enhanced consciousness of their mother, making it visible and audible. They also infuse this conscious state in certain humans who thenceforth are able to compose and perform verbal artifacts that in turn induce that heightened state in all those who see and hear them. To those they favor they give the power to know "that which is, that which shall be, and that which came before" (*Theogony* 38), to see beyond the here, learning what now is hidden, and also to time travel into the past and even into the future (Ferretti and Consentino 2013).

The nine Muses originally were not assigned to different arts and sciences. That was a division of labor devised in a later age of advanced literacy and learning, a practice not unlike that of the Catholic Church when it declares certain saints patrons of particular crafts and professions.[2] According to Hesiod, each of the nine Muses had her own name, but when Homer addressed his divine patroness, he did not do so by name. Apparently any one of the Muses would be able to assist a poet or singer, for, as Hesiod put it, they were nine maidens with the same mind—they were *homophrones*, literally sharing the same *phrên* (*Theogony* 60).

The *phrên* (often in the plural, *phrenes*) was believed to be a physical organ located somewhere in the thoracic cavity and somehow implicated in thinking. As it has been variously identified with the diaphragm and lungs, it seems to have been linked with the power of speech to form and articulate propositions. It is frequently mentioned in Homer and Hesiod as a place where spoken words are stored and retrieved in remembered utterances, so, if the Muses were *homophrones*, it meant they shared the same repertoire of words and phrases, the same verbal consciousness. *Phrên* would therefore correspond to the mind, or *Geist*, that Bruno Snell (1953) said archaic Greeks had not yet discovered—perhaps for that reason he does not mention *phrên* in his *Discovery of the Mind*. In Aristotelian terms, *phrên* would be the seat of *mnêmê*. But without recollection, *anamnêsis*, this stored information is useless. If collective memory is a people's *mnêmê* of available narratives, then collective recollection is an *anamnêsis* that occurs whenever a group comes together to attend or participate in a traditional narration. It is reperformance that, in an oral society, is the principal means by which cultural information is retrieved and a people becomes *homophrôn*.

In the last chapter we saw how narratives used specific times, places, and persons as mnemonics to connect events into meaningful narrative sequences. Since these nine singing goddesses offer mortal singers the knowledge of past, present, and future, one would expect that the early Greeks would have emphasized the Muses' role as patronesses of memory power. But this is not the case (Stoddard 2004). I am not suggesting that the Muses did not help humans remember their collective past but that perhaps their method of doing so was through some less obvious mnemonic technique. Singing songs that extol the deeds of gods and of men of ages past would have served a useful purpose by reinforcing social cohesion. But usefulness is not how Homer and Hesiod speak of Muses and those they inspire—they sing stories because this produces *delight* in the form of cessation of sorrow. Writing of the birth of the gods, Hesiod tells us the purpose for which the Muses came into being. Father Zeus lay with Mnemosyne for nine successive nights and eventually

> Memory, mistress of the Eleutherian Hills, gave birth to them [as a] forgetfulness of troubles and a respite from worries.
>
> (*Theogony* 54–55)

Hesiod placed Memory (*Mnêmosunê*) as the first word in line 54 and forgetfulness (*lêsmosunên*) as the first word in the following line. Thus Memory, through the agency of her daughters, produces her logical opposite, forgetfulness. Here Hesiod also chose to characterize the Muses' purpose not in positive terms, e.g., pleasure or the contemplation of beauty, but in negative terms: words sung to music cause a temporary forgetfulness of what otherwise might burden the mind with fear.[3]

Self-focused thought can be an unhealthy state if, when we are self-aware, we are obsessed with anxious thoughts. These mental states range through all three categories of time: regrets, guilt, and grief from our past; shame, resentment, and pain in the present; and all the anxious thoughts we project into the unknowable future. Reflections like these tend to generate recurrent images, the sort that, when they wake us at night, we cannot easily drift back to sleep. When humans construct an ideal person, they imagine some all-powerful, all-knowing immortal living in comfort and security, but even the gods seem in need of music therapy. A significant feature of many divine abodes is music: the Olympians delight in the singing and dancing of the Muses, the Persian heaven of Ahura Mazda was called the "House of Song," and the Christian heaven has as its main activity the singing to God of unending hymns of praise.

When the Greeks referred to *mousikê*, the Muses' art, they meant rhythmical words sung to the accompaniment of a lyre (*kithara*) or flute (*aulos*) and visually enhanced by the movements of dancers. This traditional art form, also called *molpê*, was exemplified by the nine Muses, singing and dancing as Apollo played upon the lyre (*Iliad* 1.603–604). This was the model upon which human *mousikê* was practiced at the courts of Menelaus (*Odyssey* 4.17–19) and of Alcinous, where Demodocus chanted while playing upon the *kithara* and young men danced (*Odyssey* 8.250–265). These two events were portrayed as expert performances put on for the pleasure of assembled guests and dignitaries, but the Greeks also enjoyed less professional *molpai*, such as were depicted on the shield of Achilles (*Iliad* 18.494–496, 565–572, 590–605).

The Daughters of Memory, who know all time, past, present, and future, are the goddesses who paradoxically offer their devotees the gift of timelessness. How is it that the Muses can provide a respite from the anxieties of time? Perhaps they can by inducing in mortals an experience that wholly occupies their memory systems, thus blocking intrusive personal memories. If the verbal artifact is a traditional composition, many in the audience will know it from start to finish and at every point along the way know "what comes next." Suspense and surprise have no place in reperformed songs and narratives, which in this respect resemble narrativized rituals, e.g., the Christian Nativity and Easter or the Jewish Chanukah and Passover. Once one fully learns a certain sequence of events, one need not concern oneself with following the plot's twists and turns or anticipating a range of possible outcomes; one can now afford to meditate on the implicit meanings of the story. Accordingly, few if any members of the Greek audience would have wondered whether Oedipus would save his city and his crown or, attending a recitation of the *Iliad*, ask themselves whether Achilles would ever return home or Hector's small son grow to manhood.

Unlike the audience, who know the story in advance, the persons portrayed by a narrator or by actors on the stage have no idea what will happen in the next moment. They share the same vulnerability and uncertainty of every human being on earth *except* the members of the audience who, for the duration of the performance, through the Muses' magic, are permitted to gaze down like gods upon the mysteries of mortality. Addressing this question of time perception in actual performances of epic, Egbert Bakker (1997) maintains that the singer did not take his audience into the past but rather brought the past into the present. Moreover, because the audience knew in advance what would befall each character, implicit in this present of the reperformance lay a future, "a future from which the epic is perceived with the knowledge and understanding of the present" (17).[4]

This ability on the part of an audience to remember the future of the person represented in the performance is grounded on their ability to access their own past experiences stored in *episodic*, or autobiographical, memory. Accordingly they take the gapped, sequential, time- and place-specific structure of this memory system as a template onto which they project the known events of the character's life and, when necessary, fill in those gaps by supplying facts and general knowledge that they have stored in *semantic* memory (Tulving 1983). These two activated memory systems are accompanied by yet another memory system, *working memory*, using which the audience would follow the unfolding performance of epic or drama both as speech and as music. As speech, the audience member would anticipate the syntactical sequence of an utterance, which in an oral artifact is usually enhanced by repetition (Collins

2013, 185–186). As music, the hearer would also anticipate metrical units, such as feet and lines. Both speech and musical form may thus be shaped to facilitate the smooth operation of short-term working memory by creating expectations that are regularly fulfilled (Huron 2006). Thus, as the hearer follows the words of the speaker, singer, or actor and empathetically shares that person's perspective, he or she enjoys that overarching mastery of time that is the gift of the Muses.

Finally, we should bear in mind that, in any culture that prizes oral performance, the practice of storytelling, singing, and rhythmic movement are not restricted to a professional class. In Greek culture, for example, every educated person was expected to have been trained in those various expressive skills classified as *mousikê*, the Muses' art. Just as spectators at a sporting event follow the action at a deeper, more satisfying level if they have themselves engaged in that particular sport, persons trained in *mousikê* would have retained in long-term procedural memory the ability to simulate mentally the complex vocal and kinesic routines involved in a particular performance. Moreover, actions and emotions referred to in the narrative, especially when reflected in the performer's delivery, are also meant to be covertly simulated by the hearer. The whole narration, assuming that it is familiar to the audience, will then unfold as a single, quasi-spatial sequence, all its components integrated within what Merlin Donald (2007) has called "intermediate-term memory." A performance such as this, sung or chanted by a soloist or by a chorus in rhythmical motion, with or without instrumental accompaniment, thus brings into play three long-term memory systems: the episodic, the semantic, and the procedural, as well as the short-term working memory and the intermediate-term systems. The effect produced by the simultaneous activation of these five memory systems is an intense state of outwardly projected attention that, as Hesiod said, grants hearers as well as participants a "forgetfulness of troubles and a respite from worries."

The Coevolution of Music and Language

The *Theogony*, like the first eleven chapters of the biblical book of Genesis, contains etiological myths that attempt to account for the way things are by recounting how they first came into being. Hesiod seems to have thought that *mousikê*, this performance that combined singing, dancing, and musical instruments, started only when the Muses first appeared. Since *mousikê* emerged at some particular point in time, it follows that there was a time before it was ever performed. Hesiod, we still believe, was right about that. To understand this composite art form, we might now ask what the human world was like before that first performance.

Since no human community has yet been found to lack music, dance, and grammatical speech, it is safe to assume that these traits coevolved before the migration of *Homo sapiens sapiens* out of Africa (circa 60,000 BCE). The relation of music to language has been a hotly debated topic in recent cognitive science since, despite the brain areas and networks they share, these two traits reveal significant differences. Music and language serve different social functions: music is used to create social bonding, whereas language is principally used to convey information (Cross et al. 2013). On the other hand, because these two behaviors are controlled by two separate neural networks, they are mutually noninterferent. Consequently, one can engage in music and language at the same time simply by singing words in rhythm and melody. "As is the case for so many cognitive skills, the exquisite unity of vocal music emerges from the concerted activity of separate processes" (Besson et al. 1998, 497).

We have a strong indication that these now separate processes, music and speech, were adapted from a common skill set of communicative behaviors. As I proposed in chapter 1, human evolution may be viewed as a series of stages marked by a progressive mastery of semiotic skills, ranging from expressive vocalizations and motor displays, interpreted as behavioral *indices*, to intentional mouth sounds and hand gestures interpreted as referential *icons*, and then to arbitrary referential *symbols*, perhaps beginning as a kind of sign language but eventually evolving into a vocal system of words in syntax. At the onset of each stage, an innovation occurred that altered the preceding communicative mode but did not obliterate it. Thus, developing the early Paleolithic skill to imitate an animal's cries or move one's hands to portray a human action did not mean ceasing to display emotional indices of, say, anger, tenderness, or fear. Likewise, the late Paleolithic ability to convey symbolic signs did not entail ceasing to communicate through indices or icons. Everyone understands that in face-to-face discourse, *visual* indicators such as facial expression, posture, and hand gestures together with *vocal* indicators such as intonation, volume, and rhythm continue to convey important information supplementary to verbal discourse.

While these two prelinguistic means of communication, vocalization and gestural display, have formed the basis for the paralanguage that now accompanies referential speech, this was not their only function. These older auditory/vocal and visual/motor elements also became the bases of music and dance—music as a redeployment of the tones and durations of expressive vocal sounds and dance as a redeployment of expressive and referential motor behaviors. This does not mean that what we now recognize as music preceded what we now recognize as language. It rather suggests that, as language with its innovative feature, the symbolic sign became the dominant mode of human communication, and the older modes were enlisted as prosodic and gestural accompaniments to speech and, perhaps concurrently, generated their own structures as music and dance.

The series of semiotic stages I outlined above correlates quite well with the first three of Merlin Donald's (1991, 2001) cognitive-evolutionary stages. His first stage, the *episodic*, is not associated with retrievable memories but with the capacity on the part of primate apes to perceive and process whole events by integrating hundreds of separate percepts, "batched together in coherent chunks" (Donald 2001, 201). Able to organize a wide array of information into a single "event perception," this animal no longer relied solely on instinct and conditioned reflexes but could now assess and manage novel situations. This grasp of whole episodes enhanced its understanding that other conspecifics also have conscious thoughts (theory of mind) as well as its ability to observe their behavior as indices of their intentions (mind reading).

Consider the following scenario: it is four million years ago, and a clan of Australopithecines has reassembled after a successful scavenging expedition. Suddenly one of them, a male of middling status, begins to howl, raise his arms, stamp, fixate his eyes, and roll back his upper lip. This behavior continues for a while. The youngsters are perplexed, but the adults have observed this in others, and the particular sounds and movements he is now making they remember his having displayed on other occasions—in this case their episodic consciousness of the moment would be informed by the early glimmerings of episodic *memory*.

At Donald's *episodic stage* we now already have the raw materials of song and dance, the *very* raw materials: vocal sound that varies in pitch, intensity, and duration together with the energetic movements of limbs and facial muscles. As emotive indices, they are intended to communicate inner states or perceived outer circumstances. This performer's mind-reading audience will have to decide whether the signs he conveys are fake, but, since he is expending a considerable effort to execute them, most will probably deem them honest. Yet, whatever they think he means, they will interpret them in the context of both the immediate episode and previous, remembered episodes.

This vocalized display is, however, neither song nor dance. What it lacks is melody and rhythm. There is nothing predictable about his vocalizations and movements. Some of his kin may be moved out of empathy to react in similar ways, but their sounds and gesticulations would not be coordinated in time either to his or to one another's. In this respect, the Australopithecine's vocalizing brain is like that of a ten-month-old modern human, an infant in the babbling phase. Something in the circuitry of the Australopithecine brain is not yet in place. For the necessary regulatory controls, we have to revisit hominid evolution some two million years later. That is, we need to view our ancestors at what Donald calls the *mimetic stage*.

Donald's second stage, starting roughly 2.5 million years ago and associated with toolmaking, represents a further socialization of our early ancestors, who now supplemented mind-readable expressive indices with deliberately planned communicative gestures. "[Mimesis] manifests itself in pantomime, imitation, gesturing, sharing attention, ritualized behaviors, and many games. It is also the basis of skilled rehearsal, in which a previous act is mimed, over and over, to improve it" (Donald 2001, 240). Such repetition served "as a mode of cultural expression and solidified a group mentality, creating a cultural style that we can still recognize as typically human" (261).

This mimesis is iconic behavior on several levels. It entails self-assessment, the capacity to measure the degree to which one matches the skills of others. Boys strive to resemble their fathers, girls their mothers, not simply on an instinctual or preconscious level but by watching intently and deliberately reproducing the actions of their elders, some of which involve precise sequences of steps. This ability to translate visual input into finely controlled motor output may have been built upon the mirror-neuron system that in nonhuman primates is associated with competitive reaching. If so, its human modification was selected to serve our more social-mimetic collaborative nature by helping us learn new manual skills, which we could then teach to others. The manufacture of tools and the use of tools to shape new artifacts were themselves iconic enterprises in that the finished objects were meant to be facsimiles of prototypes.

Mimesis also involved translating auditory signals into motor output. Since the vocal organs are not part of the musculature needed in work routines, workers could use vocal signals—rudimentary work songs—to synchronize arm and leg exertions. Those groups that had gotten the knack of using such signals to coordinate motor efforts, e.g., cutting trees and moving large stones, held an evolutionary advantage over those groups less able to keep time. Moreover, for reasons not yet entirely clear, humans who follow rhythmic auditory pulses can work longer and more efficiently than those who do not. This is equally true for weavers at their looms, for boatmen plying oars, for prisoners in work gangs, and for persons on exercise bikes with earphones clamped to their heads.

This ability of individual humans to synchronize motor output, known as rhythmic entrainment, may be regarded as an externalization of internal rhythms such as heartbeats, breathing, and brain waves, those complex synchronies responsible for coordinating every vertebrate's circulatory and nervous system. Developing the capacity to coordinate *other* bodies in group synchrony must have been a remarkable achievement for our human ancestors and may, in fact, be a key to the evolutionary emergence of genus *Homo* (ca. 2.5 mya). Beyond its utility in technical learning and group effort, this social adaptation inspired in participants a sense of belonging to a strong and protective community. At this stage, singing would take the form of rhythmic vocalizations by groups of persons moving their bodies in time to a common pulse or *tactus* (Jordania 2006).[5]

Donald's *mythic stage* began when our ancestors developed a syntax-governed speech code. Fully grammatical language has been associated with an enlarged capacity of the human brain to input and manage longer and more complex interpersonal events (Donald 1991, Dunbar 1996). Self and others, even when they do share objects of attention, even when they join in rhythmically coordinated action, find they can have different motives for doing so, different feelings, differently remembered experiences, and it is language that discriminates among those different states of mind. It not only communicates thoughts to others—it also layers and embeds them within oneself, so that now, "under the right circumstances, we can maintain several parallel lines of thought, each in a different mode. . . . Running frames within frames concurrently is routine for our species.... Our human cognitive style is linked to this multifocal consciousness, and language, in particular, is highly dependent on this feature" (Donald 2001, 258–259). At the onset of the mythic stage we may assume that longer sound structures, ones more varied than simple rhythmic repetitions, became possible—melodies and sentences and the combination of these two as *song*.

Music and language, as manifested in song and speech, incorporate a mixture of predictable and unpredictable features. The formulation of a sentence and the composition of a melody are both rule-governed transactions. Each has a kind of syntax that allows the receiver to predict *to some extent* the information that will come next. In English, for example, an article or an adjective would prompt the hearer to expect a noun, a transitive verb would be followed by an object receiving action, a preposition by an object in spatial relation to other objects. The intonation contour of a declarative sentence is also generally predictable: it rises and accelerates before lowering and slowing down, a contour that helps the hearer follow its unfolding meanings and interpret them as a completed thought. Musical styles, just as dependent on cultural differences as are natural languages, have scale structures that allow hearers to anticipate tones, a predictability further enhanced by melodic repetitions. Like spoken sentences, sung melodies also tend to produce rising, accelerating, swelling sounds before lowering pitch, lessening intensity, and increasing duration (Patel et al. 1998, Huron 2006).

The differences between these two universal human behaviors are equally important. In fact, the ability to distinguish speech from singing is itself a human universal. W. Tecumseh Fitch (2006, 179–182) enumerates the essential differences. The music of every culture has *discrete pitches* and a recognized scale from which melodies are built, whereas speech, in all but a few tonal languages, allows for continuously variable, i.e., sliding, pitches. Music tends to organize these pitches (notes) according to an underlying *isochronous rhythm* of pulses (a beat, or *tactus*), whereas speech is irregularly paced. Musical pieces, e.g., lullabies, work songs, symphonies, operas, hymns, wedding songs, and laments, exist in particular *performative contexts* and so belong to formal genres, whereas conversational speech is spontaneously generated and genreless. Insofar as verbal artifacts are employed in specific recurrent contexts, they are preserved to be *reperformed*, whereas speech is composed of ad hoc utterances, typically said once and soon forgotten.

There are practical reasons why speech is less regularized and predictable than song. As a means of sharing information, speech is an evolutionary innovation of primate vocalization. The latter sounds, when used to alert others of the presence of food or danger, did not represent states of being that could be pleasurably intensified by rhythmic entrainment; they were urgent one-way communications made to attract attention and produce specific reactions (Juslin and Laukka 2003). For the same reason, when we speak to convey information, we do not do so in song or metrical verse. If we want our message to be taken seriously, we use phrases and clauses of different lengths—we raise pitch and intensity, we shift tempo (Brown and Weishaar 2010).

Some of the neural networks our remote ancestors depended on for communication still operate within us. Because our brain has separate areas dedicated to visual and auditory processing, we can see and hear simultaneously. Thus, while we attend to speech sounds, we also see the speaker's arm and hand gestures that accompanied these sounds. Our brain is also divided into anatomically symmetrical hemispheres, each specialized to process different aspects of particular functions. The rewiring that made human speech production and comprehension possible in the left hemisphere also tweaked the circuitry of the right hemisphere, consigning to it different but complementary functions. So, as speech production became relatively centered in Broca's area, in the left frontal cortex, and speech comprehension in Wernicke's area, in the left temporal cortex, speech prosody and rhythmic patterns settled in correspondent areas in the right cortex (Hyde et al. 2008).

While speech comprehension requires a narrow focus on rapid sequences of minute phonemic differences, to process their meanings we need a broader, overarching level of attention. As the rapid, narrowly focused-upon flow of phonemes resolves itself into words, the longer arcs of rising and lowering pitch highlight the grammatical relations of these words. As the left-hemisphere speech centers became adept at managing rapid series of segmented sounds, the right was there to organize these events into suprasegmental intonation contours. We intently *listen* to phonemic sequences while simultaneously we *hear* the intonation contours that lend to these sound-segmented words meaningfully shaped structures.

Our bihemispheric brain's remarkable capacity to do two things at once, as long as one is broad and holistic and the other is narrow and analytic, I have called the "dyadic pattern" (Collins 2013). Familiar examples include: *figure-ground perception,* the ability to zoom in on one stimulus, e.g., visual object, sound, smell, etc., while keeping aware of the ambient context; *central and peripheral vision,* which enable us to focus on details while monitoring the broader optical field for other relevant objects of interest; *serial and parallel processes,* our ability to perceive and perform things one at a time while concurrently engaged in other perceptions or actions. The brain's complex bilateral divisions of labor have evolved to accommodate speech as well as song and are essential to our understanding of these two uniquely human behaviors.[6]

Of all the senses, hearing and sight became especially refined in primates and increasingly so in our own hominid branch. Our ancestors not only became adept at both receiving auditory and visual information essential to their survival but also learned to communicate that information to one another through vocal sounds and visual movements. Figure 3.1 begins at the top with the division between these two communicative channels. It is also meant to summarize the preceding account of the evolution of signs leading to language and to the singing/dancing performing art that has gone by so many names throughout the world but in early Greece was known as *molpê.*

The upper center of figure 3.1 presents the progression of Merlin Donald's first three stages, each with its requisite semiotic function, a gradual process that should be understood as cumulative, every major innovation that came along supplying *additional* means of communicating. Therefore, when full language (lexicon and syntax) came on the scene in the mythic stage, it arrived with an entourage of older communicative resources both vocal and gestural. On the vocal side, language retained aspects of indexical primate contact and alarm calls in the form of vocatives and interjections; iconic reference in words that sound or "feel" like their referents; and, finally, purely symbolic elements, e.g., arbitrary mouth sounds denoting distinct meanings. These three vocal-sign functions were modulated by overarching intonation contours that, as speech prosody, added intentional and affective nuance to the words of an utterance (Bowling 2013).

On the movement side, spoken language retained visual display in the form of indexical facial expressions, finger pointing, and body language; iconic gestures in the form of manual "air pictures" that visually

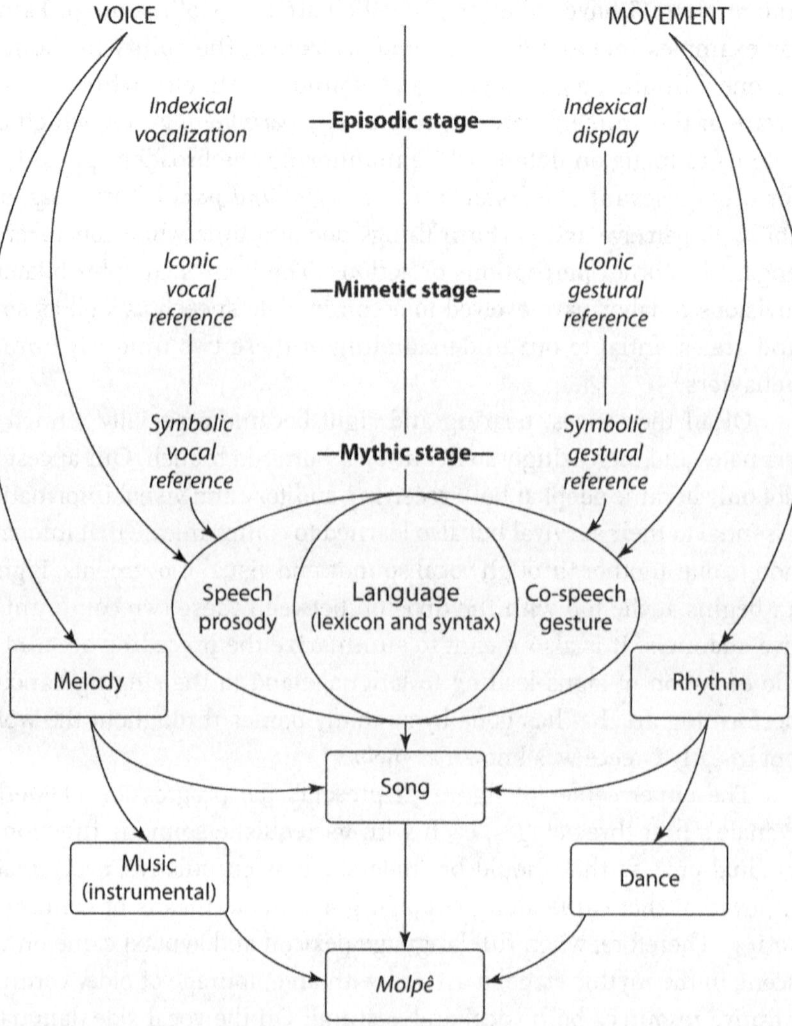

Figure 3.1 *Mousikê* and the evolution of language and paralanguage.

represent objects and actions; and symbolic gestures in the form of conventional hand signs, such as "thumbs up" and "V" for victory. The latter category of signs may have once constituted a protolanguage possessing some of the complexity of a modern sign language (see chapter 1). Like speech prosody, co-speech gestures such as these now operate in the periphery of our attention. When we are listening to someone speak, we fix our central focus on the sound stream that conveys verbal meanings, but, even as we do, we are also absorbing whatever prosodic and gestural information accompanies it.[7]

Beneath "LANGUAGE" with its two peripheral accompaniments I have arranged the elements that together constitute the performing art that the Greeks referred to collectively as *mousikê*. As the outside arcing arrows indicate, *melody* and *rhythm* derived respectively from the vocal/auditory and the motor/visual modalities and, combined with language and with each other, generated *song*. They also developed independently of language as instrumental music and dance, art forms that in turn combined with song and with each other to form that hybrid art form of instrumentally accompanied song and dance, *molpê*, which the Greeks cherished as the supreme delight of gods and mortals.[8]

Nietzsche in his *Birth of Tragedy* captured the complementary opposition of rhythm and melody in his contrast of Apollo and Dionysus.

> Music had long been familiar to the Greeks as an Apollonian art, as a regular beat like that of waves lapping the shore, a plastic *rhythm* expressly developed for the portrayal of Apollonian conditions. Apollo's music was a Doric architecture of sound—of barely hinted sounds such as are proper to the cithara. Those very elements which characterize Dionysiac music and, after it, music quite generally: the heart-shaking power of tone, the uniform stream of *melody*, the incomparable resources of harmony—all those elements had been carefully kept at a distance as being inconsonant with the Apollonian norm. . . . The virgins who, carrying laurel branches and singing a processional chant, move solemnly toward the temple of Apollo, retain their identities and their civic names. The dithyrambic chorus on the other hand is a chorus of the transformed, who have forgotten their civic past and social rank, who have become timeless servants of their god and live outside all social spheres.
>
> ([1872] 1999, 27, 56, emphasis added)

We next will consider the emotional impact of rhythm and melody combined.

Music and the Emotional Brain

When, in discussing the differences between music and language, I cited Tecumseh Fitch, I did not mention his final distinction, the fact that, whereas language is a means of referring to persons, objects, events, and ideas, music is *non*referential communication. We are confronted therefore with a

> design feature . . . easily the most difficult to pinpoint, and the topic of a vast, ancient and controversial literature: the question of meaning in music. On the one hand . . . music is clearly not meaningful in the way language is (able to convey an unlimited number of propositional thoughts or "meanings" with arbitrary specificity). On the other hand, music is not meaningless: music is expressive in some different, hard-to-define sense. It is often said that music "expresses the emotions."
>
> (Fitch 2006, 180)

Before I explore the premise that musical meaning is nonreferential and that therefore the emotions it evokes are sometimes difficult to identify, I want to observe that not all music fits that description. Like any other human artifact, a piece of music may refer to aspects of the culture in which it was produced and may do so by deploying indexical, iconic, and symbolic signs.[9] Indexical signs are operative whenever the style of the piece, or the piece itself, is so closely associated with a particular social activity that the latter supplies its affective frame. For examples of social context, consider Elgar's "Pomp and Circumstance" and Wagner's "Bridal Chorus" from *Tannhäuser.* No one wonders what range of affect each is intended to evoke. Insofar as they are associated, one with academic, the other with nuptial, rituals, they are not what some musicologists term "absolute music." To assess their referentiality, try to imagine the effect on the audience of playing the graduation march at a wedding and the wedding march at a graduation. Music can also incorporate iconic signs, sounds that resemble, for example, birdsong, wind, lapping water, thunder, galloping horses, and explosions, imitative effects that characterize "program music."

While sonic indices and icons signify by associations, the later-evolved symbolic code of language signifies more directly and, as an intrinsic feature of song, is most undeniably referential. Unlike dance music and instrumental music in the Western concert-hall tradition of the past three centuries, most world music, now and over time, has regularly included words. Songs have a right to be considered as ancient and "musical" as instrumental compositions. For the ancient Greeks, *mousikê* was a perfectly natural blend of words, instrumental music, and dance.

Moreover, a song's verbal content, not its musical structure, is most responsible for establishing its emotional meaning. Two folk ballads, for example, may use the same tune but tell quite different stories; a Baptist hymn, a romantic maritime ballad, and a miners' union organizing song from Kentucky ("Which Side Are You On?") are all set to the same melody. Not to mention that wry English drinking song from the 1770s, "Anacreon in Heaven," that got itself reworded as a national anthem, the instrumental version of which can all by itself stir deep patriotic emotion. Even when a song's precise words are forgotten, its verbal message may be subconsciously recalled. As Freud remarked, "Whoever takes the trouble . . . to note the melodies he finds himself humming, unintentionally and often without noticing he is doing so, will quite regularly be able to discover the connection between the text of the song and a subject [*Thema*] that is occupying his mind" ([1904] 1917, 174), an idea that his student, Theodor Reik, was to elaborate in his study of spontaneous music-and-word recall, *The Haunting Melody* (1953).

The function of indexical, iconic, and symbolic signs, when musically conveyed, is to construct an imaginary outer setting within which an inner series of emotional events may be enacted. These events are indeed nonreferential, but, before venturing into the controversial topic of musical emotions, we need to acquaint ourselves with some of the terms of its debate. If we are trying to locate the ultimate meaning of music in "emotions," we first need to agree upon what we mean by "emotions." Do we mean those innate responses to stimulus events that quickly mobilize the body to act in certain ways, e.g., to fight, freeze, flee, mate, eat, etc.? Do we mean somewhat longer-lasting affects, e.g., happiness, sadness, anxiety, resentment, etc.? Or do we mean less definable states of "mixed emotion," including feelings and moods, and, if there are "mixed emotions," does that mean that emotions, like primary colors, are the simple elements out of which all other affects are derived by a process of blending (Trost et al. 2012)?

However we define them, we are likely to agree that the emotions we feel when listening to music are not exactly the same as those we feel when encountering and interacting with objects and persons in our environment. Unlike those latter affects that we involuntarily experience, music offers certain emotional cues and then invites us to *choose* to *imagine* we are feeling a particular emotion. Here another distinction is crucial, that between *perceived* and *induced* emotion, i.e., between an emotion that we recognize as being expressed by the composer through the performer and an emotion that we feel we are now experiencing within ourselves through the music. This is evidently an instance of alternate perspectives: when we perceive emotion in the music, we do so from our detached vantage point, but when we feel it within us, we do so by embodying the emotional patterns represented in the music and by taking the perspective of some imagined other, be that the composer, the performer, or some idealized subject (cf. MacWhinney 1999).

After exposing a wide population of listeners to a wide range of musical genres and then polling them, researchers at the University of Geneva's Emotion and Music Lab arrived at sixty-six music-induced affects. These they then reduced to nine categories, ranking them from the most- to the least-frequently reported, and published the results as the Geneva Emotional Music Scales (GEMS):

Wonder (Filled with wonder, amazed, allured, dazzled, admiring, moved)
Transcendence (Inspired, feeling of transcendence, feeling of spirituality, overwhelmed, thrills)
Tenderness (In love, sensual, affectionate, tender, mellowed)
Nostalgia (nostalgic, melancholic, dreamy, sentimental)
Peacefulness (Calm, relaxed, serene, soothed, meditative)
Power (Energetic, triumphant, fiery, strong, heroic)
Joyful activation (Stimulated, joyful, animated, feel like dancing, amused)
Tension (Agitated, nervous, tense, impatient, irritated)
Sadness (Sad, sorrowful, blue)

(Zentner and Eerola 2011, 206)

According to these researchers, affective states are normally experienced in terms of valence, ranging from extremely positive to extremely negative values. A positive affective state is produced by a circumstance seeming to favor the perceiver and evoking an impulse to approach the stimulus object. A negative affect is produced by some circumstance seeming to thwart or endanger the perceiver and evoking an impulse to avoid, withdraw, or counteract that stimulus object. But music represents valences, negative and positive, *without* objective circumstances. "The perceptions of negative emotional characteristics do not readily translate into felt negative emotion because the listener in most music-listening contexts is safely removed from threats, dangers, or the possibility of losses" (Zentner et al. 2008, 501). Listeners become "somewhat detached from everyday concerns. A clear expression of this detachment is that *dreamy* was among the most frequent emotive responses to music in the current studies. . . . As people move into a mental state in which self-interest and threats from the real world are no longer relevant, negative emotions lose their scope" (513). As the GEMS list indicates, the predominant function of music appears to be to evoke the sort of affects that, as Hesiod said, give listeners "forgetfulness of troubles and a respite from worries" (*Theogony* 55).

The propositions that psychological phenomena are the doings of an embodied mind and that this embodiment, being the brain and the several nervous systems, has evolved over eons are now generally accepted. Most contemporary investigators therefore assume that what philosophers have termed "aesthetic emotions," including most of the affects on the GEMS list, must be grounded somehow in everyday, basic emotions and that the midbrain, the limbic system, must be where these reactions take place. The most-often-listed basic emotions are anger, fear, happiness, sadness, and disgust, but psychologists of music have opted for a somewhat modified list. Precisely how music gains access to our embodied minds has become a concern of neuroscientists over the past several decades, some of whom have continued to focus on those basic emotions.

In 2003, Patrik Juslin and Petri Laukka published a paper closely aligning those acoustic properties of speech and music that prompt listeners' emotional responses. The authors regarded five emotions, anger, fear, happiness, sadness, and tenderness (in place of disgust), as sufficiently different to be designated "discrete emotions." These are the "basic emotions" that both music and speech can elicit. (In the following list, note that the forward slashes separate vocal speech expression, to the left, from music performance, to the right; acoustic cues without slashes are simply shared phenomena; and Fo means fundamental frequency):

> **Anger.** Fast speech rate/tempo, high voice intensity/sound level, much voice intensity/sound level variability, much high-frequency energy, high Fo/pitch level, much Fo/pitch variability, rising Fo/pitch contour, fast voice onsets/tone attacks, and microstructural irregularity
>
> **Fear.** Fast speech rate/tempo, low voice intensity/sound level (except in panic fear), much voice intensity/sound level variability, little high-frequency energy, high Fo/pitch level, little Fo/pitch variability, rising Fo/pitch contour, and a lot of microstructural irregularity
>
> **Happiness.** Fast speech rate/tempo, medium–high voice intensity/sound level, medium high-frequency energy, high Fo/pitch level, much Fo/pitch variability, rising Fo/pitch contour, fast voice onsets/tone attacks, and very little microstructural regularity
>
> **Sadness.** Slow speech rate/tempo, low voice intensity/sound level, little voice intensity/sound level variability, little high-frequency energy, low Fo/pitch level, little Fo/pitch variability, falling Fo/pitch contour, slow voice onsets/tone attacks, and microstructural irregularity
>
> **Tenderness.** Slow speech rate/tempo, low voice intensity/sound level, little voice intensity/sound level variability, little high-frequency energy, low Fo/pitch level, little Fo/pitch variability, falling Fo/pitch contours, slow voice onsets/tone attacks, and microstructural regularity.
>
> (Juslin and Laukka 2003, 802)

Juslin (2010) went on to propose for further research in the affective neuroscience of music a framework that he has called BRECVEM, an acronym for seven factors, or mechanisms, activated by the hearing of music. They are: *brainstem reflex, rhythmic entrainment, evaluative conditioning, contagion, visual imagery, episodic memory,* and *musical expectancy.* The fact that these seven operate at various levels below the threshold of consciousness "leads to the intriguing scenario that *you* may know that what you hear is 'just music,' but the mechanisms [in the brain] that evoke your emotions do not. . . . At least some mechanisms do not necessarily treat musical stimuli as different from other stimuli" (Juslin 2013a, 239). Juslin's seven affect-inducing mechanisms operate as follows:

1. *Brainstem reflex*. This premammalian capacity to respond quickly to danger is activated by any relatively sudden, loud, accelerated, or dissonant sound. The least "musical" of Juslin's musical mechanisms, this alarm reflex has more in common with what I have characterized as the rhythmical irregularity of language (Brown and Weishaar 2010).

2. *Rhythmical entrainment*. Juslin's second mechanism is the opposite of the "brainstem reflex." It is the special way humans can respond to a regularized beat—a repetitive pulse rather than one that is single and sudden—by regularizing their own heartbeats and inducing in them a common motor periodicity. When this happens, they "lock in" to the rhythm, tap their feet, bob their heads, and may even move about in dance patterns. This mechanism creates that sense of oneness with others that Ian Cross (1999) believes is the primary social benefit of music.

3. *Evaluative conditioning*. This mechanism recognizes valence distinctions, i.e., positive and negative affects prompted by a given musical signal. Juslin regards this as conditioned learning, sound-cued associations linked to emotionally charged moments in one's past, and compares this effect with that produced by Wagner's recurrent leitmotifs.

4. *Contagion*. By this Juslin means the empathetic process by which perceived emotion becomes felt emotion. Music induces this contagion, he proposes, by imitating certain vocal pitches and timbres iconically associated with certain emotions. This appears to be a mirror-neuron-like effect, our premotor cortex recognizing some emotions when vocally conveyed and the limbic system automatically replicating them (Scherer and Zentner 2001, Juslin and Laukka 2003, Koelsch et al. 2006).

5. *Visual imagery*. Juslin suggests that music can prompt the listener to visualize spatial settings, landscapes, seascapes, etc. While this does not seem to me a very persuasive point, the auditory and visual cortexes do communicate with each other. Our sense of hearing can locate where sound is coming from, its speed and direction of movement, and perhaps the size of its source and thus provide us with a rough-and-ready spatial map. Moreover, if what we hear requires an immediate reaction, our visual system rapidly forms a mental image of its source object (Koelsch et al. 2013).

6. *Episodic memory*. Here a piece of music evokes an emotion associated with one's autobiographical past and, with that emotion, a sense of nostalgia and its attendant moods. This would seem to overlap to some extent with Juslin's *evaluative conditioning*. While "episodic memory"

may occasionally be cued by music, it seems, like "visual imagery," more epiphenomenal than intrinsic to musical experience.

7. *Musical expectancy*. When a musical feature, e.g., a note, a melodic phrase, or an entire melody, either confirms, postpones, or violates a listener's expectation, it makes an affective effect. This is indeed an essential factor in musical reception. Dependent on an audience's familiarity with a piece of music, expectation may be produced by several features: (1) a familiar style with a familiar "syntax"; (2) a use of repetitive structures, rhythmic, melodic, and (if sung) verbal; and (3) its status as an artifact that is performed often enough that listeners are able to anticipate its progressive variations (Huron 2006). Expectancy, when confirmed, gives listeners the illusion that they are participating, singing along with the music and sharing the emotions it represents (cf. Juslin's "*contagion*"; see also Cross 2010). Expectancy, when not immediately confirmed, produces in us a suspenseful anticipation that teases and primes our reward circuitry, and, when the withheld phrase is finally sounded, we experience an affective climax (Schultz 1998, Salimpoor and Zatorre 2013, Salimpoor et al. 2015). Expectancy, when it is occasionally violated, produces a sense of surprise that converts the emotion from one that is *felt* to one that is *perceived*. But, if one becomes familiar with the piece, even a wholly unfulfilled expectancy becomes predictable. Even a loud dissonance that jars us and triggers a "brainstem-reflex" effect can thus be pleasurably anticipated.

"The BRECVEM framework . . . proceeds from the idea that many of the psychological mechanisms do not have access to, or take into consideration, information about whether the object of attention is 'music' or not—the mechanisms respond to certain information, wherever it occurs" (Juslin and Sloboda 2013, 613–614). This sounds a lot like Steven Pinker's ([1997] 2009, 525) definition of the arts, especially music, as technologies we use to "push our pleasure buttons."[10] Be that as it may, Juslin's initial BRECVEM framework was generally well received by researchers determined to prove that music, like language, is a window into the human mind.

Affective responses to music can be quite strong, but if we define a basic emotion as a response to something in our environment that evokes a withdrawal or approach behavior, we find in music no such objective stimuli. Moreover, much of what we classify as music evokes in us much more nuanced affects (Zentner et al. 2008, Marin and Bhattacharya 2010). But since the stronger and simpler the data, the more persuasive their interpretation, researchers have a tendency to test for responses that are clear and unambiguous. This reminds one of the joke about the man searching for something under a streetlight who, when asked by a passerby what he is doing, explains he is looking for his keys. "So you think you dropped them here?" "Oh no," the searcher answers. "I dropped them somewhere up the block. It's just that it's easier to see under this light."

As brain-imaging technology has improved over the past twenty years, casting an ever broader and sharper illumination, researchers have redoubled their efforts in searching the brain for keys to the emotional meanings of music. As Zentner and his colleagues have shown, most listeners to music do so because, however we choose to phrase it, music does push our pleasure buttons. How it does so—how it maneuvers this massively complex network of neurons into "feeling good"—is itself a massively complex problem. We know that music activates circuits in both hemispheres. Some exploit holistic networks in the right hemisphere dedicated to tone resolution and regularized rhythm; others, related to the serial-processing centers in the left cortex, partly overlap with those dedicated to speech production and perception.

When Anne Blood and Robert Zatorre published the results of a brain-imaging experiment in 2001, their title served as a mini-abstract: "Intensely pleasurable responses to music correlate with activities in brain regions implicated in reward and emotion." Positron-emission tomography (PET scans) was used to monitor changes in the limbic system in response to musical selections as chosen by subjects who had in the past experienced euphoric "shivers down the spine," or "chills," while listening to them. During the experiment, when listeners reported this emotional experience, certain brain regions were seen to become activated or deactivated. Most significantly, blood flow to the amygdala decreased while flow to the nearby ventral striatum increased. This was seen possibly to

> indicate gating between behaviorally antagonistic "approach" and "withdrawal" systems. The amygdala is known to be involved in fear and other aversive emotions, as well as evaluative processes associated with socially and biologically relevant emotions, whereas

> ventral striatum mediates evaluative processes associated with reward and motivation approach behavior. Thus, activation of the reward system by music may maximize pleasure, not only by activating the reward system but also by simultaneously decreasing activity in brain structures associated with negative emotions.
>
> (Blood and Zatorre 2001, 11822–11823)

The authors also noted a concurrent deactivation of the hippocampus, an organ associated with the retrieval of stressful episodic memories.

"Gating" is an interesting concept. In electric circuitry, from which it is borrowed, gating is governed by a transistor that turns an electrical signal on or off. One might think of this as a more refined and selective version of the toggle switch we use to turn a light on or off. As a brain function, gating involves the inhibiting of one brain area while at the same time exciting another area. The process Blood and Zatorre refer to is *synaptic gating*, the mechanism by which neurons—in this case, populations of neurons—are either deactivated or activated, thereby selectively blocking or opening up the transmission of signals from one brain region to another. As they interpret their data, the regions of the amygdala associated with anxiety shut down proportionately as the ventral striatum, associated with pleasurable rewards, turns on. This is not an all-or-nothing process, a bistable flip-flop like the toggling of an electrical switch, but is rather like a seesaw: as one rises, the other descends.

We observe this in other reciprocal processes. For example, when we move our arms and legs, we depend on the synergy of opposing pairs of muscles, e.g., the biceps and the triceps to control the flexions and extensions of the elbow and the quadriceps and hamstrings to control the similar movements of the knee. Climbing, lifting, and walking would be impossible if the brain could not synchronize these complementary opposites, contracting one to the same degree that it relaxes the other. This action is mentally represented as a kinesthetic dyad. When, for example, we lift a weight, we are *focally* attentive of our biceps tensing and shortening and *peripherally* aware that, opposite to it, our triceps is relaxing and lengthening.

According to Hesiod's prescientific formulation, the function of the Muses' art, *mousikê*, is to adjust the relation of memory to forgetting. If we are to solve this Hesiodic enigma with the help of modern cognitive science, we must first identify *mnêmosunê* with declarative memory, then distinguish within that faculty two systems: *semantic* and *episodic* memory. The Muses embody the semantic archive of knowledge exemplified by traditional tales of gods, heroes, and sages. When, under their musical spell, episodic memory shuts down, our autobiographical narrative, with all its "troubles and worries" (*kaka* and *mermêra*), ceases to agitate us (*Theogony* 55).

In neuropsychological terms, strongly pleasurable music induces a flow of dopamine that excites the ventral striatum while the activities of the amygdala and hippocampus become inhibited (Blood and Zatorre 2001; Koelsch et al. 2006, 2013). When this happens, episodic memory as a continual preconscious rehearsal of *kaka*, troubles, mediated by the hippocampus, and *mermêra*, worry, mediated by the amygdala, succumbs to *lêsmosunê*, momentary amnesia. While these two structures of the limbic brain become relatively inactive, the ventral striatum, the primary pleasure center, awakens to what Hesiod would recognize as *to terpnon*, delight. "Thus the pleasure of music may be due both to positive engagement of brain areas related to reward and inhibition of areas mediating negative affective states" (Zald and Zatorre 2011, 411–412).

Though, twenty-seven centuries ago, he had no notion that the brain is where all this happens, Hesiod seems to have grasped the basic principle that a musical performance produces its effects through the simultaneous reaction of two complementary opposites combined into a multimodal dyad.

Dancing and the Inner Dance

No other species of animal uses rhythm to synchronize sounds and movements in order to achieve common objectives. One intriguing explanation of this universal human behavior, put forward by Joseph Jordania (2006), is that it is a very early hunter-gatherer, perhaps prehuman, adaptation. Those bands of men and women that survived on the grassy plains of East Africa did so by seeming larger than they individually were. That is, by rhythmically moving their bodies, a group of ten to twenty persons would seem to potential predators to be a single huge animal and, by raising their voices in loud rhythmic unison, this megabeast might represent an intimidating force. If, on the other hand, their movements and sounds were uncoordinated, this force might soon dissolve into a rout of harvestable prey.[11]

When we hear a number of persons rhythmically vocalizing in unison, this attracts our attention. When they are both singing and moving in rhythmically timed action, they seem strong and purposeful. Even nonparticipants thus can feel themselves drawn in and incorporated into a larger, invulnerable, invincible self. Rhythmic entrainment is a powerful hook, but once the hearers are synchronized with the performer(s), the patterns they see and hear induce motor responses within them. Viewers internally mirror the specific actions of the performers, a simulation produced by what has been called the "action-observation network," or AON (Calvo-Merino et al. 2005). This bilateral network, located in the premotor and parietal cortices, apparently corresponds in humans to the mirror-neuron system recorded in nonhuman primates.[12]

While brain-imaging tests have proven that trained dancers and athletes show strong activation in motor regions of their brains while viewing the performances of similarly trained persons, it is not entirely clear why viewers who have never executed those movements have similar, though somewhat weaker, reactions. While it is obviously true that spectators prefer to watch highly skilled performers rather than amateurs, the great majority of persons willing to buy tickets to view, for example, ballet concerts and professional basketball games would never be able to do the moves they are so excited to watch. When such viewers react pleasurably to the motor routines of expert performers, is it because they must put greater effort into simulating those movements (E. Cross et al. 2011)? That is one interpretation. Another is that their kinesthetic memory, attenuated over time as it might be, retains some of the motor patterns of childhood and adolescence, that period during which their proprioception of their own strength, balance, lightness, speed, flexibility, and coordination first imprinted itself in their brains.

Perception is perception for action, as James J. Gibson (1950) used to say. Something "out there," if important enough to perceive, may be important enough for us to convert our resting state into one of movement. Accordingly, of the two pathways that project forward from our visual center in the back of the brain, the *dorsal* pathway interprets not only the "where" of a stimulus but, equally important, the "how," i.e., the manner in which we should move in response to it (Ungerleider and Mishkin 1982, Milner and Goodale 1995). I will discuss this visual architecture further in the next chapter, but for now I will note that researchers into hearing have discovered two analogous pathways that project from the auditory cortex in each temporal cortex. Paired with a ventral auditory pathway, which identifies the "what" that has made a sound, a dorsal pathway performs two functions: it locates the spatial "where" and simultaneously selects the motor "how" in order to react appropriately to that acoustic source (Warren, Wise, and Warren 2005; Zatorre, Chen, and Penhune 2007).

In humans, as in most mammals, the emotions serve to register this close connection between auditory input and motor output. Zatorre, Chen, and Penhune neatly summarize how dance and music modulate these affects:

> The acoustical features of typically sad or subdued music (containing slow tempo, lower pitched sounds and smooth transitions between sounds) are compatible with the physical expression of sadness, which involves slow, low-intensity movements. The reverse applies to music typically associated with happiness or excitement, which tends to be loud, fast and high-pitched, and is hence associated with rapid, high-energy movements, such as can be observed in spontaneous dancing to music. Auditory–motor interactions . . . may therefore in part mediate music-induced emotion, perhaps providing the link between listening and moving. The psychophysiological changes that are associated with listening to music might also be a byproduct of the engagement of the motor system, and therefore would also provide afferent feedback enhancing the affective state.
>
> (Zatorre, Chen, and Penhune 2007, 555).

Just as the brain has evidently evolved the means to convert visual perceptions into motor representations, which it does when we observe dancers or athletes in action, it also has the means to convert *auditory* perceptions into motor representations, which it does when we hear instrumental or vocal music. I used this perception-action distinction in figure 3.1, when I proposed an interactive relationship between voice and movement in the evolution of language and music. While this perception-action dichotomy has explanatory power at certain levels of analysis, at deeper levels it reveals itself as merely a useful fiction. The basic condition of life is action, i.e., movement in space produced by discharges of energy. Perception is the action animals use to recognize moving and stationary objects in their environment. Vision, for example, which picks up light-mediated information as colors and shapes, is made possible by the action of eye muscles and bioelectrical pulses, "action potentials," within the blood-nourished brain. Hearing, which registers air-mediated information represented as sound, is made possible by minute movements of membranes and hairs that electrically transmit other information to other regions of the brain. Perceptions in their various modalities are thus actions or, more precisely, cognitive *re*actions.

Even when they do not result in overt actions, mental representations are also actions insofar as they involve measurable increases in blood flow, hormonal secretion, and electrical firing. Imagination is simulation, which, as mental mimesis, is cerebral action. When we imagine that we are perceiving sights or sounds or that we are engaged in doing something, we are indeed doing something. Likewise, one of the brain's most significant actions is imaginary speech, i.e., imagining it is hearing or producing speech sounds.

Inner (subvocal, or covert) speech is a uniquely human trait, one possibly as old as language. Homer may have been referring to it when he recounted seven brief inner dialogues between characters and their internal psychical interlocutors (Collins 1996). Socrates in Plato's *Theaetetus*, when asked to define "thought," replied that thought (*dianoia*) is "talk (*logon*) that the soul (*psuchê*) has with itself about whatever topic it is considering. . . . When it thinks, it is merely holding a conversation (*dialegesthai*), asking itself questions and answering, asserting and denying" (189e–190a). For the purposes of philosophical discourse, later writers, e.g., Epictetus and Seneca, could easily internalize Plato's overt dialogic form, but the person most responsible for transforming it into a literary format was Augustine of Hippo, who is known to have coined the word "soliloquy" (*soliloquium*) (Stock 2010). This internalized form of colloquy that he used as a means of self-examination and confession became a staple of Renaissance drama, and it was also appropriated by essayists such as Montaigne. (I will come back to the topic of soliloquial thought in my final chapter.)

In 1881, the psychologist Victor Egger (1848–1909) published his doctoral thesis, *La parole intérieure. Essai de psychologie descriptive*, a study that his contemporary, the linguist Ferdinand de Saussure, closely read and intensely annotated. Egger's book, together with William James's *Principles of Psychology* ([1890] 1950), soon became recommended reading for the "stream-of-consciousness" writers and surrealist artists of the early twentieth century as well as for Freud, Bergson, and literary theorists such as Mikhail Bakhtin (Pouzoulet 2012).

Less introspection based and "descriptive" than Egger's work, that of Lev Vygotsky (1896–1934) promoted inner speech as an important topic for empirical psychology to grapple with. Vygotsky ([1934] 1986) maintained that the early stage during which a child solves problems by overt self-talk does not vanish when silent thought begins at five to six years of age, as Jean Piaget had claimed. Instead, the child internalizes such speech and thereafter, all through adolescence and adulthood, thinks and self-regulates largely by means of subvocal speech. These silent utterances, sometimes dialogic, sometimes monologic, were not, however, replicas of outwardly articulated speech. Inner speech is speech for oneself, not for others, so its form can afford to be abbreviated. Its syntax, for example, regularly omits the subject. So, instead of "My brother, Edward, had to get out of town," one inwardly says ". . . had to get out of town" because one knows the subject of one's thought and need only state the predicate. As for its semantics, its words mean what they subjectively arouse in the inner speaker and are not constrained by dictionary definitions. Inner speech, in Freudian terms, would be semantically "overdetermined," its phrases condensed, its words agglutinated, and its meanings blended.[13]

In the 1950s Alexandr Sokolov began using electromyography to record minute muscular contractions in the throat, tongue, and lips occurring while subjects engage in language-mediated thought. The results of his experiments, which appeared in English translation as *Inner Speech and Thought* (1972), strongly supported Vygotsky's claims that most of what we experience as thought is indeed inner speech. Sokolov, moreover, proved that when we listen to another's speech, those same speech muscles are activated, a process he called "co-enunciation." Citing the work of B. M. Teplov, he also noted that similar reactions, "inner singing," accompany listening to music (Sokolov 1972, 51).

As neural-monitoring technology improved and the conscious brain itself could be examined for evidence of inner speech, researchers saw music as an important window into the mind. I have already discussed Blood and Zatorre's 2003 paper and briefly cited Koelsch and colleagues' 2006 paper. In the latter fMRI study, it was found that a particular region of the brain that represents the larynx reacts in mirror-neuron fashion when one hears musical melodies, a finding supported in more detail by Steven Brown and colleagues in 2009.[14] A year later, Cross observed that, in light of the insights neuroscience was now providing, "a more or less reflexive and involuntary response to music . . . is to participate by at least initiating, though not necessarily carrying through, vocal behaviours. In other words, it is as though the automatic and appropriate response to active listening engagement with music is to prepare to participate however one can—to prepare to sing" (Cross 2010, 8).

It was also during that first decade of the new millennium that Oxford University Press published a controversial study, entitled *Dance of the Muses: Choral Theory and Ancient Greek Poetics* (David 2006). Its author, the young American classicist Amirthanayagam P. David, had proposed that the reason why metrical units were called "feet" in a number of languages, including Greek and Latin, was no accident or metaphor: poetry and song evolved out of dance—not as a general activity, but as *particular* dances. For example, Homer's epic meter, dactylic hexameter, was originally a choreographic measure, one that is still used in the traditional Greek dance, the *surtós*. According to David, Homer's "ring composition," in which a series of topics progresses to a central point, then reverses itself (a-b-c-d-e-f-e-d-c-b-a), replicates in language the forward and reverse movement of the ancient performers who, their arms linked, danced in a circle or semicircle to celebrate the deeds of the gods and the ancestors.[15]

The Muses did indeed dance as well as sing verses. The human singers who appear in Homer's epics accompany their words on the *kithara* but do not dance. Instead, they are accompanied by dancers, whose movements are presumably synchronized with the singer's vocal and instrumental rhythm. In classical times the rhapsodes recited Homeric passages without dance or musical instruments. Literates able to afford written texts of poetry would read them either aloud or in a murmuring voice but would not have the visual presence of a vocally and histrionically skilled interpreter to focus on. In time, when scribal standards facilitated silent reading, persons in the privacy of their chambers could reenact those anciently staged verbal performances without overtly voicing the words.

The tradition that A. P. David has endeavored to reconstruct as the dance of the Muses, however, has never really ceased. How could it? As long as the human brain hears or reads words in a temporal sequence as phonological as well as semantic events, as long as rhythm and meter organize that sequence, the sounds of poetry—even the *imagined* sounds—will continue to induce motor responses. The ultimate dance floor for the Daughters of Memory has always been situated in the human brain. Even when its motor impulses quiver in the muscles of the chest, larynx, and tongue and issue forth as song, even when its energy radiates outward in the synchronous movements of the dancers' bodies, it is the brain that dances.

Individual Differences and the Pleasures of Music

"What does music do?" is a fair question. Those who venture to answer it must, of course, consider what music does within the organs of consciousness and, to do so, must observe musical phenomena as conscious experiences. Reductionism is motivated by the reasonable desire to understand and predict how things work by first discerning the essential phenomena and then bracketing off the merely incidental variables. On the physical, chemical, and biological levels, individual humans display quite similar properties, but, thanks to epigenetic factors and the plasticity of the brain, their affects and behaviors exhibit irreducible psychological differences. If we choose to regard these differences as extrinsic to the circumstances that elicit them, i.e., a given musical performance, and step back from them, these individual differences may seem to blur into a manageable handful of traits and eventually sort themselves out into "personality types." We *may* do this, but there is a cost involved. As Albert Einstein reportedly remarked, "Everything should be made as simple as possible, but not simpler."

Emotions have proved notoriously difficult to study in any systematic way. For one thing, researchers need to agree on the terms they use to designate these elusive states and events. But even when they do agree, specifying the emotions evoked by music presents additional problems: listeners to music tend to agree on what emotions they perceive in a given composition, but on the emotions that the music induces in them, the ones they actually feel, there is much less agreement (Juslin and Sloboda 2013). Some researchers have decided not to preselect a set of musical samples when testing subjects for emotional responses (except for control purposes, for which the experiments might include "neutral" samples). Acknowledging individual differences in advance, they insert those differences into their testing protocol by allowing individual subjects to choose the pieces that have in the past evoked in them certain emotions and levels of intensity (Blood and Zatorre 2001, Zald and Zatorre 2011, Salimpoor and Zatorre 2013).

The specter of "individual differences" haunts the halls of affective musicology, not unlike the ghostly presences of the affective and intentional fallacies (Wimsatt and Beardsley 1954) that are still sometimes heard stalking through English department offices late at night. In an effort to lay the musicological ghost to rest, Ian Cross (2008) has proposed a simple solution: let's admit, in the matter of induced musical emotion, that individual differences do exist but then agree that what music offers is "floating intentionality." That is, whatever combination or sequence of particular affects we feel as we listen to music, this specific experience, though personally important, has always and everywhere been less important than its power to strengthen social cohesion, and, to that end, music and dancing have always accompanied religious rituals, informal celebrations, sporting events, and other public performances.

By 2013, Juslin realized that his BRECVEM framework, based as it was on innate, universal features of the human brain and nervous systems, did not adequately account for receptive differences. The old debate between those, like himself, who emphasize the role of "basic emotions" in musical reception and those who emphasize "aesthetic emotions" seems to have been troubling him. His solution was to add onto his list another factor, *aesthetic judgment*, i.e., the "subjective evaluation of a piece of music as art based on an individual set of subjective criteria" (2013a, 236), thereupon renaming his acronym "BRECVEMA." As an informal sampling of these criteria, he listed such concepts as beauty, complexity, sublimity, expressiveness, originality, good taste, artistic skill, audience affect, and message. Most listeners to any given piece of music will select a certain set of these criteria and weigh some of these more heavily than others.

> Thus when two hear the same piece of music, their aesthetic judgments could produce highly different outcomes—hence the common expression that there is "no accounting for taste." Given such individual differences with respect to aesthetic criteria, it appears pertinent to ask whether one could ever hope to *predict* a listener's aesthetic judgments of pieces of music.
>
> (Juslin 2013a, 250–251)

It seems to me that Juslin's addition of aesthetic judgment improves the descriptiveness of his overall framework, but, if his goal is to discover a set of predictive principles, this revision is unhelpful. If anything, aesthetic judgment adds an extra level of unpredictability. My answer to his question would be: No, one can never hope to predict in any detail a listener's aesthetic judgments of pieces of music, nor can one ever hope to predict that listener's response in terms of the original seven BRECVEM factors. The new BRECVEMA list is, nevertheless, pertinent to our understanding of how music and, I would add, poetry give pleasure. Juslin's carefully researched and reasoned framework is useful precisely because it does *not* help us answer that irksome little question: Why does the emotional pleasure that music and poetry evoke seem different from one individual to another and even within one individual from one hearing or reading of a piece to a subsequent hearing or reading?

Despite sharing an estimated 99.5 percent of the human genome, we do seem to differ significantly in our affective responses to music and indeed to many other aesthetic patterns of stimuli. How shall we account for this? In what respect are we all different "persons?" Even identical twins raised in the same household and the same town, educated in the same schools, the same religion, the same politics, will have their differences. Even if personality tests find no appreciable differences in their attitudinal patterns, these twins will have had different experiences simply because they have been in different places at different times and reacted to different circumstances. The episodic memories they accumulate will therefore be unique, as they are for all the rest of us, and, when they review the narrative of their personal past and contemplate what the future may bring, they will do so as separate individuals. In this respect, Mnemosyne presides over the realm of individual differences.

Those singing, dancing daughters of memory were created to help us cope with the distressful consequences of our separate episodic memories and their projection into the future as hopes and fears. On the one hand we all share the distress of a remembered past we wish we could change and the dread of a future we wish we could avert, a future that may include impoverishment, sickness, our loved ones' death, and finally our own. (Not a very pleasant prospect, if one stops to think about it. And, yes, one does stop to think about it.) Perhaps our consciousness of separation is the most poignant trait that unites us all.

Current clinical psychology distinguishes two depression-related symptoms: unregulated rumination and obsessive worry. Hesiod also seemed to be aware of these two. Looking back from the vantage point of the present, we ruminate on our troubles, which Hesiod simply called the "bad stuff," *kaka*. As if that bad stuff were not bad enough, when we look forward, we contemplate what Hesiod called our worries, *mermêra*. Emily Vermeule, commenting on this latter word, said it signifies

> a kind of floating anxiety and indecision, a sense of not knowing the right course of action because the future is unclear. . . . Uncertainty and worry are also a drain on the powers. When men are extremely depressed by *mermêra*, they can wish for death to end a life of misery, but the gods are not so lucky. Zeus cannot always control events or protect his children; he is caught in perpetual feasting among women who dislike him and break his spirit with mocking words, and sons who boast and rebel against him, until the whole blessed pantheon is filled with grief and trembling.
>
> (1979, 123–124)

Hence the Muses' performances in the Olympian palace of the gods, where they "bring delight to the great mind of their father Zeus by singing that which is, that which shall be, and that which came before, all with concordant voices" (*Theogony* 36–39).

If, as Vermeule suggested, even the immortal gods suffer from some form of time anxiety, the Muses' singing about past, present, and future gladdens their father's mind perhaps because, true or not, it cancels out his depressive thoughts about events that even he cannot avert. Perhaps there is something about musicalized speech—song—that induces in gods, as well as humans, a "forgetfulness of troubles and a respite from worries" (*Theogony* 55). It is also possible that, as Chanda and Levitin (2013, 189) suggested, the beneficial effects of music are "not due to the music itself, but to embedded or ancillary factors, such as distraction, mood induction, locus of control, and perceptual-cognitive stimulation." Hesiod anticipated this too, when he spoke of the power of the singer:

> Happy is he whom the Muses love.
> Sweet is the voice that flows forth from his mouth.
> Though a man holding sorrow close in his fresh-grieving spirit

may be suffering dread with a heavy heart, if but a singer,
a servant of the Muses, intone the glories of the men of old
and of the blessed gods who dwell on Olympus,
at once he forgets his anxiety and no more remembers his griefs,
for quickly the goddesses' gifts distract him.

(*Theogony* 96–103)

What I would propose is that singing, especially when enhanced by dancing and the playing of instruments, has the power to block rumination and worry by interfering with the inner speech that mediates those depressive feelings and thoughts. This is what the singing dancers do for themselves as well as for those audience members whose covert speech rhythms become entrained to that of the words and movements of the performers. When we hear singing, two things happen: (1) we stop our own verbally mediated thoughts, and (2) we start covertly articulating the words of the song. So, as the first effect is inhibitory, the second is substitutory, a process of "gating," as Blood and Zatorre (2001) described.

The first effect is produced by imposing a rhythmic pattern on our inner speech mechanisms, thereby inhibiting speech-mediated intrusive thoughts. This has always been one of the goals of meditation, one method of which has been the repetition of a mantra, a single, often meaningless syllable or a monosyllabic word that in the process of repetition becomes meaningless (Baars 2013). The same effect can be produced by repeating a phrase, a sentence, or a set of sentences, e.g., an invocation or prayer.[16]

The second effect is produced by musicalized speech, which actively induces another set of inner speech movements. In place of our inner speech, song offers a structure of vocal sounds that have lexical meaning but are produced in ways that are quite distinct from speech (Peretz 2009, Patel 2012). Song, with isochronous rhythmic pulses, is organized in metrical units of two or more discrete tones that build melodic structures that are repeated, sometimes with variations. Speech, on the other hand, has a limited tonal range (normally, "high," "middle," and "low"), is tonally irregular, and rarely repeats intonation contours (Fitch 2006). Inner speech does repeat itself but with little or no tonal variation. Sung tones are phonatory events; i.e., they are produced in the larynx by the vocal folds and are sustained as vowels. Spoken words are phonemic events, powered by phonation, to be sure, but articulated in the mouth, which adds on those consonantal details that characterize different words. Song thus recruits the phonatory tones of our prelinguistic past to construct from them an artifact that casts a spell over our anxious inner voice.

Once our own inner speech is quieted and is replaced by the specific motor routine dictated by the music, we hear another set of words and simulate the affects they suggest. This also allows us to exit our personal timeline and enter the present as represented by the ongoing musical sound. Our past is now the sound of the last series of notes we hear, and our future becomes the anticipated course of the next series of notes. This is the way we feel when we are engaged in any purposeful action. We feel "in the flow." We ride Bergson's "pure duration" as a surfer rides a wave.

We also sense we are no longer alone, afflicted with our alienating differences. This is the social reassurance that music universally provides and is probably responsible for that cliché: "Music is the universal language." Yet, cross-linguistic fellow-feeling notwithstanding, if music, instrumental and vocal, seems to induce different emotional responses in listeners, are these differences mere epiphenomena? And what of the fact that a single individual, when reencountering a given piece of music, tends to respond emotionally to it in somewhat different ways?

This latter point brings us back to Juslin's question: Can we ever predict our or others' emotional responses to music? If each time a composition is encountered it can only evoke the same response in a listener, there would be no incentive to reexperience it. That sort of predictability would have a deadening effect. If a piece has no emergent properties to reveal—"elevator music" comes readily to mind—we have no wish to encounter it again. But it may also be the case that, when a piece fails to awaken new blends of thoughts and feelings in us, it is we ourselves that have become predictable.

Before I propose a partial answer to the questions raised by individual differences, I will state two uncontroversial assumptions. First, music is a universal human behavior because it confers a benefit in the form of hedonic rewards on those who participate either as performers or listeners. Second, to the extent that the affects accompanying these rewards vary, they must derive from individual differences in the auditor.

Aristotle's doctrine of the mean indicates that the Greeks, too, were concerned with the issue of individual differences. In his *Nicomachean*

Ethics he proposed that virtue lies at a midpoint between ethical extremes, e.g., courage as the mean between rashness and timidity. It is difficult to strike that midpoint, he says, but easy to miss it. Hence "men are good in one way, but bad in many ways" (2.6.15). Tolstoy modified this axiom in the first sentence of *Anna Karenina*: "Happy families are all alike; every unhappy family is unhappy in its own way." What I would now propose is that every listener to a musical composition brings to that experience a unique configuration of unresolved, and therefore stressful, thoughts mediated by inner speech. Then, as music shuts down the latter process, the experience of relief, that Muse-induced "forgetfulness of troubles" and "respite from worries," is experienced as a unique pleasure. In other words, the *particularity* of our pleasurable experience is a function of the cessation of *particular* intrusive thoughts. This restorative effect may stay with us for a while, for, in Shelley's words, "music, when soft voices die, / Vibrates in the memory." But eventually the memory of every soft voice dies away. As his nightingale's "plaintive anthem fades," John Keats feels "forlorn," a word that, "like a bell," tolls him back to his "sole self." Thereupon the inner debate resumes.[17]

If musical structure enhances the mnemonic of performance, it is apparently not because a musical setting per se makes a song's lyrics more easily retrievable, for, as one study (Racette and Peretz 2007) rather surprisingly concluded, sentences when spoken were more easily remembered than sentences when sung. Ancient sources and contemporary cognitive neuroscience offer converging evidence suggesting that the musical structures of songs *indirectly* enhance the memorability of their verbal content by inhibiting inner speech, specifically the subvocal chatter that episodic memory supplies with anxiety-toned references. Vocal music's inhibitory action has three mnemonic consequences: (1) by first stilling inner speech and shutting down episodic memory recall, it allows short-term working memory to process the words and send their meaning to long-term semantic memory (Baddeley and Hitch 1974); (2) by relieving the mind of anxious private thoughts and drawing listeners into rhythmic synchrony with others, a musical verbal artifact becomes associated with the pleasure that accompanies it (Cross 1999); and (3) it is therefore reperformed, thereby further impressing its content, narrative or persuasive, in the collective semantic memory, the cultural archive of a people.

Summary

Language and rhythm, specializations of the brain's left hemisphere in about 95 percent of humans, are associated with the evolution of motor controls for increasingly rapid and accurate serial actions. Tonality and melody are older, largely right-hemisphere functions that seem derived from emotionally expressive primate vocalizations. In language, the left hemisphere has the capacity to produce and distinguish fast-paced series of phonemic sound segments, whereas the right hemisphere has evolved the capacity to produce and evaluate the suprasegmental intonations that represent a speaker's emotions.

As a mode of musicalized language, singing engages the special strengths of both hemispheres, but seldom equally. A singing voice, one that produces pure, discrete tones, is most speechlike when it chants monophonically without instrumental accompaniment or verbal repetition. One might call this musically heightened speech. As we move toward the purely musical pole, language gradually loses its lexical function, vowels become extended, single syllables occupy several notes, words and phrases are repeated, instruments accompany the voice, and harmony and polyphony may begin to be heard. The further we move from the speech pole, the verbalized sentiments become less nuanced and the sentences less complex, but as long as the words remain audible, their intentionality is understood. It is only when the music overwhelms the lyrics or when the music is wholly instrumental that the performance conveys what Ian Cross calls "floating intentionality."

Across the spectrum from musically heightened speech with its metered rhythm, through song with a recurrent melody organizing its strophes, to more complex choral and instrumental compositions, music has always been regarded as a skill that powerfully benefits both performers and audience. This was the Muses' purpose, as Hesiod claimed. Somehow these goddesses' melodic voices and the rhythmic movements of their dancing bodies had the power to relieve anxiety. Though they were born the daughters of Memory, their gift to humanity, he said, was *forgetfulness*, a puzzling paradox. The approach I took toward unraveling this paradox was, first, to identify the personal troubles and worries that the Muses induce us to forget with the phenomenon of inner speech, for inner speech, alongside its beneficial effects, can also mediate recurrently intrusive thoughts—regrets, resentments, and debilitating fears. Then I suggested that the process of forgetting was one in which instrumental music and the musicalized speech of song can block a listener's alienating inner speech and in its place substitute an inward dance of rhythm and melody, which together induce a state of oneness with others.

Most musical emotions, those perceived in the music and those induced in the listener, are reported as "positive." Even when we perceive the expression of fear or sadness in, say, Mussorgsky's *Night on Bald Mountain* or Tchaikovsky's *Pathétique*, most of us do not become personally fearful or sad. Somehow, following a musical piece and anticipating its progression of sounds stimulates the brain's reward centers while suppressing the areas responsible for "negative" emotions. When, however, we look more closely at the positive moods that a piece of music actually induces in us, not merely those conventionally associated with it, we find that we do not all agree on what these emotions are. For example, does a given piece induce a predominant mood of confidence, awe, nostalgia, tenderness, or peacefulness? Different listeners will report different effects. Why? If my hypothesis is correct, we are each burdened with different private concerns, and when music momentarily erases them, the pleasure that replaces them reflects in its unique particularity the *absence* of these unique particulars. The daughters of Memory are not teachers of mnemonics but relievers of our burdens of rumination and worry. The joy they awaken in us is simply the default state of joyful, physical vitality, the "flow" that we have allowed to be covered over with dissatisfactions and anxieties.

As for the mnemonic function of musical structure, it does not directly preserve a verbal artifact. It only does so indirectly by packaging the semantic content of the piece in a form that is reusable, i.e., reperformable. We do not remember the words of a song because of its melodic structure but because that structure provides us with certain pleasurable rewards that induce us to rehear the words. It is therefore repetition that ensures the preservation of verbal content.

Four

Visual Instruments of Memory

In the previous two chapters, I have explored four mnemonic methods: *time, place, person,* and *performance,* each as a means by which oral cultures preserve narrative information. Time, place, and person are the named elements that anchor the action of the narrative; performance is the narration itself. As proposed in the last chapter, skillfully performed narration, enhanced by musical elements, induces pleasurable emotions, which in turn increase the likelihood that the piece will be reperformed and its narrative content preserved.

The final two mnemonics exist neither in narrative nor in the act of narration. Instead they are displacements of narrative information into soundless and motionless visual artifacts, either icons or symbols. Natural vistas, such as the night sky, mountains, rivers, and caves, which are indexical of ancient mythic events and the persons central to those events, first become transformed into iconic displays, such as buildings, paintings, and statuary, and then into symbolic displays, i.e., writing. In short, time becomes frozen, place becomes architecture, person becomes a carved or painted effigy, and performance becomes reseeing and rereading.

Oral cultures, dependent as they are on memory to archive information, construct a peculiarly foreshortened view of their narratable past. It is only when writing becomes a common *aide-mémoire* that a people's ongoing timeline begins to expand and gaps in the narrative begin to vanish. Writing also eventually makes it easier to store non-narrative information, from inventories and history to philosophy and lyric poetry. It, too, depends on visual mnemonics, but now the visual instrument takes the form of symbolic (graphic) signs. As a way to introduce the practice of "literature," I will conclude with a discussion of two very broad literary genres: public writing (inscriptions) and private writing (letters).

The Mnemonics of Iconism

Even though war and commerce would always require travel and mapping, whenever and wherever Neolithic farming and animal husbandry created the conditions for sedentism, the world seemed to shrink for most persons. The local landscape became sacred, and a growing labor force spurred the building of sacred architecture. Gods and goddesses were still believed to reside in the depths of the earth, sea, and sky and atop rugged mountain peaks, but conveniently located temples were now erected to invite their peaceful visitations. Artificial mountains were built on plains in the form of pyramids and ziggurats. The messages that the gods had inscribed on land and sky in the form of indexical signs were now reproduced as, or converted into, iconic signs, i.e., statuary, carvings, and paintings that could be stored in sacred places to be ritually displayed and explained to the people.

Visual mnemonic systems using iconic representations could reduce the awful enormity of specific natural indices down to a more manageable size. A lightning-wielding, storm-cloud-riding god could be rendered as a twenty-foot statue or as a one-inch amulet and still represent Catequil, Ba'al, Enlil, Indra, or Zeus. Something of the intense devotion experienced during long pilgrimages to distant caves or peaks could now be achieved by half-day processions to local shrines. The index-based mnemonics of place might still be important, but, just as the adventure of the hunt could be replaced by the sacrifice of herded livestock, in Neolithic and Bronze Age cultures iconic representations proved a more efficient means of storing and accessing the past, in particular, the genealogical and national past.

This iconic innovation did, however, present certain problems. As I suggested in my first chapter, iconic gestures as a means of prelinguistic communication would have tended toward conventionalized and abbreviated forms—a sort of gestural shorthand. The same tendency toward stylization and abstraction is evident in iconic artifacts and, as a result, can baffle viewers unfamiliar with their conventions. Then there is also the fact that they are keyed to verbal information, to cultural knowledge, and, often, to specific narratives. The relation between an iconic artifact and its verbal accompaniment is therefore like the relation between co-speech gesture and vocal speech: we may comprehend some things from watching a speaker's animated gestures, but, if we do not know the language she is speaking, our comprehension of her meanings will be woefully insufficient.

A good example of these problems is placed at the very beginning of a second-century CE Greek tale, the pastoral romance *Daphnis and Chloê*, attributed to the Greco-Roman author Longus. "On the island of Lesbos once, while hunting, I saw in a grove of the Nymphs something very beautiful, a painted image, a story of love." The author says he found it even more beautiful than the miniaturized landscape of this sacred grove with its trees, flowers, and rivulets. There were other people, too, native and foreign, who had come to this spot to view this painting, which displayed "women giving birth and swaddling babies; babies abandoned and suckled by ewes; shepherds taking up foundlings; young persons pledging their love; a pirate raid; a military invasion" (Longus 1972, 23).

This was not all the author saw. There were "many other things, all of an erotic nature," and he immediately felt inspired "to write something in response to that painting" (ibid.). What he had seen was apparently a single large mural that would now be termed a piece of "narrative art" depicting moments in the lives of two central characters. These moments would probably not have been arranged in chronological order but in the order of thematic importance: moments of greater consequence would have been depicted closer to the center of the mural and slightly enlarged. Jocelyn Small (1999) has called this "hierarchical time." All these disjunct figural moments would have been harmoniously arranged on the two-dimensional pictorial plane and probably placed on a featureless monochrome ground. Longus's job now was to convert these spatially copresent episodes into a temporal sequence.

Since narrative art presupposes narrative knowledge, Longus needed to find a person able to supply an explanation, an exegete of the icon (*exêgetên tês eikonos*). He could have invented the narrative sequence himself, but, out of respect for the painter, he wanted to learn the traditional narrative. He inquired and eventually found a local able to perform that exegesis, that "leading out" of meaning from the painted images. The result, a four-book narrativization of a single work of visual art, became the most extended ekphrasis in literary history.

The encoding of information in iconic signs is a kind of writing that requires a kind of reading, i.e., an ability both to decode it and to convey that coded information to others through language. This holds true for visual depictions, such as those that Longus learned to read at the shrine in Lesbos, and also for visual diagrams, which Charles Peirce would classify as iconic signs. Examples of the latter we find in various preliterate cultures, such as the Luba of central Africa and the Quechua-speaking peoples of the Andes. The Luba developed the *lukasa,* rectangular wooden "memory boards" that recorded tribal history and genealogy in the form of pins and beads of different sizes and colors, affixed to the surface in intricate patterns. Resembling schematic urban-transit maps, these *lukasa* were readable only by carefully trained persons. The Andeans developed the *quipu,* a similarly abstract form of iconism in which knotted yarn of different lengths and colors preserved a record of the local past. Abstract as lukasa and quipu were, the stories they preserved were, like Longus's composition, tales of flesh-and-blood people, conflicts, marriages, and alliances and similarly needed expert interpreters, exegetes, to translate the visual into the verbal. But this involved no one-to-one verbatim translation: every transfer of information from the iconic medium of the image to the symbolic medium of the word is paraphrastic and will to some extent vary from one reading session to another.

Whether it is formed as a set of separate emblems or as a single emblem composed of multiple iconographic details, this display needs to be sequenced before it can be spoken. In narrative art, i.e., representational depiction, there have been several ways to construct sequences. One is the *continuous narrative,* e.g., the Column of Trajan, a spiral bas-relief of moments in the Roman conquest of Dacia, and the Bayeux Tapestry, a long, embroidered account of the Norman conquest of England meant to be read like a text from left to right (captions, or *tituli,* added). Another similar format is the *sequential narrative,* e.g., a medieval stained-glass window, depicting a biblical or saint's story in panels to be read left to right and bottom to top, and the modern comic strip, read left to right and top to bottom. Iconic displays such as these, when they cue narrative responses, are converted into symbolic signs, i.e., spoken words and sentences.

Before continuing this exploration into visual mnemonics, I think some brief comments on visual cognition are in order. The virtual reality we contemplate when we use a verbal artifact such as a narrative depends on our ability to imagine. In many respects, visual imaging replicates visual perception (Paivio 1971, 1990; Kosslyn 1980; Collins 1991a). It follows that the more we discover about the processes the visual brain uses to analyze the features of light patterns and their sources, the more we may hope to understand the corresponding functions of imagination, specifically the ways by which language evokes mental images. Fortunately for us, the recent decades have seen some notable advances in visual neuroscience.

In the early 1980s, brain research indicated that two neural streams, or paths, project forward from the primary visual cortex in the occipital lobe (Ungerleider and Mishkin 1982). One of these, the dorsal stream, transmits impulses into the parietal lobe; the other, the ventral stream, transmits impulses into the temporal lobe. Subsequent studies determined that the dorsal stream calculates the location, distance, and speed of objects relative to the perceiver; the ventral stream selects discrete objects from the visual array, then identifies what they are (Milner and Goodale 1995, Jeannerod 2006). The dorsal stream projects an egocentric frame of reference allowing us, as subjects immersed in our environment, to find our way visually through its obstacles to achieve our goals. The ventral stream projects an allocentric frame in which we distinguish objects in their broad spatial relationship with other objects. (For a more extensive survey of this research, see Collins 2013, 88–100.)

Though these two streams function in complementary ways—in *Paleopoetics* (Collins 2013), I referred to them as constituting a unity of two, a dyad—in most circumstances one of them will be dominant, the other working in the background. For example, an athlete competing in a contact sport such as football or basketball will usually depend on the dorsal stream and project an egocentric visual frame, whereas a spectator in the stands, perceiving the action from a distance and relying on the functions of the ventral stream, calculates the ebb and flow of the two teams and their individual players by using a wide-angled allocentric frame. Of course, just as viewers of a Greek tragedy can shift perspective from themselves to a dramatis persona, viewers of a sports agon may become excited enough by the action to shift into motor-simulation mode and imagine themselves evading opponents on all sides and rushing toward the goal line or the basket.

These issues are crucially important to the functioning of visual imagination. The reason why indexical images (natural objects on sky, sea, and land) and iconic images (manmade likenesses) function successfully is because they directly conform to the two pathways of visual perception. The indexical place-and-person imagery simulates the way the dorsal stream sights landmarks and obstacles as travelers move toward their destinations while iconic visual imagery simulates the way the ventral stream selects and recognizes objects in a visual array. In both situations, e.g., a landscape or a temple complex, a center of consciousness progresses in an ordinal sequence from one locus to another, a sequence that a performer may represent in a narrative as discourse time, story time, or both.

As we have seen, these two methods were complementary ways of accessing verbal information. The journey represented indexical information stored at places along a one-dimensional pathway. An iconic display could represent that same information as a set of details arranged either in a three-dimensional gallery of statuary or upon a two-dimensional surface, a visual field that viewers could traverse by purely ocular movements. When Greco-Roman orators needed a way to remember the order of their topics, they did so by first visually memorizing a set of architectural details—a street, a garden, a temple, or their own home. Next they created mental icons associated with each topic, placing these cues in sequential order along that architecturally marked route. Then, when they came to deliver their speech, they would imagine themselves walking from one topic-cueing icon to another. This "mental walk," or "method of loci," thus combined the one-dimensional path with the two-dimensional emblematic display and then, by imagining themselves walking this path, simulated a three-dimensional experience.

Whether a narrative is presented in strict chronological order or with flashbacks and flashforwards, as linear discourse it must have its own sequence. This perhaps explains the curious linguistic coincidence of the two concepts *numeration* and *narration*. In the Romance languages the Latin verb *computare* ("reckon") was the etymon used to form verbs meaning both to *count* (objects) and to *tell* (a story). Sometimes the spelling of the two was identical, as in the Portuguese and Spanish verb *contar*. In Modern French the verbs *compter* and *conter* started out as identical forms and are still homophones, but in the seventeenth century the *mp* was introduced as a learned orthographic change to distinguish the two meanings (enumerate and narrate) and to link the word meaning "enumerate" with *computare.* In Italian the verbs are *contare* and *raccontare* (cf. French *raconter*), a pair of cognates that in English became *count* and *recount.* In the Germanic languages, another root is used: "count" and "narrate" are in German *zählen* and *erzählen*; in Danish, *tælle* and *fortælle*; and in Old English *tellan* combines the meanings of both activities. Modern English *tell* also includes its older meaning, "count," in such words as (bank) "teller" and in phrases such as "to tell time" and "all told," as in "There were fifty of them, all told."

Clearly there was a point in European history when the concepts of narration and numeration were so close that a single root, sometimes a single word, could signify these two activities that now seem so different. On this lexical phenomenon Eve Sweetser (1991, 9) in her study of the valid and invalid uses of etymology wrote: "No historical shift of meaning can take place without an intervening stage of polysemy. If a word once meant A and now means B, we can be fairly certain that speakers did not just wake up and switch meanings on June 14, 1066. Rather there was a stage when the word meant both A and B, and the earlier meaning of A eventually was lost." This is an apt description of how the English verb *tell* (A), meaning "count," faded into the English countryside, leaving *tell* (B), "narrate," as the dominant usage. The natural semantic affinity of those two activities is preserved, however, in the English noun *account,* which can still mean either a numerical or a verbal report.

So, what do numeration and narration possibly have in common? When shepherds *told* (counted) their sheep, they were doing something quite different from what they did when they *told* (narrated) stories. They counted those sheep to arrive at a sum of integers, i.e., the same number at evening as at dawn. This sum of cardinal numbers (one, two, three, etc.) was an adding up of similar items, a necessary activity but one so routine that it became a mental exercise insomniacs use for falling asleep. On the other hand, when shepherds or anyone else told *stories,* they narrated a sequence of unique incidents, a sequence that corresponded to ordinal numeration (first, second, third, etc.), each following another and preceding yet another. Thus an eighth item is understood as following a seventh and preceding a ninth, a progression that in narrative can imply a causal chain. Ordinal numbers can thus be visualized spatially as points along a line, as on a strip map, or as figures juxtaposed on a pictorial plane.

Tellers of stories, oral and written, have also combined both models within their narrative: *action* sequence, typical of the journey narrative, plays on the indexical mnemonics of place, and the *description* of objects, persons, and settings plays on the iconic mnemonics of visual displays. Together they create a virtual three-dimensional space that, with the addition of narrative time, becomes a four-dimensional imagined past performed within the theater of the mind.

Imagining the Past

There was a certain stone at Delphi. According to legend it had once been fed to Kronos. This Titan, then absolute lord of the world, was determined to prevent any child of his from ever dethroning him, as he had done to his own father, Uranus (Sky), so as each was born, he devoured it. When his wife Rhea gave birth to a son, whom she named Zeus, she hid the child away and in his place brought Kronos a stone wrapped in swaddling bands, which the Titan promptly swallowed. Eventually he vomited it up, and it fell on Mount Parnassus near Delphi, but by that time Zeus had grown up and soon did indeed dethrone his father. Zeus then took that very stone and "set it fast in the wide-pathed earth at goodly Pytho [Delphi] under the glens of Parnassus, to be a sign [*sêma*] forever after and a marvel to mortals" (*Theogony* 498–500).

So sang Hesiod. When, eight and a half centuries later, the travel writer Pausanias arrived (circa 150 CE), the stone was still there, though to him it did not seem so awe inspiring:

> Leaving the temple and turning to the left, you will come to an enclosure in which is the grave of Neoptolemus, the son of Achilles. Every year the Delphians sacrifice to him as to a hero. If you keep walking up the slope from that monument [*mnêmatos*], you will come to a stone of no large size. Over it every day they pour olive oil, and at each feast place on it unworked wool. There is also a belief [*doxa*], that it was given to Kronos in place of his child, and that Kronos vomited it up again.[1]

As I pointed out in chapter 2, oral narratives resemble episodic memories in that they situate events in both time and space, but the frequency of place names in orally transmitted narrative suggests that, to be remembered successfully, such tales rely on hearers' knowledge of regional geography. In his "Catalogue of Ships" (*Iliad* 2.484–785) Homer demonstrates the mnemonics of place by conducting us on a tour of the major towns and cities of archaic Greece, first from the east to the west, then from the Aegean islands in the southeast up to the mountain bastions in the northwest. Along the way, he identifies places by brief reference to familiar landmarks (walls, springs, rivers, and jutting peaks) and economic details (viniculture, sheepherding, dove keeping). We continue to find similar geographic allusions in Pindar, Aeschylus, Sophocles, and Euripides.

As a predominantly oral culture, archaic and classical Greeks narrowly enclosed their narrated past within the confines of their spatial present. Delphi, for example, said by Hesiod to be "wide pathed," was an important central shrine to which visitors journeyed from all over Greece, making it a hub, a cartographic landmark. It was even said to be the *omphalos*, the navel, of Mother Earth, a designation that made Greece the central nation of the world, their land the chosen stage upon which all the crucial actions of the gods had been performed, first by the family of the Titans, then by that of the Olympians. The specific indexical markers of this past were not, however, arranged spatially to correspond to any chronological sequence. As Pausanias notes, the stone that saved the life of Zeus and made the rule of the Olympians possible lay only a short distance from the tomb of the youngest participant in the Trojan War, Neoptolemus, the vengeful son of Achilles, who lived to witness the disastrous finale of the Heroic Age and the grim onset of the Age of Iron.

During the archaic and early classical period, the past was calculated in patrilinear lists of generations. In book 20 (115–140) of the *Iliad*, Homer gives a hint of how this worked. When Achilles confronts Aeneas and tries to intimidate him, Aeneas finds the time to list his pedigree. Starting with his father, Anchises, he goes back five more generations to Dardanus. Generational reckoning is not a very accurate way to draw up a chronology, but if we adopt Herodotus's generational measure of 33.3 years, we can estimate that Dardanus would have reached thirty-three years of age 233 years before Achilles and Aeneas confront each other on the battlefield. Aeneas declares that Zeus himself was the father of Dardanus but does not mention Dardanus's mother. She was traditionally identified as Elektra, one of the Pleiades, the seven daughters born of

Pleione and Atlas, a leader of the Titanic insurgency that had tried and failed to storm the bastion of the new gods atop Mount Olympus. If any early Greek chronicler wanted to use Homer to tally up the generations from Zeus's succession to cosmic kingship to the end of the Heroic Age, he might have reckoned the interval as an even three hundred years. If we take 1200 BCE as the approximate date for the fall of Troy, this makes 1500 BCE as the date of the cosmic battle, the Titanomachy, that ended the Kronian Golden Age and ushered in the Olympian regime. Needless to say, this view of universal history is as truncated as it is naïve.[2]

The early Greeks had preserved legends of their own local prehistory and of Troy, Crete, Phoenicia, and Egypt, but only after their collision with Persia did they come to realize that a larger, older, nonlegendary world lay beyond their borders. Even then, they had only the vaguest notions of the Anatolian past, of the Hittite dynasties, and the rise and fall of empires in the Levant and Mesopotamia.

The Greeks' sense of their privileged centrality and antiquity was shaken by their actual contacts with the kingdom of Egypt. One early instance was recorded by Herodotus (*History*, book 2). Another appears in the first part of Plato's dialogue *Timaeus* as derived from ancient oral lore. The Athenian sage, Solon (ca. 638–558 BCE), so the story goes, after serving a year as archon (594) reforming the laws of Athens, spent the next ten years traveling abroad in search of knowledge and wisdom. Having first arrived in Egypt at Sais in the western Delta, he endeavored to learn what the temple priests there could tell him about human origins. To get the conversation started, he volunteered to tell them what he, as a Greek, knew about humankind before and after the great flood that only Deucalion and his wife Pyrrha survived. When he finished his tale, an old priest exclaimed: "Solon, Solon, you Greeks are forever children!" The Greeks, the priest said, know nothing of the past. Their beliefs are not based on ancient traditions, and their genealogies are little better than children's stories (22b, 23b). Unlike the Egyptians, whose temperate climate and beneficent Nile have allowed them to accumulate written learning generation after generation, all other peoples, the Greeks included, have periodically lost written records because they have suffered a series of natural disasters, not just one flood but many, and not only floods but also general conflagrations. When these occur, society collapses and, as though he (and Plato) had known about the Mycenaean Linear B, writing is lost and must be subsequently reestablished. They are children, therefore, who have only oral tales—naïve myths—to represent their past.[3]

One participant in this dialogue says this is a tale (*logon*) derived from ancient oral tradition (*ek palaias akoês*). The noun *akoê* could mean "sound," "hearing," "report," or (mere) "hearsay." Plato evidently regarded an *akoê* as credible when the speaker was a creditable source, and, as he argued in the *Phaedrus,* he did not value written records as necessarily more credible than orally transmitted memories. A culture wholly or largely dependent on oral tradition tends to measure time in generations, not in centuries or millennia, so if an aged grandparent tells a grandchild a story and that grandchild lives long enough before retelling it, first-person transmission might span as many as 150 years.

However, oral transmission is, for obvious reasons, fragile and liable to gapping—to *ellipsis* and *summary*, typical of "discourse time" (see chapter 2). The anthropologist Jan Vansina, doing fieldwork in central Africa, noted that in oral cultures there were two kinds of remembered past, the recent (up to about eighty years) and the remote, the age of the originating gods and the founding rulers. Between the two there lies a gap of unknown duration and unknown activity, a "floating gap" that ceaselessly moves forward, erasing the once-recent past to make space for new persons and events (Vansina 1985).

While the recent past loses its historicity as the gap washes over it, persons and events sometimes survive, though they suffer something of a sea change and seem indeed transformed into things rich and strange indeed. If "traditions are memories of memories" (Vansina 1985, 160), then they are second-order recollections, reconstructions of already constructed pasts, and, because tradition is by definition dynamic, this is an ongoing process. Each time a traditional narrative is performed, a little bit of the recent past that has since fallen into the gap may reappear, attired now in the grandeur of the mythically remote past. When the latter happens, formerly clear distinctions between separate persons and separate events break down. Persons flow into one another, and events become conflated. Time collapses in on itself. What may have been a thousand years now seems a matter of three or four generations, an effect known as "telescoping" (Henige 1974).

Since a preliterate society is likely to exist in loosely connected regional centers, each of them develops its own oral tradition with its own heroic narratives. So, as time shortens in depth, space enlarges in breadth. If these regions are brought within a larger cultural ambit, their traditions need to be assimilated within a similarly compressed timeframe (Thomas 1989, Grethlein 2008). The remembered words of Homer and Hesiod framed an imagined past of the Hellenes drawn from many local sources: from family genealogies; from cultic practices associated with particular tombs, shrines, and oracles; and from speculations concerning the men and women who once inhabited those palaces and walled cities that now lay in ruins scattered across the landscape. Local lore presumably attached particular royal names to particular sites, e.g., Nestor with Pylos, Agamemnon with Mycenae, Menelaus with Sparta, Cadmus with Thebes, Theseus with Athens, and so forth. These ruins provided clear evidence that a powerful people had once ruled the land, a race of ancestors mightier than themselves, that had somehow perished—a race of heroes, half-human and half-divine.

One of the disasters the old Egyptian priest in the *Timaeus* may have alluded to began around 1150 BCE. Whatever triggered it—earthquake, climate change, social revolution, conquest, epidemic disease, or a mix of several factors—what followed was a series of political upheavals in the eastern Mediterranean. Populations went on the move, and once stable Bronze Age kingdoms were devastated. The inland empire of the Hittites was destroyed, as was the coastal Canaanite kingdom of Ugarit. Though Egypt eventually repelled the "Sea Peoples," it was itself severely weakened. In the Aegean, the maritime kingdom of Crete was reduced to isolated pastoral villages, and the palatial cities of mainland Greece, Iolcos, Thebes, Mycenae, Tiryns, and Pylos were burned and reduced to magnificent rubble.

Economic depression and social instability descended upon the Levant, Anatolia, and southeastern Europe. Political institutions based on agriculture, manufacture, and trade crumbled, as they would in Western Europe during the centuries following the collapse of the Roman Empire in the sixth century CE. But, like this later period, not all regions were equally affected: just as some pockets of European stability would remain, thanks to the efforts of the still-literate Catholic Church, this earlier downswing was relatively brief in the northern Levant and inland Anatolia, as the Assyrian Empire moved in to fill the vacuum. Not so in the Aegean areas, where what is now called the Greek Dark Age lasted a full three centuries. For almost two centuries after the Phoenician alphabet was adapted for use (ca. 800 BCE), the Greeks' knowledge of their past still had to depend on oral traditions centered on the ruins of former settlements, and, as the construction of a new social order required a legitimating ideology, it required the construction of a past. It was especially important to explain the transition from the glorious Bronze Age to the hardscrabble Iron Age. The gods, it was believed, had watched the fates of the ancient rulers play out in Thebes, in Athens, and in the royal citadels along the coast of the Peloponnesus and Crete. High atop Olympus, they had watched as the crew of the *Argo* sailed forth into the Black Sea and as Heracles roamed the Mediterranean coastline. But Zeus had grown weary of the presumptuous demigods he and his fellow Olympians had generated. It was he who had brought that race to a close by engineering the Trojan War, that final catastrophe in which the mortal children of the gods rushed together and destroyed themselves.

Local traditions honored local heroes whose burial sites, as Pausanias says of the grave of Neoptolemus, continued to be deemed places of spiritual power. Perhaps because these shrines and cults coexisted throughout the land, a large number of worthies seemed to have lived as contemporaries in a compressed period of time. Consider, for example, the generation of Heracles, which included Tiresias, Oedipus, Theseus, Pollux and Castor (the Dioscuri), Orpheus, Jason, and Nestor. Then we have the next generation, that of Achilles, Agamemnon, Menelaus, Ajax, Odysseus, and Diomedes, plus old Nestor, whose military career overlapped those two generations. What better way to preserve these heroes' names in a preliterate or semiliterate culture than to link them in narrative structures, namely, the Theban wars of succession, the voyage of the *Argo*, and the Trojan War?

Through a process we might term "mythic aggregation," legendary kings and warriors that, if they ever existed, may have lived many centuries apart are portrayed as joining together in momentous enterprises. Imagine, if you would, an American epic that featured a dream team of such superheroes as Davy Crockett, George Washington, General Pershing, Spiderman, Daniel Boone, Columbus, Billy Graham, and the Hulk, a band of brothers arrayed against some alien power of evil. As we well know, by the end of the sixth century the isolated city-states of Greece had to mobilize themselves to resist the encroachment of eastern empires, a fact that may in part account for the popularity of these heroic pan-Hellenic sagas that celebrated heroes from all the major regions of the land.

Without written documents to get in the way, oral cultures continually reshape their own past in order to achieve what the shapers regard as a suitable prologue to their own present. This shaping may be motivated by a desire to justify a particular social policy, to legitimize inherited privilege, to perpetuate a sense of grievance toward certain outsiders, or to portray their people as uniquely valiant and virtuous. Jack Goody and Ian Watt (1968, 30–31) referred to "the social function of memory—and the forgetting—[as] the homeostatic organization of the cultural tradition in non-literate society." The body social, they suggested, was like a living organism that takes in, reinterprets, and assimilates traditional tales, and expels whatever it cannot use—all in order to "maintain its present condition of life."

One way the rulers of the reviving Greek kingdoms legitimized their privileges was by appropriating the heroes of the mythicized Mycenaean Age. Sometimes this involved physically transporting these men's bones to, or "discovering" them in, their own territory, but more often it was done by creating a genealogical list that linked the current ruling family to an ancient hero, an effect that Henige (1974) termed "lengthening." Though it might seem the opposite of "telescoping," it is actually dependent on a telescoped past for its supply of mythic forebears. What is lengthened is the ancestral list that now must have enough names inserted in it to reach back to a plausible antiquity (Patterson 2010).

Perhaps the most remarkable example of Greek mythopoeia was the way some ruling houses interpreted the invasions that had toppled the Mycenaean states. According to this myth, the warriors that had destroyed the Bronze Age sites and brought with them the Dorian dialect were not Balkan, Thracian, or Anatolian hordes but instead genuine Hellenes, the exiled sons of Heracles. The end of what they called the "Age of Heroes" was simply the doing of another band of heroes, the Heraklids, to whom a number of new ruling houses now traced their origins, foremost among them the ruling dynasties of Sparta.[4]

When the names of the warriors, kings, and sages that had lived a mere three or four generations earlier had fallen into the "floating gap" and faded from the collective memory, their exemplary actions might still lend vigor to those remote heroes into whose myths they were now assimilated. Moses Finley ([1954] 2002, 160) felt certain that Homer's epics did not depict Mycenaean culture at all but instead was a "picture of the early Dark Age, the tenth and ninth centuries B.C., distorted here and there by misunderstandings and by anachronisms." The Age of Heroes, when gods and demigods walked the earth, would take its final form when the works of Homer and Hesiod were committed to writing. Traditional tales continued to be revered, of course, but oral culture could no longer be the sole determiner of truth or falsity. For literate Greeks the mythopoeic process was finished.

The Mnemonics of Writing

Writing, i.e., symbolic notation, had been used for millennia to keep track of multiple *things* long before it was ever used to record a narrative. The grammatical concept of the plural noun is based on the realization that things can be thought of as individual members of classes. This woman, this man, this ewe, this ax, this day are all simply tokens, specific instances of types indicated as plurals, namely, "women," "men," "sheep," "axes," and "days." The particular number of tokens in a given aggregate may be counted in cardinal numbers—one, two, three, etc., a process of addition that might use the five fingers of one hand plus the ten fingers of both hands. With the addition of the toes, the sum could reach fifteen, then twenty. When a count exceeded twenty or when an external record needed to be established, a piece of bone or wood could be scored with notches to indicate the sum. The English word "score," meaning "twenty," derives from the vigesimal system of counting. A special "score" was made to signify a sum represented by one's ten fingers and ten toes.

The earliest-known means of recordkeeping was the tally stick (or bone, shell, or stone). If we have a tally stick with, say, thirty-nine notches on one side, we have an artifact on which thirty-nine actual tokens of some type can be mapped. In that respect the signs in the stick retain a minimal iconicity. There is nothing special about an individual notch, a scratch, or any other tally mark. Like any true symbol, it is an arbitrary sign that in no way is caused by or resembles its referent, and, like an algebraic letter, its value is whatever we agree to assign it.

Marks such as these therefore require some additional sign to indicate the type to which it and its other thirty-eight accompanying tokens belong, e.g., axes, sheep, or loaves of bread. An external mnemonic for types cannot use numerical signs. Only a mark as disambiguating as a spoken word can make sense of a numerical series. An external recording device such as a tally stick needs to have inscribed upon it a special mark that designates what type of individual items has been here enumerated.

The earliest visual designations of types were probably iconic images, pictograms, such as a horned head for cattle, an oval for a loaf of bread, or a pronged staff for a mattock. Over time, some of these would be pared down into emblematic symbols, a process of abbreviation we see in the development of Chinese characters. The semasiographic marks that first came to be inscribed on wood, stone, or leather were, like that ancient system of gestural communication, manually shaped icons. The difference now was that instead of transient "air pictures," these hand-drawn images were permanently fixed in space (Ingold 2007). Persons able to understand these marks could translate them into speech, that system of vocal signs that had always maintained clear phonological distinctions between nouns. The notion that visual signs need not be pictographs, that they could bypass iconicity altogether and directly signify those vocal signs, was slow to catch on. The mediation of one system of arbitrary signs, language, by another system of equally arbitrary signs, writing, was a formidable conceptual feat. It is no wonder, therefore, that glottography, or speech-sound inscription, would be an expertise that for millennia was relegated to highly trained specialists.

Writing emerged in the cities of the Tigris and Euphrates, the Nile, and the Indus valleys in the fourth and third millennia BCE, apparently under the social pressure to manage the exchange of food and manufactured goods and to regulate debt and credit (Mullins, Whitehouse, and Atkinson 2013). Writing is quite different from the earlier mnemonics of time, place, person, performance, and iconic display. Although those five help in the retrieval of information stored in long-term memory, writing simply assumes the function of long-term memory (episodic and semantic) for the data it transcribes. It does not exercise those memory systems but activates instead our *short*-term working memory, thereby obliging us to exert much more concentrated attention on its serial elements.

When any cultural innovation arrives, words need to be found to name and describe it. In the twentieth century, for example, when the internal combustion vehicle replaced the horse-drawn carriage, the new machine adapted some of the old terms, dashboard, driver, horsepower. Airplanes adapted seafaring terms, e.g., captain, pilot, navigator, aeronautics, airship, crew, passengers, boarding, cruising, rudder, hatch, and cabin. More recently the personal computer eased its way into our lives by borrowing those old-school office terms desktop, folder, file, trash, and mail. When we "wrote" by striking the lettered keys of a typewriter—now a computer keyboard—we still called the product a "manuscript," literally a *handwritten* document, and we still call it that even if it is an electronic "soft copy." Wherever and whenever writing was introduced, old terms were similarly recruited, and these words usually meant the act of scratching, e.g., Greek, *graphein;* Latin, *scribere;* German, *schreiben;* and Old English, *writan.*

Early words for reading, however, are more subtly allusive because this process is carried on within the soundless, viewless skull behind the eyes. Stephen R. Fischer (2003, 12) notes that, in various ways, "reading appears . . . coupled to such cognitive scanning processes as tracking, weaving, tool-making, berry gathering, face and gender recognition and many others, whereby flood of visual data—shapes, units, patterns, orientation, sequencing—is assessed at a glance." The five "scanning processes" that, according to Fischer, reading drew upon are each worth a closer look.

1. *Tracking.* Reading graphic symbols usually involves following a line of marks, as a hunter would follow an animal's tracks or a traveler would follow a pathway or set of landmarks. The Greeks had regarded oral narrative as a *diêgêsis,* a "leading through," and therefore the art of hearing a narrative was the ability to follow the voice of the speaker (see chapter 2). The ability to follow single-file horizontal lines of clearly drafted characters, a skill essential to literate *diêgêsis,* required readers to scan the row and pick out the significant details.[5] The Latin verb *legere,* which meant both "to read" and "to pick (out)," also meant to track, i.e., to trace a set of footprints (*vestigia*) or to sail along the coast within sight of specific towns and natural landmarks. In English we still have the idiom "to pick one's way" (through some uncertain environment). This link of reading with tracking also suggests a link to the mnemonics of place.

2. *Weaving.* The loom, a Neolithic invention that preceded writing by several centuries, was in simplest form a wooden mechanism on which are strung a set of vertical threads (the warp) through which is interwoven, back and forth, a single horizontal thread (the weft, or woof). The Romans called the warp threads *stamina* ("standers"), a row of them an *ordo,* and referred to the act of setting up an *ordo* and starting a new

weaving project by the verb *ordiri*, which could also mean "begin to speak." The word *ordo* in all its later Latinate cognates, e.g., *ordine, orden, ordre, order*, retains those implications of planning and attention to detail that it took from the craft of weaving.[6] Handling the weft, in Latin *trama* ("crosser"), by passing the shuttle back and forth was the tricky part of the weaving process and, ever since Homeric times, has been a metaphor for strategizing and conspiracy, possibly influenced by the spider's web. In the Romance languages *trama* and its derivatives came to mean a plot, including a story plot.

Textile analogies to writing have had a long history, one based on a set of common features. Both a written surface and a woven cloth are two-dimensional planes formed by one-dimensional lines, i.e., pen strokes and threads. Then there was the material used: at a time when consumers were much more aware than we are now of how their goods were made, readers would be regularly reminded that, for example, papyrus was manufactured by combining vertical and horizontal fibers and that, when stored or transported, both cloth and written documents were rolled up. Moreover, textiles such as linen and silk were in some cultures the preferred writing material: they could either be inscribed with letters that might be inked or embroidered as well as interwoven with images. The latter, a highly technical art form, recalls the preliterate mnemonics of iconic display.[7]

3. *Toolmaking.* While this term narrowly refers to production, not to reception, if we broaden it to include the *using* of made tools, we have here an apt analogue to reading. Tools are made to extend the power of their users. Some nonhuman animals have been observed to use found tools, such as sticks, leaves, and stones, but humans transform the shape of natural objects, customize them for particular tasks, vary their style, and preserve them. As I have argued (Collins 2013), we may speak of verbal artifacts—songs, stories, dramas, and poems—as tools that are made and preserved in order to be reused over and over again. Each verbal artifact, oral or written, extends a different set of human powers, a set that includes social intelligence, object and environmental knowledge, perspective taking, as-if play, thought as mediated by inner speech, long-term memory, and visual imagination. The special feature of *written* verbal artifacts is their stability. They are no longer mediated by a storytelling performer: the erstwhile audience member has now become the performer of the work.

4. *Berry gathering.* By this term we mean any foraging activity involving wide-angled visual scanning for the sake of picking out some sort of object, approaching it, and then manually collecting it. When writing was first introduced, some cultures, e.g., the Greek, the Latin, and the Germanic, regarded reading as an activity so closely related to foraging that it was given the same word. The earliest recorded meaning of the Greek verb *legein* was "to pick out," "to count," and, somewhat later, "to recount." In the fifth century it was associated with the noun *logos* (discourse) and came to mean more exclusively to "speak," "say," and "recite," i.e., read aloud from a written text. The Latin *legere* and the German *lesen* each had two main meanings, namely, "to pick out" and "to read" and, because reading also implied speaking aloud from a text, *legere* and *lesen*, like the Greek *legein*, were long associated with recitation.

Reading is a learned adaptation of the visual search routines but one in which the visual field has been reduced to the size of a page, the path to a line of writing, and the act of walking to the movement of the eyes from one word to another. As a cultural innovation, reading is a visual activity that recruits both the dorsal stream, to pursue the optic flow of graphic characters, and the ventral stream, to select significant shapes and combinations of shapes and then identify their meanings.

5. *Face and gender recognition.* In visual perception, one social function of the ventral stream is associated with face and gender recognition. The brain invests much of its resources in face recognition because individual members of tightly knit social units such as families, clans, and tribes cannot afford to be anonymous or unrecognized by others, and, to be held responsible for their actions, they all need visual identities.[8] As we have seen, gossip and more formal storytelling have always needed to "name names"—and so the theme of hidden and revealed identity has always been a major narrative motif. In Aristotle's view of plot (*muthos*), the recognition scene (the *anagnôrisis*) was a powerful event. The verb *anagignôskein*, literally to "know again," or "re-cognize," appears in the *Odyssey* (11.144) when the hero calls up the souls of the dead and is shocked to find his own mother among the shades.[9] He asks his guide, Tiresias, if she could look at his face and recognize him as her son. He is told that she can, but only after she tastes some of the blood of sacrifice he has poured out. This she does and is soon able to recognize him. When the Greeks imported the Phoenician alphabet, it was highly sig-

nificant that they regarded the act of reading writing as so close to the act of recognizing that they extended that verb *anagignôskein* to mean "read."

Our English word "read" derives from the verb *rædan*, which meant "interpret" or "advise." This implied not only a picking out of particular words (cf. Latin *legere* and German *lesen*) and a verbatim recitation of its words but also a general enunciation of its meaning. The notion that writing is covert speech and reading is overt speech is very ancient: the Sumerian and Egyptian words for reading and writing were synonyms for "speaking" (Fischer 2003).

It is important to note that the five activities that Fischer lists, tracking, weaving, toolmaking, berry gathering, and face and gender recognition, all predate the emergence of writing, some by thousands, others by millions of years. They illustrate how certain activities, having ostensibly nothing to do with language, could be enlisted in that new activity, the reading of writing (Anderson 2014). Yet, like these "cognitive scanning processes," reading also enlists visuomotor processes: the visual recognition of letters, i.e., sight reading, and motor simulation, i.e., subvocal articulation. Thanks to our brain's visuomotor coordination, we are quite capable of reading words as visual structures (gestalts) and as articulatory sequences. We recognize them by their shapes and simultaneously support our interpretation by subvocally sounding them.

Writing, we say, puts our thoughts "on the record," whereas the speech that we exchange with others seems to us to be confidential. This, however, was not always the case. In cultures in which only a small percentage of a population is literate, speech is the way to publish one's thoughts, and writing is the way to conceal them. Only in recent centuries and only in particular societies has literacy been promoted as a skill that all persons should learn. Prior to that, writing was regarded by most as a mysterious code. The word "glamour" derives from the word "grammar," the field of learning that, as "grammarie," was regarded by the unlettered as conferring occult powers on the lettered. "Spelling" and "casting spells" suggest the same connection. The modern French noun *grimoire* means unintelligible gibberish or unreadable scribbling, but in medieval French it meant a conjurer's book, seen as a book of *grammaire*. The ancient North Germanic script we know as "runes" derives from a word that meant "secret."

Since writing is a symbolic system superimposed on that other symbolic system, speech, it introduces an extra layer of encryption. If, as Voltaire said, language was invented to conceal thoughts, then writing may have been invented to conceal language. But of course, secret codes are invented not only to exclude the uninitiated but also to share information with a trusted few. On the individual level, sharing personal information means sharing secrets, that precious social coinage. Just as we value our own secrets, we have a natural desire to expropriate the secrets of others, and the more power these persons seem to wield, the more power we derive from divulging their secrets. As I suggested earlier, one motive for oral narrative is the revelation of the intimate speech and actions of powerful personages, of gods, kings, queens, and heroes. Writing, that second layer of encryption, served as a secret channel for the transmission of personal secrets.

Public Verse and the Epitaph

The earliest purpose for writing was, as we know, to keep an accounting of goods, such as grain, livestock, and objects of trade, and of social obligations, such as debts, taxes, and military service. As economic centers grew into royal and, later, imperial cities, literacy spread beyond the circle of priestly accountants to that class of civil servants known as scribes and then to children, mostly boys, whose families could afford to have them trained in this useful discipline. The literate class, which in the ancient world constituted a professional urban elite, would be able to read publicly displayed information and exchange privately addressed information in letter form. Long before anything resembling book publishing was possible, these two kinds of writing were practiced and, as I will argue, significantly shaped all later literary genres.

Public writing, when it was meant to be permanent, was an expensive undertaking. If it was an epitaph carved in stone, the proper material had to be quarried, dressed, transported, the letters incised, perhaps a bas-relief or a statue added, a proper site acquired, and the monument erected. While those who paid for it might have preferred a more detailed accounting of the virtues or exploits of the person celebrated, they knew that words were costly, and, even when expense was no consideration, they also knew that literate passersby would be more likely to stop and peruse the inscription if it was short and to the point. Another factor was tone: since the implied speaker was not personally present to supply gesture and vocal nuance to the words, the text would necessarily assume an impersonal distance that was increased by the fact that this monologist did not require any verbal response on the part of the reader or could even know who this passerby was. Early monumental texts from Egypt, Mesopotamia, Persia, and Greece all shared what we now think of as the classical style—brief, sententious, and impersonal—traits that may have derived from the practical constraints imposed by their medium.

Writing is a highly restrictive channel of communication when compared to speech. This was especially true of public writing, i.e., inscriptions, or epigraphy. It did, however, invite readers to supply certain minimal speech features, namely, intention and voice. In a given culture a literate traveler would be able to identify quickly the genre of an inscription, e.g., historical memorial, honorific plaque, epitaph, or cenotaph, and process it accordingly. Whatever message they conveyed, these concise texts tended to be read as though someone were audibly intoning the words. It was as though every piece of publicly viewable writing had been enclosed in invisible quotation marks. Many inscriptions were accompanied by images of the speaker, bas-reliefs with writing carved across them or statues that required readers to read the words inscribed upon their bases. On tombs and headstones, we also find images of the dead and words that represent their message to the living. We also occasionally find this in even smaller forms, e.g., on Greek ceramics depicting figures of myth with letters flowing from their lips, a representation of speech events that is still the way comic strips are drawn.[10]

The Greek word *epigraphê*, like its Latin translation *inscription*, means letters *written on* something, a stone wall or a bronze plaque, often accompanied by an iconic display, a painting, a relief, or a freestanding statue. An inscribed monument builds upon that older mnemotechnic in which an icon of a person becomes associated by viewers with certain events and places. The difference writing makes is that now a text, including that person's name and remarks on his or her life and character, is incorporated to support or establish this shareable memory. When the lyric poet Simonides (ca. 556–468) said that "painting [*zôgraphia*] is silent poetry [*poiêsis*], and poetry is painting that speaks," he said so in the cultural context in which the two media were already intimately combined in monumental art. This form of epigraphic mixed media thus externalized what Allan Paivio (1971, 1990) called *dual coding*, i.e., the cognitive process by which information is stored in two distinctive forms, as mental imagery and as language, each reinforcing the other.

If we examine the various genres of early public writing in terms of speech-act theory, we can begin by noting that expressives and declaratives seem ill-suited to this form of display. Stone or bronze inscriptions are meant to stand forever, so expressives such as "what a lovely day," "I wish I were home," or "I love Julia" would be unlikely texts, though expressions of grief are common on epitaphs when such feelings seem permanent. Declaratives, such as "I sentence X to banishment" or "I pronounce you man and wife," effective as oral speech events, are unsuited to public writing, though such events might be referred to. Simonides was also credited with inventing the "method of *loci*" (see earlier in this chapter), by which a series of mental images could cue a series of words or topics (Yates [1966] 2010).

As a means of externally storing communal memory, public writing could include commissives and assertives—commissives that acknowledge a continuing relationship, e.g., an alliance between two nations, a pledge of friendship, or a profession of religious faith, and assertives that list the accomplishments of a ruler, the good deeds of a wealthy citizen, or the virtuous and exemplary life of a person buried beneath the monument. These public statements, by and large laudatory assertives, store cultural information by preserving through time the names of certain persons whose spirits, after death, were presumably gratified when such words were read.[11]

Directives are especially important adjuncts to assertives and sometimes can even stand alone, as in the Delphic inscription "*Gnôthi seauton*" ("Know thyself") and the headstone motto "*Memento mori*" ("Remember thou must die"). We moderns are constantly barraged with public directives in written form, e.g., "stop/go"; "stop, look, and listen"; "police line—do not cross"; "keep in line"; "do not litter"; "vote for X"; "drink Y"; "drive a Z"; "eat here." As speech acts, directives may precede assertives by, in effect, calling out to a passerby and commanding attention. Epitaphs in the Greco-Roman tradition are fictive speech events that follow a formula: (1) the passerby is directly addressed in the vocative and (2) told to read the inscription that represents (3) a voice that may either belong to the deceased person, the person responsible for setting up the stone, or the stone itself; then (4) the names of the deceased and of the person responsible for the monument are mentioned; (5) some information concerning the life of the deceased is recounted; (6) a sentiment of grief is expressed; and finally (7) the passerby is told to do something with the knowledge imparted by the text. While many epitaphs include all seven features, sometimes seeming to fill in the blanks in a mechanical fashion, others select from them and shape the genre with artistic subtlety.

Imagine a traveler pausing to read an epitaph. He bends toward the stone slab and, shading his eyes from the sun with his hand, murmurs the sounds of the carved characters. The words on the stone represent themselves as the voice of a person whose life's journey has ended here and for whom, like the stone itself, time has stopped. The traveler, noting the length of his shadow, reckons he has only four or five more hours of light. The winding road beckons, and he has other things to do.

Probably the most famous epitaph in classical antiquity was that written by Simonides in honor of the three hundred Spartans who gave their lives at Thermopylae to slow the advance of the Persian army into mainland Greece. Transliterated and lineated as an elegiac couplet, it is as follows:

> *Ô xein', angellein Lakedaimoniois hoti têide*
> *keimetha tois keinôn rhêmasi peithomenoi.*

And translated in literal, unmetered English:

> Stranger, go and report to the Spartans that here
> we lie in compliance with their words.

We have here the voices of the dead directly addressing the wayfarer, requesting not only that he read the text but also act as a courier to their distant leaders in the Peloponnese. This conveys no emotional expression because it is the dead who had accepted their fate who now speak, no comment on the life of those who had died there because every Hellene would have known Thermopylae and its place in their recent history. What makes this minimalist poem so powerful is how the tone of the first line contrasts with the second. The first line suggests an urgent need to send the news of this battle to Sparta. The second line tells us that the speakers are now in the timeless realm of the dead and how they interpret their destiny.[12]

If there is any wish common to all epitaphic voices, it is the wish to be remembered, to be known by the living. In his study of classical Greek epitaphs, *Inscribing Sorrow,* Christos Tsagalis (2008) examines the four-line text identifying the grave of Hegilla: "All should hear my exact age. / At twenty-five I left the light of the sun. / With respect to my manners and temperance, / My husband knows best how to speak of them." Tsagalis finds the first line most interesting in that it does not directly address the reader but instead says that all, everyone, *pantas,* ought to hear (*acousai*) the inscribed words. This raises two questions. Who is talking, and who is hearing? Are the readers, silently scanning the stone, imagining they are hearing a voice, Hegilla's or some generic female voice? This might be possible, but another explanation is that the heard voice is that of a literate wayfarer sounding out the words or, in the case of the Hegilla epitaph, speaking them aloud so that everyone, including the unlettered, could hear them.

Jesper Svenbro as early as 1976 and in *Phrasikleia* (1993) noted the role of the literate passerby in ancient Greek inscriptions. Here he quotes an epitaph dating from the early fifth century and found on Euboea, the large coastal island northeast of Attica:

> Greetings, o passersby! [*Chairete toi pariontes*] I rest, dead, under here. / You, who draw near, read out [*ananeimai*] who the man is that is buried here: / a stranger from Aegina, Mnesitheos by name; / and my own mother Timarete set up for me, as a monument / on top of the mound, an inextinguishable stele, / where through all the days to come someone will tell [*tis erei*] passersby: / Timarete set me up for her dear deceased son.[13]
>
> (Svenbro 1993, 48, 50)

Svenbro then comments that the verb used for the command "read out" comes from the verb *ananemein,* which literally means "to distribute among others, as we might say, 'share this with others.'"

"Sharing" presupposes a prior dividing of some countable thing into parts. The relation of this action to oral information sharing is preserved in the English verb *to impart* and the German verb *mitteilen*. I suggested earlier in this chapter that *telling* (narrating) is closely connected to *telling* (counting). At this point I will offer the speculation that the activity involved in both these senses of the word *tell* (to narrate and to count) is also connected with cutting into parts, or shares, and imparting them to others, i.e., *dealing* them out. (Cf. the English "good deal" and the French "*beaucoup*," both meaning a generous "cut" of something.) If there is any usefulness in this lexical archaeology, it may be to identify a deeply buried conceptual metaphor formed in a number of different linguistic communities, a metaphor used to understand the complex process by which remembered information is transmitted and preserved in an oral culture and written information in a semi- or fully literate culture.

The purpose of these early Greek epitaphs was to preserve a name, the familiar name that a person responded to in life, associate it with the site of interment, identify the family and place of birth of the deceased, and attach perhaps one or two suggestive details that point to a larger narrative. Death was regarded as an unavoidable evil, but, after dying, oblivion was an evil that could be evaded. Inscription, as a form of public writing, enlisted every literate passerby to help the bereaved family pass on the remembrance of that name to others (Derderian 2001).

Private Prose and the Personal Letter

Millennia before the invention of writing, Neolithic agriculture and animal husbandry had created self-sufficient sedentist communities, but with the domestication of the horse, the invention of the wheel, and the forging of bronze and iron tools and weapons, humans were once again on the move, and their lines of connection were getting stretched to the breaking point. Literacy facilitated the expansion of kingdoms and the creation of empires. With writing, rulers could use couriers to convey precise instructions to commanders in the field and diplomats in foreign courts. Whereas rivers had been the swiftest means of travel, they had not always been the shortest connection between two points. In the Bronze Age systems of roads were beginning to be built, permitting armies, trade, and information to move at maximum speed between the centers of power in the Middle East. Sea lanes would also be used, necessarily so for island dwellers, but they still posed dangers, not the least of which was piracy. Nevertheless, for many citizens of the reviving cities following the Greek Dark Age (ca. 1100–750 BCE), travel was becoming a way of life.

The Iron Age saw increased displacements of persons and populations. Short-term displacements were occasioned by maritime and overland trading, military campaigns, pilgrimages, and, later, tourism. Intermediate-term sojourns were occasioned by the necessities of work, military postings, and, for the elite, education in particular centers of learning. Long-term displacements took the form of immigration, diaspora, exile, and involuntary servitude. The genetic and cultural homogeneity of populations was progressively breached as conquerors and conquered intermingled, albeit unequally, in the same geographical spaces, and various notions of foreignness emerged: for Greeks they included the foreign guest (*xenos*), the resident alien (*metoikos*), and the settler (*epoikos* or *katoikos*).

Circumstances such as these promoted letter writing as a first-millennium BCE form of maintaining social networks. Of course, just as not all public inscriptions were epitaphs, not all letters were personal. Some were military communiqués, diplomatic letters, business letters, letters of recommendation, letters of invitation, thank-you letters, and letters accompanying gifts. Nor were all letters private. Before the invention of printing made multiple copies quick and easy, it was standard practice to address a single handwritten letter to a group and have a designated reader recite it to an assembled audience. Proclamations were thus disseminated, as were teaching documents, such as the letters of Epicurus and of Paul of Tarsus. Recitation was, of course, also necessary in order to reach persons who could not read.

By "personal letter" I mean a text written or dictated by one named addresser and then sent to one other person, the named addressee. This contrasts with an epitaph, which, as a piece of public writing, addresses whoever comes along as a "passerby," "wayfarer," or "stranger." Again, unlike an epitaph, which is essentially a monologue in so far as it excludes the reader from any interlocutory role, the personal letter, though monologic, is fully meaningful only as part of a correspondence, an ongoing long-distance conversation. In that sense, it may be considered a dialogic document.[14]

That each letter in the correspondence is a single conversational utterance is a conventional fiction, however, as the interposition of days or weeks between responses hardly makes for a real conversation. While dialogue is the fictive model for the ongoing exchange of letters, the fictive model for the individual letter is the face-to-face meeting. A Greek or Roman letter typically begins with a third-person narrative tag, e.g., "Publius greets his friend Quintus" or simply "Publius to his friend

Quintus, greetings," and usually ends with "farewell" or "stay well." This one-way "conversation" includes the wish on the part of the addresser that the addressee is enjoying good health and, as though the addressee has asked, the assurance that the addresser is also quite well, thank you. What may follow are requests and news, some of which may be confidential information. The salutation on top and the valediction on the bottom also indicate to the addressee that the document is complete (Muir 2008). Epitaphs, in contrast, usually start with the salutation of a general sort but rarely end with a farewell.

If the message was written on a wax tablet, a *ptyx* or *tabula*, or on two sheets of papyrus, written sides facing each other and tied with a papyrus fiber, it would be secured by a clay seal, perhaps marked with a distinctive impression. Unlike an epigraphic script that actively invites public performance, the personal letter was writing strictly intended for literate perusal and was either read inaudibly or recited in solitude. The popularity of collections of real or fictional letters lies partly in the thought that we are reading someone's private mail and learning intimate secrets.

Several other differences are also worth noting. The epitaph is fixed in one spot; the letter, which has already perhaps traveled many miles by land or sea, remains portable in the hand of the recipient. This means it can easily be taken about, reread many times, and its meanings carefully teased out. The epitaph, on the other hand, must be recorded mentally if it is to be retained in the memory of passing readers and, for that reason, is composed in a few lines of carefully crafted verse. The personal letter, being rereadable, can afford to be in longer, more expansive conversational prose. As part of a monumental display, including perhaps architectural features, a statue, and adjacent buildings, gardens, or landscape, an epitaph can occasion an engrossing visual experience that does not invite readers to visualize other nonpresent places or persons. A letter, on the other hand, can and often does describe absent places, persons, and events for the reader to imagine.

Personal letters, moreover, are not intended to endure through the centuries. In Greco-Roman times they were normally written on readily available material, which usually meant papyrus of middle or low quality, and each was a single document. If a letter is later collected and published in multiple copies, it stood a better chance of surviving, but *if* it was collected, it was probably because the letter writer was an illustrious public figure, a philosopher like Aristotle or a statesman like Cicero, in which case its content would be less personal than didactic. Yet even celebrity was little help in preserving these texts: Aristotle's letters were entirely lost, and Cicero's went missing until the fourteenth century, when Petrarch rediscovered most of them and had them published (Rosenmeyer 2006, Poster and Mitchell 2007).

By and large, our archive of Greek and Roman personal letters has been reduced to a handful of manuscripts accidentally preserved in the dry climate of North Africa and to a number of published works that adopted the personal letter as a literary genre, e.g., the *Epistles* of Horace, the *Heroides* of Ovid, and the love letters of Philostratus II. Because of the dearth of actual examples from this period, I have drawn up a theoretical model based on the conditions that apply to any and all personal letters, then and now.

A personal letter operates in the discourse world of an addresser and an addressee, a pair of real-world participants that here depend on written monologue but might, under more desirable circumstances, communicate in oral dialogue. As such, a letter may be regarded as constituting a text world that depends for its structural elements on the spatial and temporal factors unique to the epistolary situation. Figure 4.1 represents that situation.

A/a. Prior lives. Unless the addresser is writing to a parent or a close sibling, we assume that she has had years of lived experience before meeting the addressee and that this provides some context for her personal history. Likewise her addressee will have also had a preacquaintance history. The past history of each may be partly known to the other and referred to in her letter. (Note: for the sake of convenience, I will refer to the addresser in the feminine and the addressee in the masculine. The capitalized letters in the graph are used to indicate the location and times of the "I," the addresser as the point of view, or deictic center; the lowercase letters indicate the circumstances of that absent addressee.)

B. Shared experience. Both addresser and addressee will have had some sort of mutual history, which is assumed to include face-to-face dialogue. *B* represents the principal context for the epistolary text, the shared frame of personal reference, subsuming the cultural context that the two parties are also likely to share. The top horizontal bar, extending to the right from *B*, represents the place of the addresser; the bottom bar represents the place of the addressee. These two separate bars represent their current spatial separation, a fact that may become an important

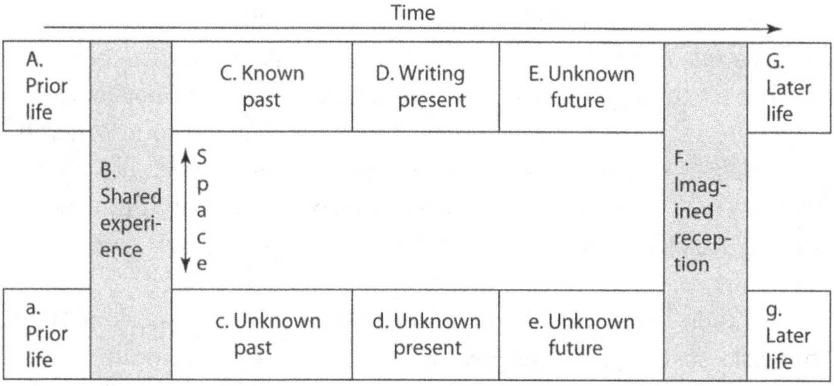

Figure 4.1 The epistolary text world.

underlying theme in the text. When referred to in the letter, *B* is customarily set in the simple past tense.

C. The known past. We will assume that the letter writer, as the deictic center of this text, knows her own prior life (*A*), may be acquainted with *a*, and can refer to *B*. *C* is that specific portion of her own immediate past that includes the events that have occurred since her last communication with him, whether in person, by letter, or by other means.[15] Whatever portions of this that she chooses to include in the text will be in the nature of news or gossip, recounted from episodic memory or hearsay and, of course, from her particular vantage point in the present. The customary tenses for *C* are the simple past and the present perfect.

c. The unknown past. This, on the lower bar, represents the same temporal interval but, because it was experienced by her absent addressee, it is a past interval as yet unknown to her. Though she can empathetically project herself into it, it must remain for her a topic of speculation, affectively toned perhaps by one or more emotions, such as envy, concern, anxiety, or other uncertainties. Its customary tenses are the simple past and the present perfect, its mood the conditional.

D. The writing present. Though a letter writer surveys all epistolary time zones, the consistently dominant tense in every letter is the present. It is understood that this verbal medium mimics face-to-face direct discourse as indicated by the formula of the salutation, which in the Greco-Roman convention included the names of the two parties as though some third party were describing their meeting one another here and now, e.g., "Secunda Publio salutem dicit" (Secunda says hello to Publius), and the valediction.[16] The writing present will sometimes be thematized in the text through references to the weather, the immediate surroundings, the mood of the writer, etc. As part of this desire to concretize this present moment, she will also style the discourse in such a way as to convey a particular recognizable voice pattern, a rhythm, tone, and diction suitable to the effect intended—cool, casual, funny, demanding, intimate, etc.

In *d, the unknown present,* the two parties are now coincident in time but dislocated in space. His present, *d,* unknown to her, coincides with her present, *D,* unknown to him, but this is not a symmetrical relationship: in *D,* the addresser is aware that her addressee is *un*aware that she is at this moment engaged in writing. This serves to frame the letter as a peculiar form of discourse. Unlike a dialogic response, which is normally determined in a present context by the principle of relevance (Wilson and Sperber 1990), a personal letter is self-initiated, self-relevant, and potentially self-revelatory. When this interval is thematized, the writer uses the present tense.

E. The addresser's unknown future. The addresser knows that her addressee cannot know what she is now verbalizing and may not even know a letter is in transit. Reflecting on this may inject a degree of anxiety into her text. Since her addressee is spatially separated from her, it will take time for the letter to arrive, an interval into which she projects plans, fears, expectations, hopes, desires, etc. that may find their way implicitly or explicitly into the text. The customary tenses used to indicate these considerations are the simple future and the future perfect.

e. The addressee's unknown future. If his lived experience in *c* and *d* are, at best, conjectural to the addresser, those of *e* are merely hypothetical and are as yet as unknown to him as they are to her. All that she can be certain about is the uncertainty they both confront. The customary tense for *e* is the simple future.[17]

To summarize: So far, the addresser has been able to imagine her addressee by accessing episodic memory data from *B,* the times and places in which they were physically copresent before separation. She has also been able to imagine him projected into three additional spatiotemporal fields, each with its unique cognitive/affective contents: *c,* the interval of separation before she begins her letter; *d,* the interval he experiences during which she composes the letter; and *e*, the interval between the writing and the arrival of the letter.

F. The imagined reception. Finally, there is *F*, the place and time in which he reads her letter and in which the two share a copresence, albeit imaginary. Here the spatiotemporal zone of the act of writing, *D*, becomes mapped onto *F* ("as you read this, understand that I am now . . .").[18] The underlying present tense of the writing here subsumes the future event, the reception/reading, and draws addresser and addressee into virtual communion. The addresser's deictic projection of herself into this event is rarely so complete as to warrant an expression such as "I am here with you"—if a projection of *F* into the future occurs, it would likely take the future perfect, as in: "When you get this letter, I will have . . . ," for even in the imagined moment of reception the writer knows that her presence to her reader is at best wraithlike and hovering. As she verbally projects herself into the present moment of the reception (her addressee's present tense), she knows that the present moment of the *writing* will at this point be already days, perhaps weeks, in the past—that even that portion of her unknown future, *E*, will have become a portion of her known past. Beyond the reception of the letter may be its relevance to her and his future lives, dimly envisioned as *G* and *g*.

As we have seen, a letter functions like a ghostly extension, a sort of astral projection of the addresser across the space-time gap that separates her from her addressee. As a medium, a letter operates under peculiar constraints. Any here-and-now communication, being dialogic, would allow for an oral exchange of questions and clarifications, all supplemented by paralinguistic information in the form of gestures, facial expressions, and intonation. A written communication is not so manageable. Its interpretive context is not grounded in a perceptually shared present but is instead constructed out of fourteen distinct temporal intervals, or zones, only four of which (*A*, *B*, *C*, and *D*) the writer can know. Though we probably all take this communicative format for granted, this piecing together of four knowns and ten hypotheticals is, when one stops to reflect on it, an extraordinary cognitive achievement.[19]

From the addressee's point of view, the text can be a palimpsest of multiple intentions and emotions. Reading between and underneath the lines, he may discern or construe conflicting, unresolved attitudes on the part of the writer. These may, of course, be unfounded assumptions on his part, but a monologue written and read in silence and solitude can indeed reflect the state of mind similar to what John Keats called "negative capability, that is, when a man is capable of being in uncertainties, mysteries, doubts, without any irritable reaching after fact and reason" (Keats 1899, 277). As I will suggest in my next chapter, monologic literary genres, such as the lyric, derive from the personal letter much of their emotional complexity.

Summary

With this chapter, coming as it does midway through the book, I deal with that transitional period, which has arrived in different places at different times in history, when writing emerges in an oral culture and, as a new mode of communication, generates new kinds of information. The main purpose for communication appears to be, first, to attract the attention of others and, second, to motivate their behavior by re*mind*ing them of particular things and what to do with them and generally to *mind* the voice of the speaker. Extended verbal structures such as stories and songs, which we collectively possess and store as verbal artifacts, also need ways to remind us of what they have to say. For this they need appropriate mnemonic systems.

In previous chapters, I have focused on the mnemonic systems by which oral cultures preserve extended verbal information. This they achieve by using named places and persons to cue the unfolding of extraordinary events and by repeatedly performing those narratives to imprint them voice-to-ear in the memory of listeners. When in this chapter I took account of visual mnemonics, I first considered iconic signs fashioned to resemble persons and places and thereby bring to mind narratable events. In this way I intended to broaden the concept of pictographic script to include architectural design, paintings, and sculptures as instruments of an eye-to-voice-to-ear communicative relay that called for a kind of what we might call preliterate reading.

One of the most important purposes of iconic representation and visually cued speech in an oral culture is the preservation of collective memory as an archive of a people's past. What is the past that an oral culture imagines as its own? In the absence of written records, iconic signs and the interpretation they prompt create a version of the past that is often quite different from our own. Not only is it filled with what we recognize as mythic agents and incidents, but it is also subject to gapping and foreshortening, or "telescoping." Using Greek archaic culture as an example of a functionally oral society, I suggested how such a society came to view its past and the leading figures of that past.

Visual icons had long existed in order to be converted vocally into speech and thereupon evoke visual images in listeners' minds, but the crucial next step was to create a set of visual percepts that represented speech directly. With this innovation, visual marks lost their iconicity and became graphic symbols of sounds, at first of syllables, then of phonemes. Writing thus became arbitrary *visual* signs for speech sounds that, several hundred thousand years earlier, humans had strung together to form arbitrary *vocal* symbols for perceptible referents.

The two early genres of writing, the epitaph and the personal letter, I regard as key to the development of literate art forms because both seem driven by that primary motive of communication, the personal desire to be remembered. In different formats, each embodies those two wishes: be mindful of me, and be motivated by my wishes. The epitaph does so by preserving a name in the face of death, the personal letter by maintaining presence in the face of absence. Exercises in wishful thinking they may be, yet both reflect that deeply human instinct to be with others or at least not to be permanently alone, and, like music and dance, they are means of affirming social solidarity. The fact that literary texts are written in silence and solitude and, in the course of cultural evolution, came also to be read in silence and solitude may account in part for the themes of nostalgia, loss, and yearning we associate with literary prose and poetry.

Five

Poets' Play and Plato's Poetics

Singing is a universal human behavior combining words that awaken the memory, melodies that arouse the emotions, and rhythms that move the body. The singer, who might also play a drum or stringed instrument, therefore always occupied a prominent place in social and religious gatherings, which often would be held during times of leisure and feasting after other group activities, such as hunting, planting, harvesting, and herding.

In ancient Greece, these performers, the *aoidoi* (singers), were the essential agents of the Muses' art, or *mousikê*. Singers in an oral culture, though prized as archives of communal information, were also adept at improvisation, which would lead to nontraditional, innovative compositions. It was writing, though, that facilitated the preservation of such performance pieces, e.g., the songs of Sappho, Alcaeus, Anacreon, and other lyricists of the seventh and sixth centuries BCE. By the beginning of the fifth century, many, perhaps most, singers were also songwriters or, as we now use the term, authors. To designate this activity, the Greeks chose the common verb *poein* (to build or make) and began calling them *poiêtai*, poets.

In this chapter I will first review the emergence of this new performing art not as I did in chapter 3, as the play of music and dance, but as the play of language, the medium of *mousikê* that writing foregrounded. When Plato viewed poetry, he focused especially on that one element of the Muses' art, the words. As we shall see, he rejected not only publicly performed discourse but also, and, for somewhat different reasons, privately read texts. The latter rejection is especially strange given that he was one of the most prolific writers of the classical era. As a philosopher deeply concerned with verbal practice and theory, Plato has cast his shadow forward over twenty-four centuries of literary thought.

The spread of literacy among the Greeks from Plato's Athens to the third-century Hellenistic kingdoms held a significance that was difficult for those then living through it to appreciate fully. We might compare this two-century transition to the spread of the Internet over the last two decades. Greeks by the end of the fifth century BCE, as we by the end of the last century, had to have been aware that something big was underway. Some had a notion of what was coming, but few, if any, could assess the extent to which this innovation would change the way humans interact with one another and perceive the world or how the new technology would transform the older verbal genres.

Enter the Poet

We have a natural tendency to assume that our own cultural categories may, with a little effort, be found in other cultures both contemporary and past. This has led us to read back into classical Greco-Roman cultures the generic concepts we now regard as axiomatic, e.g., prose as distinct from poetry, oratory as distinct from fiction, and lyric as distinct from narrative poetry. Since one of the principal values of literary genres is that they frame readers' expectations, such assumptions may become barriers to an understanding of these classical texts.

For example, the ancients seemed not to have made a sharp distinction, as we do, between epic and drama. Except for the formal differences between a solo performance that alternates narration with speeches and an ensemble performance by multiple masked actors, Plato and his contemporaries regarded both these genres as narratives (*diêgêseis*) publicly presented by vocally and histrionically trained performers. As mentioned in the last chapter, both Plato and Aristotle regarded Homer as the first tragedian, Aristotle adding that the mock-epic *Margites*, also attributed to Homer, was the first comedy (*Poetics* 1459a30–b17). As for their plots (*muthoi*) and characters, epic and tragedy drew from the same, predominantly Homeric, cultural archive, as did hymn, dithyramb, choral ode, and lyric.

Then there is that even more fundamental problem of what to call that class of persons responsible for composing works in any of those varied genres. The history of professional titles can be highly significant, for whenever the name for a given artisan changes, it is likely that the art itself has changed. For example, the name for the person who would materially produce books in the early fifteenth century CE was "scrivener," or "scribe." By the end of that century, some of those artisans began to be called "printers." We know why that nomenclature changed and need to inquire now why an analogous name shift occurred in Greece in the fifth century BCE.

Before the fifth century, composers and performers of verbal artifacts, as distinct from forensic and deliberative discourses, were commonly referred to as *aoidoi* (singers). Neither Homer, nor Hesiod, nor the early lyricists called them "poets." The earliest representation of the archaic singer appears in the *Odyssey* in the persons of Demodokos and Phemios. Of Demodokos, King Alcinoos of Phaeacia says: "On him, above all others, some god has bestowed the gift of singing to give delight through whatever his spirit (*thumos*) moves him to sing" (8.44–45). Later, the Ithacan bard Phemios, hoping the returned Odysseus would spare his life, tells him, "I am self-taught (*autodidaktos*). The god put all kinds of passages (*oimas*) into my mind (*phresin*)" (22.347–348). In these statements Homer accounts for the improvisatory skills of a singer as a collaboration of two autonomous agents: first, an external god and, second, an internal spirit. The latter, the *thumos*, was believed to be located within the chest region, the *phrenes* (translated above as "mind"), in which personal memories and general knowledge were stored. As the seat of intuitive judgment and instinctual energy, the *thumos* is perhaps what was understood here as the *autos* in *autodidaktos*: "the god put the passages in my *phrenes*, where my inner self, my *thumos*, could learn them." The divine and the human sources were different, yet they had one feature in common: they functioned without fully conscious control on the part of the singer. This aptitude may be enhanced by practice but is not really extraordinary. After all, every time we open our mouths in spontaneous speech, the sources of the thoughts we express and the words we choose are nonconscious. When we feel an impulse to address a topic, we can usually do so with unpremeditated fluency.[1]

During the fifth century, a new professional title begins to appear: "poets" enter the scene as "makers" but not necessarily as singers of their own compositions. *Poiêmata*, consequently, became "verbal *objets d'art*, the almost tangible products of an artisanal process" (Ford 2009, 132). The first evidence of this word as applied to verbal artifacture appears in Herodotus together with the first of a series of nouns formed with the related suffix *-poios*, as in *epopoios* (verse, or speech, maker, i.e., epic poet). Later in that century new words were coined, such as *melopoios* (lyric maker) and *tragôidopoios* (tragedy maker). By the fourth century the suffix was used for other genres, too, such as *iambopoios* (maker of iambic verses), *elegeiopoios* (maker of elegiac couplets), and *dithurambopoios* (maker of dithyrambs). (Cf. the English suffix *-wright* as in "wheelwright," "shipwright," and, by extension, "playwright.")

When Aristotle set out to define the *technê* of poetry, he meant to include all matters pertaining to *poiêsis*, which for him meant the authorial "making" of epic, tragedy, comedy, dithyramb, flute playing, and lyre playing. Like the visual arts that employ the colors and shapes, these six performing arts are imitative media. That is, these six "poetic" arts imitate human feelings and character through rhythm, speech, and harmony, singly or in combination (*Poetics* 1447a23–28). His inclusion of lyre and flute playing as media of poetry seems strange to us. We do not now think of musical compositions as poetry and so should ask ourselves why Aristotle did. One answer is that, having declared the overarching principle of *poiêsis* to be imitation, he considered nonverbal music, using rhythm and harmony (melodic structure), to be imitating human emotion. But this leaves us with yet another question: if a "poet" is a skilled craftsman who produces "poems" for skilled performers to enact, who are the persons who make the compositions that lyrists and flutists perform? That is, if *epopoioi* were makers of verse narratives (epics) and *melopoioi* were song makers who set words to melodies, who were those responsible for making wordless, instrumental pieces? When we search Greek dictionaries for those *-poios* words that might denote the composers of music for stringed and wind instruments, we find that, while the words *luropoios* and *aulopoios* do indeed exist, they mean *makers* of actual stringed and wind instruments, not composers of musical pieces.

This implies that Aristotle considered the essential function of a poet, whatever his or her generic medium, to be the artful crafting of a specific instrument to be used in a mimetic performance by another artist, such as a rhapsode, actor, singer, or musician. In its introductory sections the *Poetics* suggests that a *poiêtês*, broadly defined, is any maker of a useful artifact, i.e., a tool, or instrument. As Aristotle proceeds, he narrows this meaning to that of a maker of *verbal* artifacts. But, broad or narrow, his usage of this term indicates that a division of labor has emerged in which production and performance are two distinct *technai*—the one is *poiêsis*, the other *mimêsis*.

As Andrew Ford (2009) has pointed out, the primary reason for this name shift from "singer" to "poet," as well as the division of labor between composer and performer, was the increasing use of writing in the fifth century. Pindar, Aeschylus, and Sophocles still spoke of "singers" (*aoidoi*), but prose writers and, later, Aristophanes and Euripides referred to "poets" as the makers of verbal artifacts. Eventually Greeks and, somewhat later, Romans came to sense that verbal artifacts could be divided between publicly spoken or sung performance on the one hand and privately read or recited writing on the other. The latter, newer kind of verbal artifacts came to be generalized as *grammata* in Greek and *litterae* in Latin, each the plural form of a word meaning "written character," i.e., *gramma* and *littera*. Accordingly, any discourse inscribed on any surface was seen as a series of separate letters, a visual representation that sharply differentiated it from the audible stream of sounds produced by orators, actors, singers, or everyday conversationalists. *Grammata* and *litterae* came to signify what we now mean by "literature," not necessarily "imaginative literature" or "*belles lettres*" but simply anything written down, including history, law, philosophy, and natural science.

It was writing that from the seventh century, the close of the Greek Dark Age, exerted those pressures that gradually were to change how these cultural artifacts were understood and used. Those genres dependent on imaginative invention, or poetic play, though slower than history, philosophy, and the sciences to adapt to the new medium, also registered its steadily growing influence. But from the fifth to the third centuries BCE the most sensitive gauge of this literate transformation was rhetoric, which, as the principal means of conducting public affairs, pervaded all forms of verbal experience.

Plato reacted with dismay to written epideictic oratory as the Sophists practiced it. Their demonstrations of what writing could do when artfully deployed seemed to him a dubious improvement over traditional oral culture. By the mid–fourth century, Isocrates, Alcidamas, and Aristotle had become focused on two rhetorical styles, *lexis agonistikê*, long used in spontaneous forensic debate, and *lexis graphikê*, increasingly used in formal epideictic speechwriting (*Rhetoric* 3.12). The older improvised style, which Plato had extolled in the form of Socratic dialectic, was energetic and dialogically responsive; the newer, carefully prescripted style was marked by monologic argumentation embellished with epigrammatic figures, learned allusions, and sustained delivery. Yet, despite the benefits that written style had won through careful composition, it had lost the liveliness of interpersonal exchange.

In the *Poetics*, Aristotle's main interest was tragedy and its formal differences from epic, but, as a systematic thinker, he felt obliged, as I just noted, to deal with those other imitative arts at the start, if only to set them aside as lesser genres. He even felt obliged to mention another medium-defined art form, even though he says he has *no name* for it. This art form imitates simply by using speech (*logos*). Here the anonymity of this genre seems a consequence of the ambiguity of that word, *logos*, which could mean speech, written prose, reason, or language. This genre evidently perplexes him. What shall we make of those unmetered dramatic sketches, the mimes of Sophron and Xenarchus and the Socratic dialogues (*tous Sôcratikous logous*, 1447a29–b11)? (Note, by the way, that Aristotle regards these two kinds of composition as "poetic" even though they are not written in verse.) As prose pieces written in the agonistic style, both mimes and philosophical dialogues represented lively conversations designed to illuminate differences of opinion and of character. These *poiêtai* had taken this sometimes combative give-and-take, so essential to tragic and comedic conflict, and made it available on the written page to the amateur actor or imaginative reader.

Aristotle may not have had a name for this kind of writing, but he obviously knew how it worked. Even when one hears someone tell the story (*muthos*) of Oedipus, one can be deeply moved (*Poetics* 1453b3–6), and toward the end of his treatise he declares that tragedy, like epic, does not need to be performed to produce its effects: it also produces a vivid (*enarges*) experience when only read (*anagignôsei*)—though its most vivid effects are produced when it is fully staged with *mousikê* (song, musical instruments, dancing, gesture, and acting) together with visual spectacle (1462a11–13, 17–18).[2]

Theatrical pieces that could be enjoyed without having to be staged had become popular among the growing literate class, but this innovation seemed anomalous to Aristotle. One reason it might have emerged in the fifth century was the fact that by then Greek settlements had spread far beyond the Greek mainland and Aegean region, especially far from the cultural capital, Athens, an expansion that would accelerate following Alexander's conquest of Egypt and the East.

War, too, had been a factor. Since travel and trade between the Greek homeland and the cities in Sicily and southern Italy were periodically disrupted during the long Peloponnesian War (431–404 BCE), many in the West were eager for the latest productions of playwrights such as Euripides and Aristophanes. Plutarch in his *Life* of Nicias reports that the Syracusans were so starved for theater that, following Athens's naval debacle of 413, an Athenian prisoner able to recite speeches from the latest tragedies of Euripides might win his freedom. It is therefore possible that dramatic texts such as mimes may have caught on first among Greeks living abroad.

Word Play

A *poiêma*, broadly defined, is a thing made or done, a noun equivalent to the Latin *factum*. Its narrower meaning is a skillful construction, an artifact, e.g., a house, a woven garment, a hand tool, in short, any humanly useful object including a "work of art." As with other tools, we carefully preserve such works until we have occasion to use them again, and, when we do, we transform them in interesting ways. As we interact with them, they return to the status of playable instruments. Even paintings and sculptures require human intervention. A painting in a totally dark room loses its colors and contours until light is turned on it and a pair of skilled eyes contemplates it. Similarly, a dramatic script, a musical score, or a novel, when stored on a shelf, is an object that can only be fulfilled through sound, real or imagined.

A tool is not a standalone object. It is an extension of the human body, its brain, nervous systems, and musculature, and is inextricable from human intention. A made tool entails the distinction between *what* it is and what it is *for*, and what links these two is intention. The latter can be fairly straightforward when the context is apparent, but play behavior problematizes intention. We especially associate the performing arts of music and drama with play: musicians "play" instruments and "play" melodies, and actors "play" roles in "plays." Prose fiction and poems—these too must be played, either aloud or in the imagination of the reader.

Play is an activity that entails pretending that X is Y while knowing full well that X is *not* Y. This obliges us to hold onto two contradictory ideas at the same time. These persons on the stage, for example, are both the characters they play and the actors we know them to be. And these musicians, whose emotions seem revealed in those surging waves of sound we hear, seem to embody some higher power, though we may know them well and know that their lives outside the concert hall are quite ordinary. Even the visual arts involve degrees of imaginative play and call for our voluntary engagement in illusion.

This instinctive capacity for play is not uniquely human but is shared by most mammals and birds, especially young ones. Ethologists distinguish three kinds of animal play: locomotor, object, and social play (Bekoff and Byers 1998, Burghardt 2005). *Locomotor* play involves extraordinary, energetic movements. *Object* play occurs when an animal, typically a carnivore, pounces on an inert object, treating it as though it were prey.[3] Locomotor play is often combined with object play, e.g., when a dog takes a stick in its teeth and shakes it while making exuberant leaps or spinning about.

Locomotor and object play (with its subset, exploratory play) are solo activities, but *social* play is somewhat more complicated. Not only does it involve two or more participants, but it requires that these participants accept in advance that their interaction will be *play*. When Gregory Bateson (see chapter 1) observed monkeys in rough-and-tumble social play, he reasoned that they must have already exchanged what he called a metacommunication, a signal that meant that subsequently communicated signs, in the form of aggressive displays and actions, were not to be interpreted as what they ordinarily mean, namely, threats of physical attack and injury. Dogs very clearly metacommunicate play by performing the "play bow"—head back, front legs extended forward, chest touching the ground, back legs stiff and tail erect. We humans, both young and old, have our own ways to send the play signal—through body language, facial expressions, tone of voice, explicit utterances, and ritualized game settings. Social game play can be more complex than the other two behaviors. It may, for example, include elements of locomotor and object play in the form of chasing, escaping, wrestling, and celebratory

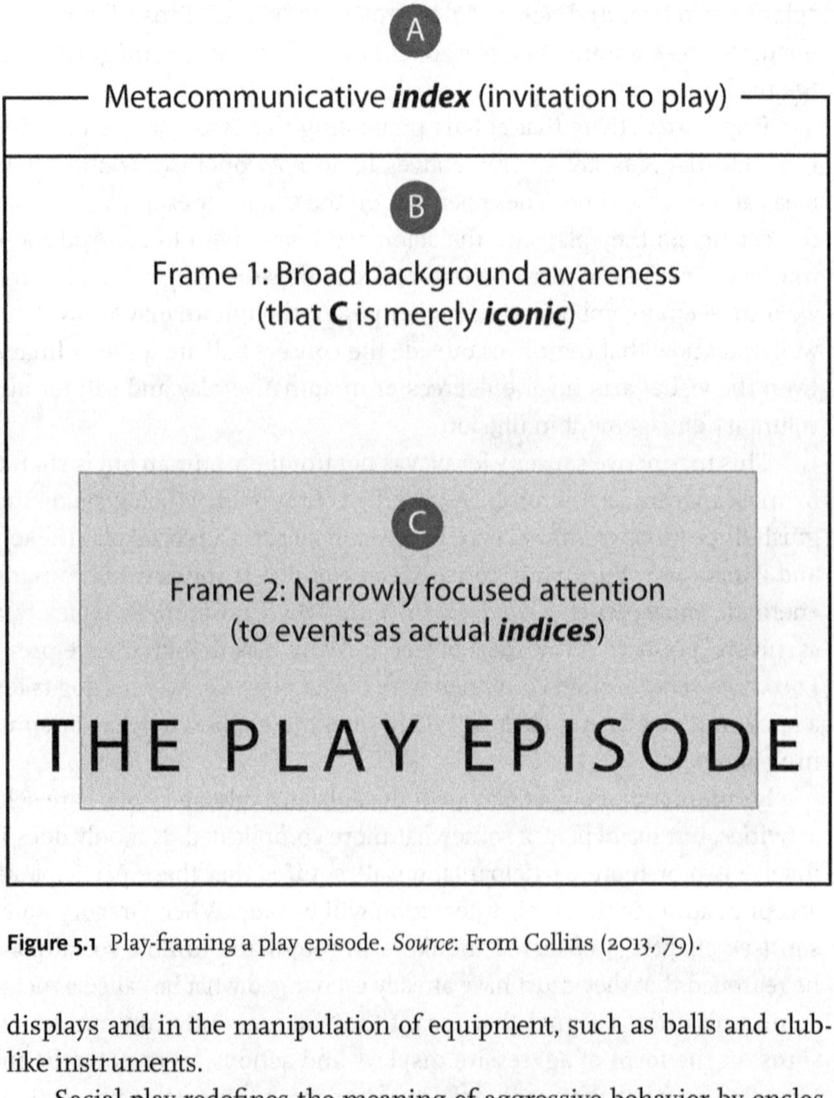

Figure 5.1 Play-framing a play episode. *Source*: From Collins (2013, 79).

displays and in the manipulation of equipment, such as balls and club-like instruments.

Social play redefines the meaning of aggressive behavior by enclosing it within what Bateson called a "play frame." In semiotic terms, this happens when a set of actions that normally serve as indexical signs (of dominance, submission, hostility, fear, violence, etc.) are framed, or recontextualized, by the mutual knowledge that these indices are actually icons, i.e., performances that only resemble an aggressive encounter. Figure 5.1 represents this play framing.

We humans can do all three kinds of animal play, but, when we perform them, we are able to do so in ways that are uniquely human. Our metacommunicative skills, enhanced by language, can initiate much more nuanced and complex forms than can any nonhuman animal. This is especially the case when we engage in social play, such as flirting or good-natured teasing and banter. Here we assume an attitude that our coplayers know we do not really have. When, however, we assume a different identity, we shift into a wholly new form of play, *imitative play*.

We learn imitative play quite early in life. Within the first year of life, babies exhibit the need to share their parents' identity by imitating their speech sounds and intonations. Culturally evolved skills are each built upon this imitative drive, rooted in the play instinct of our exceedingly social species. Our success in this presupposes the empathetic aptitude to imagine ourselves as others and assume their perspective. When, for example, we listen to music, we must first allow it to entrain our biological rhythms before we can experience the emotional patterns of the melodies in synchrony with others (Cross 2008). The social play of nonhuman animals may sometimes exhibit what we interpret as imitative play. But no species has developed the knack of playful mimicry to the same degree as humans, who regularly combine locomotor, object, and social play with imitative play especially in the form of role-playing.[4]

When that uniquely human trait, language, serves as the medium of imitative play, we might term that behavior "word play." (Note that this is a broader concept than a "play on words" such as a pun, acronym, malapropism, etc.). In this sense, however, all language is word play. The signs expressed by the sounds we use when (in English) we articulate the word "dog" or "bird" are arbitrary signs, i.e., symbols, that represent these only because English speakers agree to use them to designate these two referents. This agreement is the precondition we all accept when we first acquire speech, no matter what linguistic community we happen to be born into. Every baby's impulse to learn language and its parents' impulse to teach it, by way of exaggerated intonation, repetitions, eye contact, and other interactions, are essentially play behaviors, making language the first game we ever learn to play.

If, as I suggest, our aptitude for word play is based in animal play behavior, we might inquire whether the expressive kinds of animal play are also reflected in the human performing arts of the sort identified in chapter 3 with *mousikê*. For example, the three kinds of play, locomotor, object, and social, align in certain ways with those three complementary performing arts, dance, music, and drama. Dance qualifies as a human mode of locomotor play; the playing of musical instruments is a form of object play; and drama, presented as a conflict, or agon, among two or more participants, corresponds to social play.

The ways in which these three performing arts are dissimilar from animal play behavior involve those two uniquely human traits: mimesis and language. When human locomotor play becomes dance, it often takes the form of rhythmically coordinated patterns involving a group of dancers mimetically reproducing movements in concert with others. Likewise, the performance of instrumental music often includes verbal content and a number of instrumentalists in mimetic repetition. Drama, which can incorporate rhythmically timed and metrically organized dance and music, foregrounds speech, sung or declaimed. Finally, the performing arts accumulate a repertory of artifacts, in the form of preserved, repeatable movements, sounds, and words, which is not the case in either nonhuman play or in improvised human play.

The verbal arts have woven word-play principles into the very fabric of their artifacts. Unlike the play Bateson observed in the monkey house, which I have analyzed in semiotic terms as indexical signs enclosed in a frame of iconicity, word play is not a twisting and turning of bodies in motion but rather the twisting and turning of symbolic signs that we have come to know as tropes. Thus metonymy, i.e., the use of a noun to direct attention from a cause to an effect, from an effect to a cause, from a part to a whole, or from a whole to a part, is clearly a verbalized form of *index* perception. By asserting a partial resemblance between two objects or concepts, metaphor plays with verbalized *iconic* signs. Another major trope, irony, plays entirely within *symbolic* signs and for that reason may be regarded as the purest form of word play. Of these three tropes, metonymy seems the least playful, for, as a verbal form of indexicality based on the contiguity of two items, it requires no extraordinary leap of the mind across semantic domains. If we are looking for a good example of mental leaping, we might choose irony, the trope by which some or all of a discourse must be understood as meaning the opposite of what it seems: here we find a verbalized mode of social play, such as teasing and banter, when expressed through litotes, hyperbole, and some forms of sarcasm. However, the most exuberant variety of word play is surely metaphor, including simile and personification.

Figure 5.2 shows how metaphor may be fitted as word play into the earlier play-frame diagram. The word play of metaphor, like every other

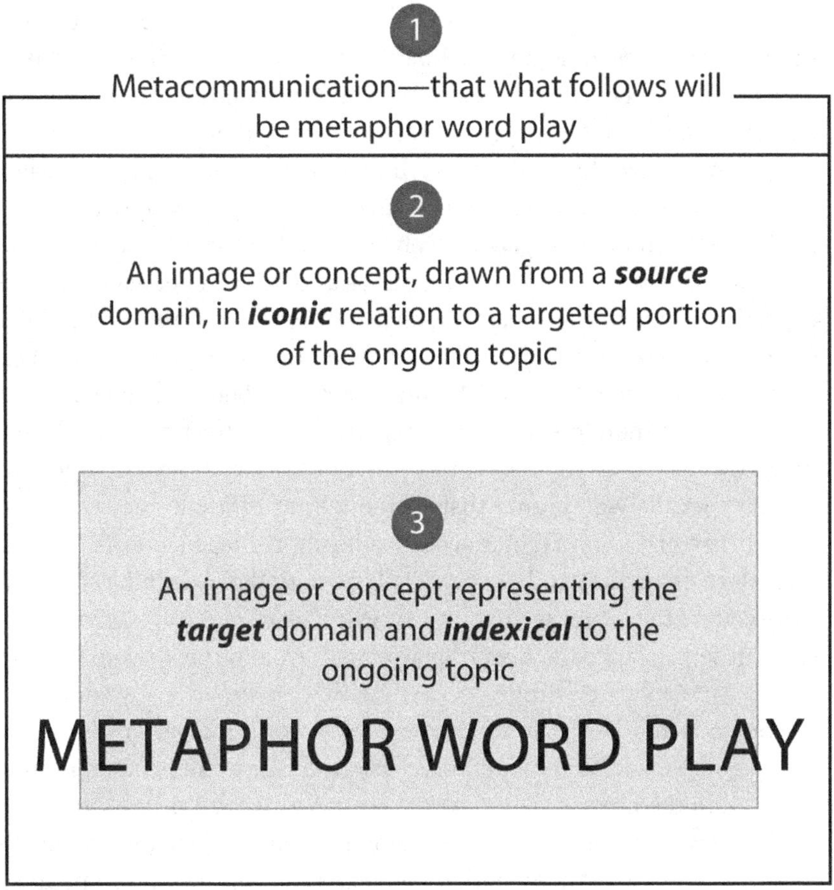

Figure 5.2 How metaphor as word play is play-framed.

kind of play behavior, works only when two disparate intentions coexist, separable only by semiotic difference. One advantage of the foregoing diagram is its emphasis on the simultaneous copresence of both components. The centrally framed component (3) is a portion of the ongoing topic of the discourse, a portion indexically connected to the elements that form the situation present for our consideration. For example, when Romeo declares, "But, soft! What light through yonder window breaks? / It is the east, and Juliet is the sun!" our focus is momentarily shifted from Juliet as an object of Romeo's passion—an indexical complex situating Juliet as a Capulet and Verona as a dangerous network of feuding families—to the image of the sun that here serves as an iconic sign. The momentarily present metaphor image, or concept, (2)—here, the sun—drawn from a relatively remote source domain, iconically signifies a portion of the ongoing topic as its target (3)—here, Juliet.

Whenever we become aware that we are about to hear or read a verbal artifact, be it prose fiction, drama, song, or poetry, we can predict that some form of word play is likely. If we value the experience of this verbal artifact, we accept the play mode as a condition of engaging in this experience and willingly suspend disbelief for the interim. As I pointed out in chapter 2, when anyone begins to tell us a story, we first of all expect that, since this tale is a *re*telling of a *re*membered episode or set of episodes, it is therefore subject to expansions, contractions, flashbacks, flashforwards, and all those other discontinuities typical of episodic-memory recall. We assume that in the telling of a tale the sequential chronology of "story time" may be radically reshaped into "discourse time" (see chapter 2) and are especially alert to the possibility of breaks in discourse time. Just as discourse time is a word-play transformation of story time, metaphor is a word-play transformation of discourse time. When a metaphor or simile occurs, the flow of situationally connected discourse is momentarily interrupted by a word, image, or digression that seems to take us momentarily away from the events and topics represented in the piece, and we alertly respond to this by shifting into play mode. Consider, for example, the Homeric simile that interrupts the narrative at the very moment that Menelaus suffers a wound from an arrow that has pierced the corselet about his waist:

> As when some woman, a woman of Maionia or Caria,
> Stains ivory purple to make a cheek piece for horses,
> This prize lies stored in a chamber and, though coveted
> By many a horsemen, is kept for a king alone,
> Reserved for his horse's adornment and his own glory,
> So, Menelaus, your fine thighs were stained with blood,
> Your calves as well and down over your ankles.
>
> (*Iliad* 4.141–147)

Figure 5.3 represents metaphor and several other varieties of word play in relation to two variables, *size* (length in words) and semantic *complexity*. The vertical y-axis represents the relative range of size typical of particular

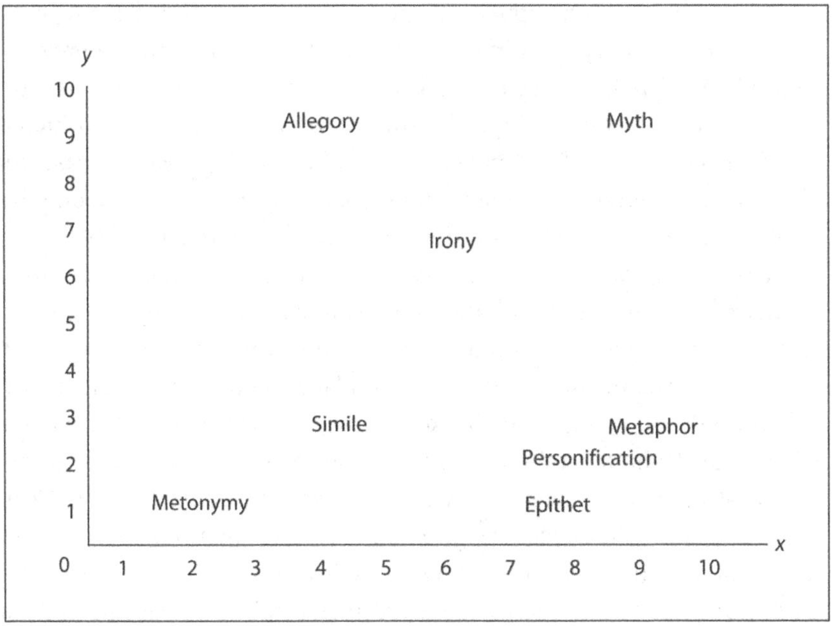

Figure 5.3 Varieties of word play in relation to two variables, size (y) and complexity (x).

verbal entities; the horizontal x-axis represents semantic structure ranging from the most simple and explicit structures (unfolded, i.e., overt and revealed) toward the left to the most complex and implicit (infolded, i.e., covert and concealed) toward the right.[5]

Metonymy is a very brief, usually one-word image that readily prompts some other image. For example, a metonymy that substitutes a word meaning "sharpened point" for the whole sword, as does the Latin noun *mucro*, simply draws attention to its deadliest detail while prompting us to imagine the entire weapon in peripheral vision.[6] By that definition, the association of part to whole, synecdoche, becomes a logical subset of metonymy. More broadly considered, any single reference to an object in a spatial field may metonymically evoke other contiguous or concomitant details, e.g., the mention of a wall evokes the image of a top and a base, a ship evokes the sea it sails on, a chariot evokes a flat surface it moves over. Having the minimal length and complexity, metonymy is the least playful of the tropes.

Simile, which may appear in either short or extended form, plays with the momentary merging of two disparate items, as in the previous example of Menelaus's wound and the dyeing of ivory, and may be considered a subset of metaphor. The simple fact that it includes an indicator, such as "like" or "as," makes it an explicit analogy. By contrast, the absence of such indicators makes metaphor an implicit analogy, one that requires more participation on the part of the hearer/reader.

Homer's epics abound in similes but contain few of what we recognize as metaphors. On the other hand, these works include many instances of formulaic metaphors, i.e., epithets. Familiar examples include "winged" (words), "wine-dark" (sea), "rosy-fingered" (dawn), and "shepherd" (of men).[7] When words are "winged," their likeness to birds is implied (infolded), but what avian traits are thereby mapped onto words? *Swiftness*, certainly, but what of *freedom*, *wildness*, *gentleness*, *loftiness*, and so forth? "Wine dark" may imply intoxication, grief, opacity, and the underworld, and a king, when termed a "shepherd" (of men), invokes a complex pastoral network of metaphors in which soldiers are sheep and sometimes sheepdogs, enemies are wolves, and the battlefield is an abattoir (Collins 1996). Personification functions as another subset of metaphor when it infolds (implicates) a conscious human agent, e.g., dawn as a goddess whose physical presence is preceded by her fingers reaching upward from below the eastern horizon. In this instance, personification also functions as that subset of metonymy, synecdoche, since the goddess's fingers are here the part that stands for her whole body.

Moving upward in the word-play chart, we come to consider examples of folded meanings operating not so much on the level of words, phrases, and clauses but instead suffusing whole discourses. Irony, for example, may indeed occur as a word or phrase, but, if it is identifiable as a speaker's or writer's *attitude*, we may decide to assign to an entire work a meaning different from, usually opposite to, the statements the author makes. In short, the implicit content is cognitively dissonant to the explicit content, and it is our job to ex-plicate this plicature.

Allegory, as explicit play on a large scale, is simile writ large. Allegorists who enter into this word play are normally obliged to share with us the key that unlocks their code, that is, if they expect us to play with them. Not to do so would be as senseless as their inviting us to engage in a game the rules of which are known only to themselves. Socrates, for example, can be ironic as a means of leading interlocutors from bafflement to enlightenment, but when he composes his allegories, he always reveals to us the implicit meanings folded into the explicit details of the narrative.

Myth, however, is a code without a key. As an orally transmitted cultural narrative, it may indeed have begun as an allegory or may be retrofitted with a key supplied by later explicators, but such keys, if they ever existed, no longer accompany narratives passed down over many generations. It is the nature of traditional myth to imply various interpretations that tellers and hearers may either dispute or hold in a state of perpetual equilibrium. The power of mythic discourse thus lies in the mind-teasing, unresolvable polysemy of its episodes.

At the risk of conceding the obvious, let me state that my analysis of word play in figure 5.3 is intentionally "fuzzy." The numbers 1–10 are not there to quantify precise degrees of length or complexity but merely to represent the "more" or "less" of each value. Though it is sometimes useful to draw fine distinctions, there is an advantage sometimes in stepping back, as I have done, and recognizing the family likeness of such forms as simile, metaphor, allegory, and myth as instances of semiotic word play.

Plato on Poetic Lies

Dancing and music making may be judged on a scale from unskillful to highly skillful, but, as soon as language inserts itself into a performance, it introduces another criterion: truth. The Greeks were evidently troubled by the concept of truth, *alêtheia*, and its logical opposite, a falsehood, *pseudos*. Whether *alêtheia* originally meant "unforgottenness" (or "unforgetfulness") or this was a folk etymology, the noun did imply the social obligation to remember what one had witnessed or had been taught and, when necessary, to share that information with others. It also implied the responsibility to remember one's promises to others. In English we have that archaic expression, to "plight one's troth," meaning to "pledge one's fidelity," to promise to be "true" to someone. This social function of memory, which I discussed in chapters 1 and 4, thus includes accurately accessing particular contents from episodic memory and committing oneself, through "prospective memory," to certain specified actions in the future.

In any society that conducts most or all of its business through face-to-face oral interchanges, truth is a precious yet fragile currency. Before writing becomes an option, one person can look another in the eyes, observe his or her body language, listen for telltale vocal inflections, and, if deemed warranted, interrogate that person further. Once an assertion is written down, however, its source no longer needs to be present. Like an affidavit that attests to its own veracity, a written document cannot be conversed with, nor does it have a face, body, or intonations that may support or cast doubt on its words, a point that Plato's Socrates makes in the *Phaedrus*.

The thought that poets are liars did not, of course, originate with Plato. Hesiod had centuries earlier acknowledged the deceptiveness of poetic language when in the *Theogony* (27–28) he represented the Muses as informing him that "we know how to tell many falsehoods that are similar (*homoia*) to real things, / but, when we wish, we also know how to utter truths." As protégés of these goddesses, do poets also willfully lie? And how can we ever tell when the Muses and the poets are lying, if falsehoods and truths look so much alike? The Greeks apparently did accept that poetic discourse, especially in narrative genres, is often at variance with the truth. Had it been a justifiable response to the national need to recover somehow an unrecoverable past that Homer and Hesiod fabricated a Greek prehistory of gods and superhuman warriors? Or is lying simply the way poets make their living?[8]

Plato's judgment on traditional cultural values, as argued by Socrates in books 2, 3, and 10 of the *Republic*, was that much of what passed for "truth" was actually falsehood and that the two men most responsible for promulgating these errors, Homer and Hesiod, had inserted into their poems numerous lies concerning gods (book 2) and men (book 3). These poets often chose to portray the actions of gods and heroes as selfish, mindless, violent, or deranged without any proof that this was so. The gods, Socrates reasoned, have altogether too much power to need to resort to deception and too much wisdom ever to descend into irrational passions. As for the heroes of the past, human though they were, it is morally irresponsible to represent these role models as acting shamefully. Imitative by nature, children must never be taught these tales.

Why, he asked, do poets represent gods and heroic ancestors in such a light? Because, he said, in a public performance it is far easier to portray extreme emotions than it is to show wise and calm states (*Republic* 10.604e–605a). By this answer, Socrates implied that false representations of gods and heroes are a function of the performance medium and motivated by the ambition of professional performers. In the absence of a courtly patronage system, poets depended for their livelihood on a paying public they were eager to please. In this respect, poets had become the equivalent of tuition-charging sophists and hired speechwriters (*logographoi*). In short, yes, lying was simply the way poets earned their living.

The pursuit of real truth needed to be clearly distinguished from the pursuit of counterfeit truth, of lies that merely imitate reality. The conflict between the two, said Socrates, was the "ancient quarrel of philosophy and poetry" (10. 607b). To illustrate his point, Socrates offered a set of verse fragments, slurs presumably written by poets against philosophers: "the bitch yelping at her own master," "that howling hound," "big in the idle forum of fools," the "conquering mob of the overly clever," and the "subtly refined analyzers" who still "live in penury" (10.607b–c). Since it is an ancient quarrel, Socrates could not justly be blamed for having started it. Early in his career, Socrates recalls (*Apology* 22a–23e), he had spent time interviewing poets, asking them to justify their truth claims. He soon found, however, that they could not do so or even tell him what their verses meant. Divinely inspired or not, they must have been out of their minds when they composed those verses. His disparagement of poets was why his chief accuser, the poet Meletus, had denounced him.[9]

Falsehoods (*pseudea*) and truth (*alêtheia*) constitute a polar duality that Plato aligns implicitly with the two presentational modes, *mimêsis* and *diêgêsis*. So, in verbal performance, *mimêsis* occurs when an actor or singer speaks in the persona of someone else, and *diêgêsis* occurs when speaking in his or her own voice. In the first instance, performers become disguised as others; in the second they appear as themselves and testify to the truth of their narratives. Spontaneous dialogue, when used by two or more persons to disentangle truth from false opinion, relied on first-person, *diêgêsis*-like speech in that each participant was expected to contribute what he, not anyone else, truly thought about the topic at hand. This is what Plato referred to as dialectic.

As narrative, first-person *diêgêsis* can, of course, be unreliable even when the narrator thinks that he or she is telling the truth. We've all known people who regularly misremember their own past lives and, deluded though they are, speak with absolute conviction. What they utter are "true lies" (*Republic* 382a), words that are most truly lies because the speaker believes them to be true. Moreover, since these speakers are convinced that the falsehoods they tell are truths, these lies are more readily believed by others who then pass them on to yet others. This deplorable state of invincible ignorance, Plato declares, is the imprisonment of the soul in a shadow world of illusions, which in book 7 he depicts in his "Allegory of the Cave." If illusion is an addiction leading to the imprisonment of the soul, then poetry, even when used recreationally, serves as its "gateway drug."

We, I think it's fair to say, find this conclusion on its face absurd. Fictionality, as a feature of literate word play, has become a concept that we now simply take for granted. As readers, we expect to assume the perspective of fictitious characters, to visualize what any one of them sees, fear what he dreads, and desire what he covets, even if that turns out to be his neighbor's wife and his neighbor's expeditious demise. In short, we submit to the illusionary power of words, imitate these characters in our imagination, and, in so doing, become ourselves the voluntary instruments upon which their authors play. If transported back to fifth- or fourth-century Greece, we might sooner agree, on this issue at least, with the sophists than with the Socratics or the Platonists. We might sense a certain aesthetic kinship with Gorgias, who was known to have composed rhetorical "playthings" for his own amusement and who posed the paradox that "through its plots and emotions, tragedy creates a deception in which the deceiver is more just than the nondeceiver and the deceived wiser than the nondeceived."[10]

While we might regard the works of Homer, Hesiod, and the tragic poets as sources of insight into the human condition, interpreted within a specifically Hellenic cultural context, Plato's fellow countrymen, obviously lacking our time-distanced perspective, tended to view them as sources of truth and reliable guides to proper behavior. Their faith in this oral archive thus resembled the faith of latter-day religious fundamentalists in their own written archives. If we are to believe Plato's assessment, even Athenians had such an undeveloped capacity for verbal play that most could not manage to take their *poiêsis* with a grain of Attic salt. For that reason, as he saw it, they were easily duped by sophistical orators and by politicians who had learned the art of playing upon their credulity.

It is perhaps precisely because we have learned to suspend disbelief and moral judgment while reading a piece of literature that we may also assume Plato's perspective on societal ills and their treatment, if we try. We might begin by recognizing that life for him could not always have been white-robed holiday excursions to the Piraeus or witty soirées in the metropolis. Plato had grown up during what was probably the most disastrous period in ancient Greek history. Soon after their victory over the invading armies of Persia in 480, Athens and Sparta, each with its allied cities, entered into a prolonged period of hostility, decades of open warfare interrupted by intervals of truce just long enough to permit each side to rearm. If traditional dating is correct and Plato was born in 428, his first experience with Athens, which Pericles had called the "Education of Hellas," would have been that of a city painfully recovering from a Spartan siege and from a plague that had killed off a third of the population, which had sought refuge within its walls. Many of the survivors blamed the democratic leadership of Pericles, himself a victim of the plague, for the expansionist policies that had led to this conflict with Sparta. Others questioned their own faith in the gods.

Over the next eight years Athens, bolstered by its navy, recovered much of its strength, and in a number of campaigns fought its rival to a standstill. But in 415, during a period of truce, it turned to the west and undertook a large-scale expedition aimed at capturing the powerful Sicilian city of Syracuse, allied with Sparta. Poorly planned and executed, this venture ended two years later in the utter destruction of that portion of its fleet and the loss of tens of thousands of soldiers and sailors. Demoralized and wracked by dissension, Athens nevertheless fought on for seven more years, until, in 404, exhausted and impoverished, it capitulated to Sparta. Dictating the conditions of peace, Sparta installed a government of oligarchs, the Thirty Tyrants, sworn to suppress Athenian ambitions abroad and citizens' rights at home. When after a year these men were ousted in a popular revolt, a new period of social unrest ensued. Conspiracy theories abounded, followed by recriminations, confiscations, and summary executions. In 399, Plato's old teacher Socrates, rumored to have been connected with Critias, one of the leaders of the Spartan-imposed government, was tried on unrelated charges and executed.

From that point in time, the twenty-nine-year-old philosopher must have looked back upon a whole generation with rage, grief, and shame. Historical events had profoundly shaken Athenians' trust in their gods, their leaders, their neighbors, and their own inner resources. As Plato saw it, the reason they had lost trust in themselves lay in the very concepts they and their fellow Hellenes had so long trusted, the orally transmitted knowledge that constituted their identity as a people. In *Preface to*

Plato, Eric Havelock (1963) termed this their "tribal encyclopedia" and observed that the verses of Homer, Hesiod, and other poets, choral, dramatic, and lyric, had been recited to the young, who were then expected to memorize them as guides to correct judgment and right actions. Acquiring and later accessing this preformulated, catechetical knowledge required a suppression of selfhood. That is, while engaged in rehearsing a given passage, one's mind could not easily step back and contemplate the meanings conveyed by this series of words.

> The poem's structure, rhythm, syntax, and plot, its very substance, have all been designed for a situation in which "I" do not exist. They provide the machinery of self-identification, the magic of the spell, the drug that hypnotises. Once I end my absorption in the poem, I have ended the poem too. . . . The contents of the encyclopedia cannot be identified by retrospective analysis . . . but in the epic story they are *implicit,* not *explicit.* They appear only as acts and events performed by important persons or happening to important persons. This was inevitable as long as the law was to live in memory. For memory could identify effectively only with acts and events.
>
> (Havelock 1963, 216–217, emphasis added)

Note that the narrative structure of this "encyclopedia" conveys *implicit* meanings. Its word play is that of myth (cf. fig. 5.3). Havelock was not suggesting here that fifth-century Greeks lacked a concept of the self, as Bruno Snell had claimed in regard to archaic Greeks, but that the "unexamined life" and the ills that accompany it were consequences of a misplaced trust in a poetic knowledge base rife with misrepresentations.

On the other hand, verbal falsehoods can sometimes be allowed, says Plato's Socrates. We can justifiably tell lies in three circumstances: to our enemies, to our friends when any of them becomes deranged and may do himself harm, and to anyone when speaking about ancient times. In the latter case, because we cannot know the precise truth about those events, the stories we make up should be as similar to the truth as possible (*Republic* 2.382c–d). Here Plato apparently has in mind etiological accounts that, unlike those in the *Theogony,* are philosophically insightful myths and allegories. Allegory, such as that of the cave in the *Republic* and the charioteer in the *Phaedrus,* were examples of explicit word play: no one who heard Socrates narrate them could have been deceived.

These, plus Plato's myths of "ancient times," such as that of the Demiurge in the *Timaeus,* Atlantis in the *Critias,* Thamus and Theuth in the *Phaedrus,* and even the spherical first humans, as Aristophanes narrates in the *Symposium,* would all be acceptable stretchings of the truth. The latter myth, we should recall, was censured neither by Socrates nor by any other of the guests because they knew very well how to receive a tale in the spirit of the teller. Plato himself must have picked up the wink and the nod from his master, who, as his disciple wrote, was not above spinning falsehoods to treat the ailments of those among his countrymen whom he regarded as perilously deranged.

The most notable, or notorious, of these falsehoods is the "noble lie" (*Republic* 3.414c–417b), the "Myth of Metals." According to this tale, all citizens of the ideal city would be taught that they were formed in the earth out of a mixture of gold, silver, bronze, and iron, a four-part division reminiscent of Solon's four nonhereditary, economic "tribes." Their relative status would be determined early in life by their response to the challenges of education. Those who were mostly gold become guardians, those mostly silver become soldiers, and the rest, a mix of bronze and iron, become farmers and craftsmen. Though he explains it to Glaucon as an allegory, it works only when it operates *as a falsehood,* i.e., when it is *not* assumed to be an allegory but rather a simple truth. Only then will it serve to inspire obedient citizens and self-sacrificing soldiers. Lies such as this Socrates regards as a dose of needful social medicine, a *pharmakon* that must be administered to the citizens of his ideal city (5.459d).

When Socrates shares this lie with Glaucon, he professes to be embarrassed by it. It is, after all, an outrageous fabrication that would certainly qualify as an entry in any tribal encyclopedia. As though to exculpate himself, Socrates remarks that it was not he who made it up—he simply borrowed it from the Phoenicians. By this he seems to allude to the myth of Cadmus, the Phoenician, who left home in search of his abducted sister, Europa. In his wandering, he entered a desolate land where he encountered a dragon, which he slew, then, on the advice of Athena, sowed its teeth in the very earth upon which the city of Thebes was destined to be built. Those teeth soon sprouted up as armed warriors that fought one another furiously until only five were left standing: these became the ancestors of the five noble families of Thebes. This, by the way, was the same sort of autochthony myth that Athenian politicians had told their own citizens: long ago their ancestor, Erechtheus, had emerged from the very soil of Attica, and that is why their right to the land was absolute and inalienable.[11]

Oral poets, as editors and publishers of tribal encyclopedias, had been well aware of their own importance as long as this traditional lore remained relevant to their audiences. They were, therefore, strongly motivated to promote this cultural archive and, with it, their own professional status. Rulers were similarly motivated to promote their own office by employing poets to sing the praises of the kingly heroes of yesteryear. Hence the epic tradition and the age-old interdependency of rulers and their bards.[12]

Homer celebrated this relationship in a number of passages, but it was Hesiod who made it explicit in the "Kings and Singers" passage in the *Theogony* (80–103). The Muses, he said, bestowed upon both classes of men the divine gift of inspired speech, kings to mediate conflicts persuasively, the poets to pacify the conflicted minds of individuals. Together, they brought peace and stability to mankind by redirecting attention both from outer and inner quarreling.[13] It seems to have been Plato's project to confront this alliance of kings and poets and replace it with a new alliance, that of kings and *philosophers*.

Plato and the Imitative Imagination

Factual misrepresentation poetically disseminated and passively absorbed was, in Plato's mind, the consequence of another, more insidious, kind of deception. This was the misrepresentation of oneself as another person and the pretense of possessing the knowledge of that person. Thus poets and their proxies, actors and rhapsodes, lie not only when they assert untruths but also whenever they imitate, i.e., impersonate, others. In tragedy, the most popular poetic genre of the time, a poet tells a story by employing mask-wearing performers to imitate personae, but in epic the rhapsodes deployed a combination of narration (*diêgêsis*) and role-playing enactment (*mimêsis*). When, for example, a rhapsode stands before an audience and shifts from speaking *about* Achilles to speaking *in the voice of* Achilles, he necessarily represents himself falsely (*Ion* 541c; *Republic* 3.393–394). Not only does he utter whatever falsehoods Homer has versified, but, as Plato implies, in the very act of role-playing he himself embodies a falsehood. We would probably say, yes, but so what? Can we not pretend for a moment that we in the audience are in the presence of Achilles? Plato's student, Aristotle, would agree with us, but Plato had a point to make: our vulnerability to deception comes from our tendency to confuse appearance with reality, a moral weakness that verbal art, *poiêsis*, shamelessly exploits.

In the *Ion*, a dialogue between Socrates and a professional performer of Homeric epics, the philosopher gradually maneuvers the rhapsode into a corner, forcing him to admit that he cannot claim to possess the expertise of any of the characters he performs, for example, generalship, when he portrays Homer's military leaders. Like the shape changer Proteus, Ion seems in a state of constant self-transformation (*Ion* 541e). How does he do it? Socrates proposes two possibilities: either Ion is a fake, or Ion is divinely inspired. The hapless rhapsode, offered a choice, opts for the latter, and on that note Socrates ends the interview.

Ion's dismissal, preceded by ironical expressions of reverence for his divine gifts, anticipates the dismissal of the similarly protean poet in the *Republic* (3.398a). But this latter passage is also preceded by a set of reasons why this sort of poet should be sent away. At issue now are not the falsehoods and pretensions to knowledge conveyed in poetry but the style (*lexis*) in which poems are performed. It was the style, after all, that drew audiences and made false and unethical statements so attractive. By *lexis*, Plato may have also meant "diction," in the sense of "word choice," but what he focused on in book 3 of the *Republic* were the paralinguistic effects of voice (*phônê*) and of body language (*schêma*), as when a performer turns from narration to speak a given character's words, he might assume a particular tone and timbre, pace his words slowly or trippingly, and choose for them a particular amplitude. He might also visually imitate that character by facial expressions, hand gestures, and posture. In short, a performer of a poem with directly reported speeches might deliver them as would an actor on a stage.

Such theatrical effects must have been electrifying to audiences. Passages of Homer would seem to come alive as the rhapsode assumed this, then that, persona before their very eyes. And not only personae, but, as Socrates notes, sometimes these performers produced a full range of sound effects, vocally mimicking thunder, wind, hail, axles, pulleys, musical instruments, dogs, sheep, and birds (*Republic* 3.397a).

The results of this mimetic display would be extreme variations in rhythms and numerous disruptions in emotional tone, but, aesthetic principles aside, this performance style would encourage young minds to similar imitative excesses. Imitation, it would seem, only breeds imitation. The future leaders of the city would be ill-served if they grew up thinking all they needed was to play a variety of impressive roles. In Socrates' ideal state, every man was to practice a single skill, a single vocation, commensurate with his natural gifts. A man skilled in composing or performing epic verse would, therefore, have no place in this city because he would be "twofold or manifold" (*diplous . . . oude pollaplous*). His skill in imitative play, involving in this case more than one identity, would render his nature two-ply, or *duplicitous*. Should such a "man, so clever as to be able to turn himself into all sorts of things and imitate everything, arrive at our city, prepared to exhibit himself and his poems [*poiêmata*]," the guardians should treat him with all due respect as a sacred and marvelous fellow but inform him that they have no use for men like him and forthwith send him on his way to some other city.[14]

Throughout his dialogues Plato compares the making up of verbal compositions (*poiêsis*) with the art of realistic painting. Visual artists, he said, emulate mirrors that automatically represent whatever they reflect: both create appearances of reality but never reality itself. If, for example, a painter wishes to depict a bed, he must depend on another technician, a carpenter, to furnish his model. The latter makes particular beds, one after the other, each slightly different, but he does not make "Bed," the Form, or Idea, the divinely created prototype that exists forever prior to any single piece of furniture identifiable as a bed. In Plato's example (*Republic* 10.596–598), the painter knows nothing of the skills of carpentry and can represent that object only in its visual properties, on a two-dimensional surface, and only from a single angle of vision.

The comparison of painting with poetry was well established before Plato placed it under philosophical scrutiny. Simonides had famously defined painting as silent poetry and poetry as painting that speaks. From Homer down to contemporary tragedians, all poets had tried holding the mirror up to nature in order to create painterly images of reality, but, as Plato decided, in claiming they knew what they were talking about, they were just as ignorant as painters concerning the true nature of what and whom they imitated. Poetry and sophistic prose, like paintings, were faulty replicas of reality, *mimêmata* that could never initiate their hearers into the contemplation of reality itself. Like the bright, lifelike colors with which painters depict their images, a poem's melody, meter, and rhythm can weave a magical spell. But, uttered without those musical "colors," these words reveal themselves as mere rhetorical bombast (*Republic* 10.601).

Considered from a semiotic perspective, the visual images of painting and the mental images evoked by a poem's words both imitate real things in the outer object world by functioning as iconic signs. But in the *Cratylus* (432b), Plato takes a more radical turn when he declares that words, the symbolic signs of language, are themselves imitative. If, as Plato's Socrates argues, naming and painting are analogous, then conventional and natural signs become linked through their one common feature, imitativeness. Plato has forthwith expanded the semantic domain of imitation to include all forms of meaning making and communication.

If speech is as imitative as painting, what about writing? If speech is an auditory imitation of visual appearances and alphabetic script is a visual imitation of speech, then writing is doubly implicated in the shadow world of illusion. Moreover, writing and painting seemed already naturally linked because the verb *graphein* was commonly used for both activities. In the *Phaedrus*, Plato isolates a particular deficiency in writing:

> Writing, Phaedrus, has an uncanny quality, quite similar to the painter's art in which creatures stand before you like living beings, but if you ask them a question, they maintain a solemn silence. And so it is with writings: you might think they spoke as though they had consciousness, but if you ask them anything because you want to know what their sayings mean, they always declare only one and the same thing.[15] And not only that, but every discourse, once it is written down, gets circulated everywhere, among those who understand it, as well as among those who have no interest in it. It has no idea to whom to speak or not to speak. When abused or unjustly berated, it always needs its father to help it, for it is powerless to protect or help itself.
>
> (275d–e)

Here Plato shifts the focus of his critique. It is not so much that speech and writing are imitative like painting but that they too may be used to produce immutable, therefore unresponsive, artifacts. For example, verbal artifacts, such as epics and tragedies, are memorized by rhapsodes and actors and, when performed, run their course uninterrupted by members of the audience—certainly not by barefoot philosophers eager to engage performers in question-and-answer sessions. Whether voiced by a rhapsode or read from a papyrus scroll, Homer's characters "always declare only one and the same thing." As Gregory Nagy has consistently maintained, the root meaning of the Greek noun *mimêsis* is "reenactment."

For Plato the serious seeker of wisdom pursues dialectic, i.e., the improvised conversational analysis of topics such as truth, justice, goodness, love, and the soul. This inquiry must not be trammeled by received opinions but strike out in whatever direction seems to promise clarification—or, as he puts it, wherever the wind of the discourse carries it (*Republic* 3. 394d). His first article of faith was his belief that everyone had deeply deposited within him or her the ultimate knowledge of reality and that learning was really a matter of remembering what had been revealed in a previous state of being. In Socrates' analogy, he, the master dialectician, served as a midwife to deliver the knowledge with which his interlocutors were already in labor to bring forth (*Theaetetus* 148–151).

In the *Phaedrus* this birth metaphor goes through several changes. First of all, a written discourse is like a bastard child whose father is not there to defend it. An unwritten discourse is later described as a seed sown in the soul of another that inevitably brings to birth a legitimate child. As the metaphor develops, the image becomes agricultural and, as the source domain shifts, so does the target: now it is no longer the opposition between legitimacy and illegitimacy but rather that between seriousness and play. Spoken dialectic is seriously productive, but writing is a playful activity meant to amuse oneself and astonish others.

On the other hand, and Plato's thought is often otherhanded, writing does have its uses, principal of which is to remind us of what we already know. For a philosopher, it is not an external memory store but a set of memoranda (*hupomnêmata*) meant to help us revisit thoughts already "inscribed in the soul" (*Phaedrus* 276, 278). Written documents thus function as mnemonic aids, in ways similar to paintings, statuary, and architecture, as I proposed in chapter 4.

For a philosopher, writing is play that yet serves a useful purpose by preserving a "treasury of reminders against the forgetfulness of old age—and not only for his own use, but for any others who travel this same path" (*Phaedrus* 276d). Is this indeed why Plato wrote his dialogues? Are they memoranda? Many of them are actually framed as recollections, some second or third hand, of Socratic conversations. Most preserve the playful, bantering tone of the master, and though they sometimes sketch out metaphysical doctrines, they do not constitute a systematic philosophy. If the document known as Plato's seventh letter is authentic, he may have transmitted to his students an esoteric system that he never entrusted to writing. Though serious inquiry demanded face-to-face dialectical engagement, there was also, evidently, room for playfully engaging with writing, that ghostly likeness (*eidôlon*) of living, breathing discourse (*Phaedrus* 276a).

To his readers, his admirers included, Plato has always appeared inconsistent on a number of issues. Speaking through his main character, Socrates, he castigates the lies of poets but recommends not only that the guardians of his ideal city lie to its citizenry but over time that the guardians themselves believe the lie as well. He decries the poets' myths as suitable only for small children, claims that reason (*logos*) is the only proper discourse for adults, but makes up numerous myths of his own. He declares mimesis an illogical cultural habit and the arts that practice it, such as tragedy, pervasive dangers to the life of the mind, all the while creating dramatic dialogues with protagonists and antagonists in specific settings. He inveighs against writing as a *technê* that immobilizes living thought yet represents that thought in a massive library of prose works.

What if Poetry were to call Plato to the bar of justice and question why she should not ban all philosophers, at least all Platonic philosophers, from her own ideal city? Why, as Shelley might ask, should poets remain forever the unacknowledged legislators of the world? Why are these players of language accused of lying? The poet "nothing *affirmeth,* and therefore *naught* lieth," Sir Philip Sidney wrote. In litigating what he called that "ancient quarrel of philosophy and poetry," had not Plato unjustly accused poetry of the very word play he himself practiced?

In his own defense, Plato might answer: "Show one instance in which I told a lie? *I* never lied, because *I*, too, affirmed nothing. In fact, I never once represented myself as a speaker in any of my dialogues. I merely represented the affirmations and denials of others."[16]

Was Plato a player after all? Three years before his death he extolled a "seriousness [*spoudê*] that is not out of tune when combined with play [*paidia*], the sister of seriousness" (*Letter* 6.323d). When he died at the age of eighty, one tradition has it that in his deathbed his disciples found a copy of Aristophanes.

Plato and the Formation of Genres

"Isn't everything said by storytellers or poets a narration [*diêgêsis*] of events past, present, or future?" "What else could it be?" answers Adeimantus (*Republic* 3.392d). So narration, as a linear succession of words told by a speaker, is for Plato the inclusive category, the genus comprising all the possible ways we have to communicate temporal phenomena. Within this genus its species can now be differentiated according to the ways by which this communication is effected. One way is by *simple narration,* when the actual teller is the only source, e.g., dithyramb; another way is mimetic narration, when the tale is told via role-playing (*dia mimêseôs*), e.g., drama; and yet another way alternates telling and role-playing, e.g. epic (394b–c).[17]

Plato's triad of narrative genres has had a long and pervasive influence on literary studies. Based on the way verbal artifacts present themselves, these "presentational modes" have for centuries seemed universal ideas as convincingly distinct and simple as the triangle, the square, and the circle. By contrast, other genres, such as lament, pastoral, epithalamion, and satire, would be defined by their culturally specific subject matter; yet others, such as elegiacs, iambics, the sonnet, and the villanelle, are defined by their verse forms. Recognizing this difference, Goethe termed the three Platonic genres "natural forms" (*Naturformen*) and sensed them operating below the surface of conventionally evolved forms of literary expression.[18]

If Plato's three genre categories are fundamental, do they constitute a cultural-evolutionary sequence? Western literary theorists have felt drawn to such speculations. Which of the three was the primal genre, and how did the next genre proceed from the first, then the third from the second? Or is the third some Hegelian synthesis of the first two? Would Homeric epic represent the primitive narrative matrix from which drama and lyric grew? Western literary theorists have traditionally associated Homer's directness of expression and innocent eye with the "childhood of the race," the bard only a notch above Rousseau's awe-struck savage. They saw in this sequence a pattern of cultural evolution later repeated in western European literature, moving from saga and romance to troubadour songs and sonnets to sixteenth- and seventeenth-century drama. Or does dithyrambic lyric, the primal *cri de coeur*, mark the protopoetic starting point? And what of tragedy—is it not linked to rituals of human sacrifice, and should it not therefore be the primordial verbal art form? It is tempting to claim Platonic authority for this or that evolutionary genre theory, but to do so, I submit, is to misinterpret Plato's intentions.

The first mentioned species of narration, simple narrative (*haplê diêgêsis*), poses a bit of a problem for later genre theorists and apparently for Plato as well. To represent it he nominated dithyramb, a genre of song associated with the cult of Dionysus, in which the principal singer narrates a story directly, without impersonation. But Plato nominates it reservedly: dithyramb best exemplifies this form of discourse, "I suppose [*pou*]." He does not identify lyric poetry with dithyramb, but, insofar as Greek lyric represented the thoughts, feelings, and life experiences of the poet in the first person, most literary historians have simply assumed that Plato would have so categorized it. But whether he indeed viewed the lyrics of such poets as Sappho, Alcaeus, and Anacreon as akin to dithyramb, in the *Republic* he betrays little interest in simple *diêgêsis*, leading us to conclude that he did not regard it as a very significant mode howsoever it manifested itself. In his late dialogue *Laws*, moreover, his main character, the Athenian Stranger, in the course of critiquing contemporary musical culture and commenting on several ritual genres of hymnody, mentions those songs that celebrate the birth of Dionysus, parenthetically adding, "I think [*ormai*] they are called 'dithyrambs'" (701b). What an offhand way Plato chose to refer to what we, over the centuries, have come to regard as the quintessential poetic genre, the lyric!

Then there was "pure mimesis," exemplified by tragedy. Whatever the first syllable of the word *tragôidia* originally signified, the rest of the word is clear: tragedy was a *sung* performance, less like modern drama than like Italian opera, which was designed to be a revival of classical drama. According to Aristotle, tragedy grew out of certain ritual celebrations of Dionysus, during which a dancing chorus sang dithyrambs to which the first dramaturge, Thespis, added a soloist. The word for this soloist was *hupokritês*, which means "answerer" or "interpreter," suggesting that the chorus either posed a question or told an enigmatic tale to which the soloist would offer a clarifying reply, which might in turn prompt another choral song. As this dramatic genre evolved, the soloist assumed a principal role in the tale that the chorus was narrating. His answers would therefore appear authoritative because he had either witnessed or participated in the events described in the narrative. As the chorus represented the narrator and the soloist (now the actor) imitated a persona in the tale, early tragedy reproduced, to some extent at least, the same alternation of *diêgêsis* and *mimêsis* that Plato noted in epic. This complicates the neat distinction of drama as "pure mimesis," for just as epic is a form of storytelling that combines narrative and embedded speech, so early tragedy may be seen as a combination of those same elements (Walton 1980).

Over time, the mimetic element came to predominate simply by the addition of onstage characters, two by Aeschylus and three by Sophocles. An actor could still be interrogated by the chorus or its leader, the *corypheus*, but now it was also possible for two actors to enter into dialogue, thus bypassing the chorus. Then, once three characters were copresent, the whole speech situation could be assembled onstage: a first-person *I*, addressing a second-person *You*, about a third-person *He* or *She*. As the actors took over the storytelling through mimesis, the term *hupokritês* shifted meaning from "answerer" to "actor" to "pretender"—and eventually to the modern word "hypocrite." The actors' lines belonged to that part of the performance originally termed the *epeisodion*, i.e., an addendum, something additional inserted into the choral narrative, but as the role of the chorus diminished and that of the actors increased, "episodes" became equivalent to "acts," evolving eventually into a five-part structure, divided by choral interludes, a form that Greco-Roman dramatists bequeathed to the European Renaissance.

Every inquiry into the evolution of classical genres inevitably takes us back to Homer and the epic form that Plato classified as narrative by means of both *diêgêsis* (narrating) and *mimêsis* (role-playing, or imitation). There is something special about epic, wherever and whenever it is composed, and there is reason to suppose that this combination of narration and reported speech is the prototype of all other verbal artifacts. The epic format represents the way social events are told—who *did* what to whom as well as who *said* what to whom and when and where all this doing and saying happened. Like the retrieval of episodic memories, in which there is a present rememberer and past events that usually include other persons, narrative necessarily includes narration plus imitation, or, as Aristotle called it, "diegematic *mimêsis*" (*Poetics* 1459b30–37). Though every individual epic has its culturally differentiated conventions, its hybrid structure simply represents the way we humans store and communicate social information and is therefore the default genre of all verbal artifacts.

Consider figure 5.4, a three-part diagram representing epic narration in a play-frame format. First, the situation is contextualized by the knowledge that the action at hand is to be a storytelling with all the culturally specific norms that may obtain. Second, the audience focuses upon a storyteller who performs a *diêgêsis*, which, as I suggested earlier, is an act of "leading [them] through" a set of imaginary scenes in which persons are portrayed not only doing actions but also engaging others in speech. Third, we hear the words they say as the narrator shifts into *mimêsis* to represent the voices of these imagined personae. At no point should an audience confuse the speech of the various personae (frame 2) with that of the narrator (frame 1), but at no point in this performance can they totally ignore the fact that they are experiencing these several voices through the single voice of the narrator.

Many of us, myself included, have over the years used the phrase "mixed mode" to designate this combination of *diêgêsis* and *mimêsis* that Plato claimed differentiates epic from other forms of storytelling. Eric Havelock referred to Homer's "mixed style" and "mixed mode" (1963, 11, 21, 24). Accordingly, we have found ourselves also using the adjective "pure," as in "pure *diêgêsis*" and "pure *mimêsis*." An example of this appears in Gérard Genette's *Narrative Discourse* (1980, 162–163), which in a footnote (162n2) he explains: "The common translation of *haplê diêgêsis* as 'simple narrative' seems to me a little off the mark. *Haplê diêgêsis* is the narrative not mixed (in 397b, Plato says: *akraton*) with mimetic elements: therefore, *pure*."[19]

Since in any opposition of "pure" and "mixed" the mixed usually connotes an adulterated or blurred state, one would assume that Plato would approve the pure and decry the mixed. We know that he came to feel that the content of Homer's epics could have a bad influence on the young, but in his treatment of style, did he reject Homer because epic combines the imitation of various characters' speech with simple narrative? If this double form of narration were an undesirable mixing of pure elements, would he not have extolled pure forms as morally superior?

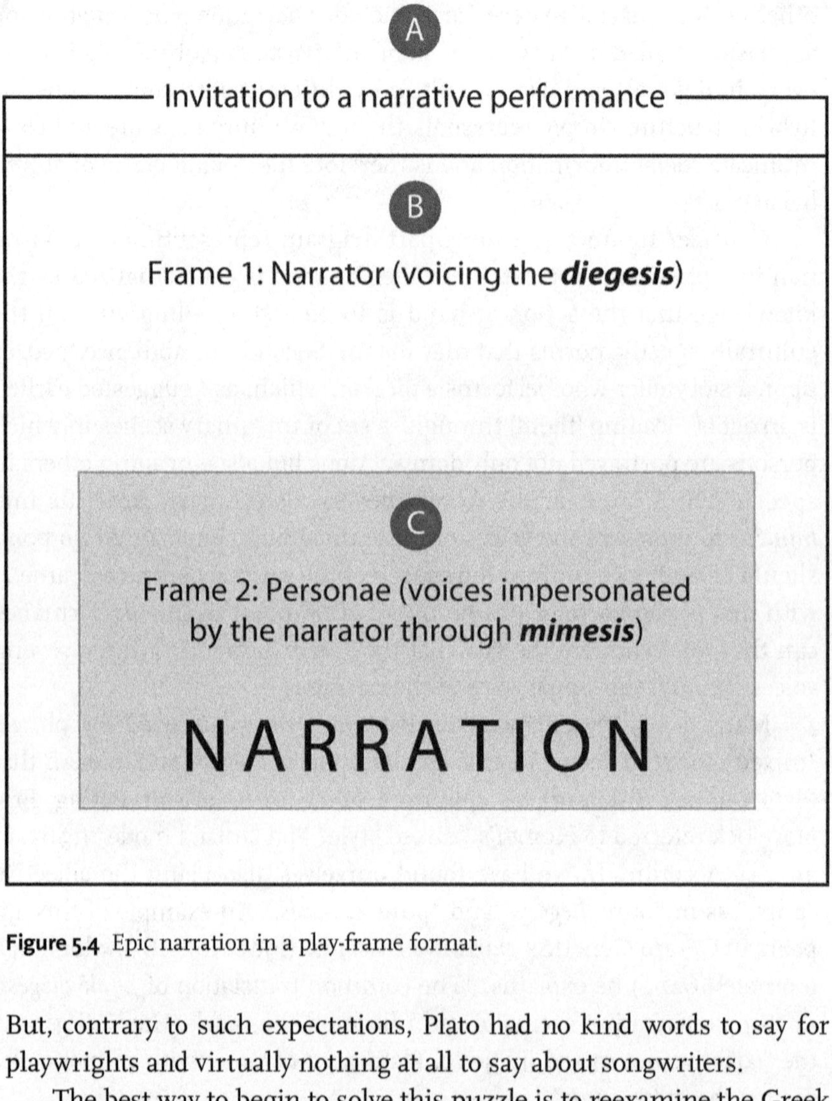

Figure 5.4 Epic narration in a play-frame format.

But, contrary to such expectations, Plato had no kind words to say for playwrights and virtually nothing at all to say about songwriters.

The best way to begin to solve this puzzle is to reexamine the Greek text, where we find that Plato never identifies a kind of *compositional* style as "mixed." What is mixed (*kekramenon*, 397d) is a certain *performance* style, as well as the performers who practice it.[20] As I indicated in the last section, Plato felt that when performers of their own or of others' works overmimeticize, they mix (up) their compositions both vocally and gesturally and as a result disrupt the musical continuity of the work. This is not only artistically wrong but is also morally detrimental because those portions that lend themselves to such imitative excess tend to represent persons in their most irrational and deranged moments.

How shall we account for this time-honored misreading? We might begin by recognizing in ourselves two cultural biases. The first is our tendency to interpret Plato's poetics in terms of book literacy. Plato did read and write, as did most poets from Hesiod on, but the normative venue for poetry was not the library or the private chamber. It was the public place where Greeks assembled to be entertained by trained singers, actors, and reciters. Style (*lexis*) pertained not only to the words but also to the vocal and gestural performance of the words, the manner in which the words (*logoi*) were conveyed to an audience. Poetry was still *mousikê*, the Muses' art. The second root of our misreading is our tendency to view all human phenomena as evolving from simpler to more complex forms. Of the two perspectives, the literate and the evolutionary, the latter is more useful to the study of oral poetics, but neither is useful in elucidating Plato's cultural critique of *poiêsis*.

Since Plato never declared epic to be a "mixed mode," we need to revise our view of the origins of the three Platonic genres. For a start, Plato did not regard the simple modes as primary forms that preexisted their combination in epic, as we might regard the primary colors red and blue preceding their mixture as purple. Instead, Plato implies that epic with its two modes of narration is the preexistent genre. By virtue of the stories he tells, Homer is not only the father of epic but, according to Plato, also the "first teacher and leader of the tragic poets" (*Republic* 10.505c).[21] In this light, the two simple genres, lyric and drama, appear to emerge as independent art forms extracted from the ancient master genre, the storytelling style that epic most closely resembles.

I am not proposing that Greek lyric and drama actually split off from Greek epic at some point in time but that all three derive from a common, full-featured storytelling antecedent that is as old as human speech itself. If indeed the simple genres of lyric and drama constitute portions selected from that full narrative spectrum most closely reflected in epic, then we should find elements of both lyric *diêgêsis* and dramatic *mimêsis* in epic. In specifically Homeric epic, however, we find little in its strictly diegetic passages that we can identify with an actual person, such as we find in Hesiod and the lyric poets. Where we do find them are in the mimetic passages, i.e., those portions in which the narrator allows personae to speak. These *mimêsis*-embedded *diêgêseis* are voiced in three distinct forms: (1) inner monologue, (2) outer monologue, and (3) dialogue.

In Homeric epic, *inner monologue* occurs seven times, five times in the *Iliad*, twice in the *Odyssey*.[22] In each instance of inner speech it is prompted by a life-threatening emergency that forces a character to consider his options. (For a discussion of inner speech, see chapter 3.) If these episodes are meant as transcripts of silent thought, they are nonetheless presented, in accordance with the conventions of oral poetry, as overt speech, as characters "speaking to themselves." Truly covert inner speech is best represented within *written* poetry, once the role of the silent solitary reader becomes the accepted convention. Eighteenth- and nineteenth-century European lyric poets were able to exploit this quasi-covert mode of speech to great advantage.

Outer monologue occurs within a narrative when a character speaks to a person or persons but receives no answer either because the addressee is not present, as in apostrophe, or for various reasons does not offer a response. Lacking the dynamic agonistic tone typical of dialogue, it seems easily assimilated into the narrator's voice, as a voice remembered from his or her past, and does not interrupt the line of address established between performer and audience. Yet, as direct speech, it exists in real time as "scene" without therefore the temporal acceleration or slowing down of normal narrative discourse (see chapter 2). Prayers and laments have this character: see, for example, Chryses' prayer to Apollo (*Iliad* 1.37–42), Penelope's prayer to Artemis (*Odyssey* 20.61–90), Thetis's lament for the imminent death of Achilles (*Iliad* 18.52–64), and the laments of Andromache, Hecabe, and Helen for the slain Hector at the conclusion of the *Iliad*. Outer monologue is also the mode of formal public speeches, which in Homer proceed uninterrupted as long as the speaker, typically male, holds a scepter passed to him by a herald, e.g., Telemachus's speech to the Ithacans (*Odyssey* 2.37–81).[23] This highly expressive speech mode is found both in lyric and, as soliloquy, in drama.

Dialogue occurs, of course, when characters are portrayed as speaking to one another. If it is a sustained exchange of words in real time (as *scene*), the audience will focus on it as though they were hearing speech that is no longer addressed to them by the narrator but as speech as it flows back and forth between these two or more onstage participants. These voices call for imitation on the part of the narrating performer because these alternating speakers may need to be distinguished from one another. The relation of epic dialogue to drama is indisputable, for "without the example of Homer, showing the heroes and heroines of myth conversing in dialogue in a high style, Attic tragedy would never have come into existence in the form that it did" (Griffin 2004, 156).

The essential elements of epic, as the Greeks understood it, are modes of speaking. The very word *epos* (epic) means "speech." Greek lyric and drama, insofar as they historically derive from Homeric epic, each drew upon a different set of speech modes. Lyric assimilated inner and outer monologue, and drama assimilated outer monologue and dialogue.

We may never know in any absolute sense whether singing, miming, or storytelling came first in human evolution. As Jeffrey Walker (2000, 166–167) remarked, asking which genre is "more fundamental" leads inevitably to the chicken-and-egg dilemma. As expressive and informational behaviors, singing and miming may have preceded speech, but once humans mastered language, the most important verbal behavior would have been oral narrative employed as a means of sharing social information. Oral culture, I submit, possesses this narrative spectrum in which various bands of discourse, some narrow and specialized, others more comprehensive, may assert themselves from time to time as prominent genres, then recede, while the entire spectrum remains to furnish the breath of narrative modes typical of epic. This spectrum of genres, in dynamic interaction over millennia of preliteracy, survived, albeit with significant adjustments, after the introduction of writing.

Summary

In the process of oral transmission, verbal artifacts became the possession of performers who edited them to suit themselves and their audiences and passed them on for later generations to continue this process. When such compositions were written down and copies circulated, standard versions were established and, when performers themselves became capable of writing, they began to compose wholly new works, authorized and identified with their names. In ancient Greece, this was marked by the emergence of a new professional class, as some "singers" came to be recognized as "poets," i.e., makers of original verbal artifacts.

The oral art of storytelling presupposes the willingness of teller and audience to imagine themselves in particular places, perceiving and engaging in actions, and assuming the perspectives and personae of others. This verbally cued imagining is a special kind of play directly dependent on our capacity to form and maintain two ordinarily contradictory interpretations of a situation at the same time. This is apparent when animals, ourselves included, engage in social play of the rough-and-tumble variety. A play episode of this sort usually proceeds without mishap because both participants understand that these aggressive displays are nonhostile in intent. In semiotic terms, those aggressive indices are understood as actually iconic signs, i.e., they *resemble* hostility without threatening the injurious consequences normally associated with hostility.

This doubleness is also the basis of verbal play at every level of complexity: from the word, as in pun and metonymy, to the phrase, as in metaphor and irony, and to the entire discourse, as in allegory and myth. It is this doubleness, or foldedness, that Plato finds problematic. As he defines the issue, appearances, i.e., iconic signs, produced by a process of imitation, are necessarily deceptive when they are mistaken for reality. The works of Homer, for example, pose a danger for any society that takes what he says of gods and heroes as literally true. In other words, when Homer becomes one's inerrant scripture, one descends into a sort of polytheistic fundamentalism. The unexamined life that is not worth living is, for Plato as it was for Socrates, a life devoted to received opinions and therefore averse to the adventure of conscious inquiry. Writing, too, is problematic, for, just as a portrait is an imitation of a person's visual presence, so writing is an imitation of a person's speech. The only purpose these soulless representations can serve is to remind us of what we already know. Presumably Plato would regard his own dialogues as mere reminders of how living, breathing thinkers interpenetrate one another's minds to interrogate the meanings of the words they use every day.

Genre theorizing ostensibly begins with Plato's *Republic* and his distinction between two presentational modes, *diêgêsis* (narration) and *mimêsis* (imitation). All storytelling and *poiêsis* is *diêgêsis*, Plato declares: some compositions are simple *diêgêsis*, some are *diêgêsis* by means of *mimêsis*. However, theories that posit simple *diêgêsis* (as dithyramb and lyric), simple *mimêsis* (as drama), and "mixed mode" (as epic) as a trichotomy force that paradigm on Plato's actual argument. While Plato, in his character Socrates, takes exception to the notion that Homer and the poets who followed him are founts of wisdom and usable knowledge, he is especially critical of the performance style of contemporary poets, especially their tendency toward extravagant histrionics. It was their presumptions of inspired wisdom combined with their flamboyant performance style that he deemed "mixed." It was this early fourth-century cultural craze that he condemned as "theatocracy," not what we encounter in the quiet pages of a book of poems.

Six

Writing for the Voice

Despite the silence of the written surface, the literate imagination has from its very beginnings been deeply engaged in restoring the immediacy of the human voice. In this chapter I will first revisit the question of why the Greeks and Romans preferred to write their words in continuous script without inserting spaces between words and suggest how speech simulation may have influenced this choice. Then I consider that other literary uncertainty, the evolution of the lyric, the genre Plato and Aristotle ignored, that nevertheless over the centuries has become recognized as the prototypical poetic genre. This is a long story, but an important chapter in it was certainly the revival of Greek lyric in late republican and early imperial Rome. It was during this period that four established genres, oratory, drama, inscription, and letter writing, exerted influence on the lyric and, each in its own way, left its lasting mark on this songlike form of writing.

In Roman lyric, the uninterrupted voice of the single speaker represented in the text may be differentiated by the physical presence or absence of the addressee. If the addressee is present, the lyric is a dramatic monologue; if absent, it is either apostrophe or epistolary speech. In the first-century Roman lyric, the personal letter became a standard model for this newly reconstituted, nonsung, speech-simulating literate genre.

As fashioned by Catullus, Horace, Tibullus, Propertius, and Ovid, these written artifacts were identified with the voices of their makers but did not serve as vehicles for inner speech or the meditative, affect-toned soliloquizing that has come to characterize the post-Renaissance lyric. There are, however, indications that, before this thinking voice emerged, poets were beginning to play the role of protagonists in a larger, autobiographical enterprise. We see that in the erotic confessions of Catullus and the elegists, in the later poems of Ovid, and, despite his ironic persona, in the odes of Horace. We also find an increased insistence that they were not mere wordsmiths but inspired prophets (*vates*) who through their works had each created a deathless self that lives on as a vocal consciousness whenever and wherever in the world his words are read.

Voces Paginarum

The popularity of written mimes, mentioned in the last chapter, should be understood in the context of the general preference among classical Greeks and Romans for reading texts aloud, especially literary texts. This cultural fact has led scholars to inquire into the extent to which the ancients practiced silent reading. How readers processed writing is a significant question, since writers would have designed their works to suit the reading habits of their public. This scholarly debate, which in the last century paralleled the orality-literacy debate, also has important implications for our inquiry into writing-mediated imagination. To what extent did literate poets see themselves as makers of vocal-performance scripts?

Literary recitation reflects the Greek and Roman love of eloquence, but it also was a practical means of advertising new compositions. In the late Republic and the early Empire, copies of works in progress, in either prose or verse, might sometimes be circulated among friends and friendly critics and their comments solicited, but an author of a new work more regularly read it aloud before a gathering of invited guests, which sometimes might include the general public. Though theater, chariot racing, and gladiatorial spectacles had become the most popular mass entertainments, public recitations also attracted a number of interested persons.

In 1927, the Hungarian scholar Josef Balogh published a sixty-four-page monograph, "*Voces Paginarum*" (voices of pages), that offered a new way of explaining why the ancients preferred to read literary works out loud: they generally *could not read silently*, and those very few who acquired this skill were regarded as objects of wonderment. Balogh's key proof text was a passage in the sixth book of Augustine's *Confessions*. In 384 CE, that thirty-five-year-old spiritual searcher traveled to Milan, where he sought the counsel of Bishop Ambrose. Augustine had hoped to discuss his spiritual struggles, but the bishop was preoccupied with busy people who needed his services. His only free time, it seemed, was devoted to

> refreshing his body with nourishment and his mind with reading (*animum lectione*). But as he was reading, his eyes would move through the pages and his heart extract their meaning, yet his voice and tongue kept still. When we were present—for no one was forbidden to walk in and it was his custom that no visitor need be announced—we saw him reading silently (*legentem . . . tacite*) and never otherwise.

Assuming that Ambrose's silent reading must have had a good purpose, Augustine and his colleagues suggested reasons to justify the bishop's behavior.

> After sitting there for a long time in silence—for who would ever dare disturb such an intensely focused man—we would leave the room and speculate that he did not wish to be distracted during the brief time he had laid aside from the noise of other people's business to refresh his mind. We also speculated that perhaps he was wary lest the author he might be reading (*quem legeret*) might have phrased something rather obscurely and that it would be necessary to explain it, or enter into a discussion concerning the more difficult questions implied, for the sake of an intent and perplexed listener, and in so doing spend less time than he wished in reading his book [literally, "in unwinding his scrolls"]. On the other hand, a more plausible cause for his reading silently could have been his need to preserve his voice, which was very easily weakened.
>
> (*Confessions* 6.3)

As Balogh interpreted it, this passage proved that even highly educated persons in Greek and Roman antiquity were unaccustomed to reading silently and that reading, *lectio*, normally meant what we mean by "lecturing." If Ambrose had been reading in a normal manner, the people in attendance would have been able to hear his every word and interrupt him by questioning the meaning of obscure expressions.[1] Yet, just two books later in the *Confessions* (8.12.29), when Augustine recounts his conversion experience, he tells of sitting one evening in his garden and hearing from a nearby house a child's sing-song voice saying over and over "pick up and read, pick up and read." He rushed back inside, where his friend, Alypius, was sitting, and impulsively opened a codex of the Epistles of Paul and read at random a verse from Romans 13 *in silentio*. Had he by then acquired the knack of silent reading, or, as Balogh proposed, was this the result of an extraordinary psychical experience? At any rate, when Augustine states that he had read it *silently*, he does so without commenting on it, as though this act of silent reading was not at all extraordinary.

Two years later George Lincoln Hendrickson (1886–1962) responded to Balogh by suggesting that Greek and Roman readers practiced a variety of reading styles, including subvocalization and auditory imagination. In regard to the latter, Hendrickson felt it significant that the verbs "to hear" (*akouein* and *audire*) could be used to refer to the act of reading. He cited the passage in which Phaedrus remarks to Socrates that anyone who thought he was a physician because he *heard* something from a book (*ek bibliou pothen akousas*) would be crazy (*Phaedrus* 268c4) and Socrates' anecdote in which King Thamus of Egypt predicts that a generation of readers will grow up as "hearers of many things (*poluêkooi*)" yet will have really learned nothing (*Phaedrus* 275a6). (See chapter 5.)

But, when one read, did one silently "hear" the phantom speech, i.e., practice auditory imagination, or did one actually hear one's own overt articulation of the text? And what did Augustine really mean when, in a dialogue with his son on the nature of words, he says that writing "is made for the eyes, but from that point on it pertains to the ears so that it may come into the mind" (*De magistro* 4.8)? When Hendrickson quoted this passage, he added, "While it might be possible from the point of view of modern psychology to interpret these words in a quasi-figurative sense of so-called inner speech and inner audition, yet for antiquity and for Augustine it may be affirmed with confidence that the word was articulated sound (*omne verbum sonat*), as he says in another place and that therefore to read was to read aloud" (Hendrickson 1929, 185).[2]

This partial capitulation to Balogh's position does not seem, however, to be supported by the *De magistro* text. Augustine does not say anything here about overt speech. The written words seem to pass from eye to ear without overt lingual mediation, as they did for Ambrose and for Augustine himself at his conversion experience. To define "word" as "sound" is not the same as saying that every word that is read must be uttered so that the reader's ears may outwardly decipher that sound. It simply means that a word is received and stored as essentially an acoustic, rather than an optical, sign. Though the mind decodes a written word first by its graphic shapes, it normally must convert those shapes into sounds, overt or covert, in order to verify that visually scanned word's meaning.

This revisionary interpretation of mine accords with several subsequent statements Hendrickson makes:

> When we speak of silent reading, it will be recognized that there are many gradations between vocal and silent reading, descending from distinct oral utterance to indistinct murmurs, to whispers, to mere lip motions, and so on through unconscious muscular movements of the tongue, throat, or larynx, to pure eye-reading unattended by any enunciatory effort. . . . The ingenious psychologist of the laboratory has devised methods to detect muscular responses of which the subject is quite unconscious. But with or without muscular reaction there is present in most readers an inner or mental enunciation of the words read, with its corollary, an inner or mental audition of them. This is the so-called *parole interieure*, which the French scholar V. Egger was the first to define.
>
> (1929, 193, 194)

In 1968, Bernard Knox published "Silent Reading in Antiquity," an essay that marked a decisive turn in this debate. While acknowledging Hendrickson's contributions, Knox assembled a considerably larger list of citations proving that the ancients, though they normally read texts aloud, were perfectly able, when need be, to read them silently. After all, this was how private, personal letters would usually be read in the company of others. After citing the excerpt from Antiphanes' *Sappho*, the riddle of the letter, which I will discuss presently, Knox examined a passage from Euripides' *Hippolytus* in which Theseus discovers tightly clutched in his dead wife's hand a tablet, in which she has falsely accused his son (her stepson) of raping her. He opens it and reads it to himself while the chorus voices their concern. Then he cries out. "Alas! What misfortune has now been added to misfortune! Luckless man that

I am! . . . This tablet shouts out, shouts out horrors! . . . Such, such is the vocal song that I see in these written words" (874–880). Theseus, as Knox points out, is actually represented on stage as reading this text silently. At the very least we can conclude that silent reading was not regarded as an impossible task; no divinity infused into his mind the words that Theseus says he "heard."

In 1997, two scholars, neither one aware of the other's research, revisited this debate. The medievalist Paul Saenger published *Space Between Words*, in which he argued that the key to the ancients' preference for reading aloud was that their texts were written in rows of words without spaces between them, i.e., in *scriptio* (or *scriptura*) *continua*.[3] Supporting Balogh's argument, Saenger asserted that readers needed to articulate the syllables serially in order to recognize the separate words. It was not until the seventh century in Ireland and the twelfth century throughout Europe that scribes inserted spaces between words, an innovation that Saenger asserted made silent reading finally a feasible practice. The speed that silent reading made possible also promoted, according to Saenger, the advancement of European philosophical and scientific thought.

In the same year that *Space Between Words* was published, the Russian philologist A. K. Gavrilov published his article "Ancient Techniques of Reading," in which he endorsed Bernard Knox's position and further buttressed it by making reference to the psychology of reading. Reading aloud obviously coordinates the speech-motor centers of the temporal lobe, including Broca's area, with the visual area of the occipital lobe. This neural hookup is involved in establishing what has been called the "eye-voice span" (EVS), the distance between the words being pronounced and the several words lying to the right that the eyes scan before articulating them. The length of this span varies, depending on the proficiency of the reader, the clarity of the lettering, and the predictability of the syntax. This visual span also allows the reader to anticipate the proper intonation, speed, and pausing of the voice in order to render the recitation as close to natural speech as possible. What this presupposes is that the visual system recognizes the words and retains them for a split second in working memory; then, by the time the voice utters them, the eyes have already shifted rightward to recognize the next chunk of words. Like a skillful guide on the pathway of graphemes, vision is therefore always several steps ahead of voice (Taylor and Taylor 1983).

To support his case, Gavrilov was able to cite an early passage from the Roman rhetorician Quintilian's *Institutio oratoria* (*On the Orator's Education*) that clearly indicates that first-century teachers were aware of EVS and viewed its efficient use as an educational goal: "To look off to the right, which is how everyone teaches reading, and look in a forward direction, is not only reasonable but also simply the way it is done. Since one must see what is coming next before one can say it, a very difficult procedure, one's attention must be divided in such a way that one function may be handled by the eyes, the other by the voice" (1.1.34). This is "very difficult" only at first, for Quintilian is dealing here with early-childhood education.

Besides the eye-voice span, Gavrilov reasoned that inner speech must also have played a role in Roman literacy by bridging the gap between overt (fully articulated) and covert (silent) reading procedures. This

> semi-conscious subvocalization . . . remaining below the threshold of audibility for other people . . . can produce on us an aesthetic impression no less than we get from reading [texts] aloud. Thus research on a number of different aspects of reading tells us, in effect, that the expressions "reading aloud" and "reading to oneself," although they seem to be opposed, actually pick out *mutually complementary* forms or facets of the reading process. Both are present, in different doses, whenever reading takes place.
>
> (Gavrilov 1997, 61)[4]

Texts written in *scriptio continua*, in contrast to our space-separated words, must have made reading a somewhat slower process, but we ought not exaggerate its difficulty. In either format, a reader's eyes move forward by quick saccades, fixating briefly on the stems of content words (verbs, nouns, adjectives, and some adverbs)—this is what Quintilian meant by looking "off to the right." In modern European languages, given their paucity of inflexions, the spaces have become essential to smooth reading. In Greek and Latin, however, content words are inflected (verbs most highly, nouns and adjectives to a lesser degree, and adverbs sometimes), while function words (prepositions, conjunctions, and many adverbs) remain, as in most languages, uninflected. The inflected endings of content words indicate the completion of words, and their prefixes, usually formed from prepositions, indicate their beginning. In *scriptio continua*, these two boundary markers, together with short function words, would have been scanned peripherally while the reader focally targeted the meaning-rich stems of content words. In short, the word boundaries where modern languages insert spaces may have been marked in Greek and Latin by prefixes and inflexions.[5]

Pauses at clausal and sentence endings and for inhalations are necessary features of speech, but, despite these exceptions, most spoken discourse is what we might call *dictio continua*, with no silences inserted between words. Ancient readers of *scriptio continua* would therefore regard our space-separated texts as unnaturally fragmented representations of speech, and speech was precisely what their unbroken script may have aimed to record. Even when a text was not recited aloud, its silent or murmured reading would necessitate a more consciously experienced subvocalizing, during which one would be aware of one's breathing, the flexions of one's tongue, and the movements of one's lips. Doing so, one would also be aware of the rhythms and phonemic values of the words. That is, one would "hear" oneself even as one silently "uttered" the text. Reading a text often enough to become familiar with its syntactical turns and cadences would be especially appropriate for literary prose or verse, the pleasures of which repaid this effort. Writing for the voice thus meant writing not only for the outer voice but also for the inner.

The Genealogy of the Lyric Voice

What Plato and Aristotle regarded as "lyric poetry," i.e., song texts, they viewed as a minor art form. Aristophanes implied that by the late fifth century BCE the young regarded lyric as hopelessly old fashioned (*Clouds* 1353–1358). When, following the Peloponnesian War, dithyrambic lyric was revived by the "New Music" movement associated with virtuosi such as Timotheus of Miletus, Phrynis of Lesbos, and Philoxenus of Cythera, its critics deemed its metaphors and newly coined epithets pretentious display. "You're singing dithyrambs" came to mean "You're talking nonsense" (Ford 2013). It was only under the pressure of writing, exerted slowly and steadily over centuries, that lyrical song eventually metamorphosed into "poem."

Sadly, the body of Greek lyric poetry is fragmentary and, except for the so-called Song of Seikilos, its music lost despite modern efforts to reconstruct it. What we are left with are texts, most of them portions of poems quoted by later prose writers. Though inner monologue has over recent centuries become the dominant form of lyric, in the ancient world it was more the exception than the rule. The often-cited lines attributed to Sappho, "The moon has set / and the Pleiades. It is the middle of the night. / Time passes / and I lie in bed alone," sounds like inner monologue, but since it is a fragment, it may originally have been set within an apostrophe to a neglectful lover or a speech put in the mouth of a mythic character, such as Penelope or Endymion. The form I will examine instead is the paradigm most typical of classical lyric, outer monologue, which on closer examination reveals four distinct stylistic types.

I take this basic lyric speech situation to be that of a specific addresser speaking a monologue to an addressee or to an assembly of addressees. The addresser may be represented either as present and orally uttering a verbal message or absent and communicating through a written text. The addressee may be represented either as immediately present to the speaker or absent, as is the case of a reader of a text. Apostrophe is a special form of word play: the addressee is actually absent or otherwise incapable of hearing the addresser's speech, but the latter is represented *as though* directly speaking to that addressee. Whereas the lyric addresser is always a single specific person, whether the author or a designated persona, the lyric addressee may either be a general or a specific person.

Figure 6.1 represents four distinct genres: epideictic oratory, drama, inscription, and the personal letter. The first two are presented as oral performances, the second two as written texts. These are not sharply demarcated stages any more than are Merlin Donald's four cognitive-evolutionary stages. Rather they mark the successive stages of maximal influence exerted by each of these ongoing genres in shaping the lyric in the classic form of outer monologue. These four, listed on the left, are placed under the addresser because, according to the lyric convention, the speaker of an outer monologue determines the genre of the speech and its addressee(s).

Epideictic Oratory

Aristotle in the *Rhetoric* (book 1, chap. 3, or 1358b) uses the three principal tenses to differentiate kinds of persuasive discourse. Speech that deals with future actions and employs hortatory or dissuasive discourse

Addresser		Addressee			
Present	Absent	Present		Absent	
(Always specific)		General	Specific	General	Specific
1. Epideictic oratory		Audience			
2. Drama			Another persona		
	3. Inscription			Any passerby	
	4. Personal letter				The intended recipient

Figure 6.1 Models for lyric discourse as outer monologue.

belongs to *deliberative* rhetoric. That which deals with past actions and employs defensive or accusatory discourse belongs to *judicial* (forensic) rhetoric. That which deals with present occasions or timeless principles and employs value distinctions, such as praise and blame, belongs to *demonstrative,* better known as *epideictic,* rhetoric. This third kind of rhetoric, epideictic, equally persuasive in intent, was not meant to lead to outward action but rather toward inward changes of mood, thought, or belief. It characterized ceremonial speech such as religious discourse and funeral orations. Like music, which has as one of its social functions the potential to draw an audience into a state of oneness through rhythmic entrainment, epideictic employs repetitions and figures aimed at creating emotional unanimity.

As a predominantly oral culture, archaic Greece (from the ninth through the eighth centuries) used publicly performed speech to achieve what later would be called "rhetorical" effects. Though Aristotle wrote separate treatises on the arts of rhetoric and poetry, in practice these two arts were not quite so disjunct. This is Jeffrey Walker's compelling central argument in his *Rhetoric and Poetics in Antiquity* (2000), which he introduces by discussing Hesiod's "Kings and Singers" passage (*Theogony* 80–103). It is sometimes supposed that kings allay social unrest through rhetoric, and singers allay inner distress through poetry. Yet Hesiod insists that both classes of men personally receive the gift of mellifluous speech from the Muses. Both, Walker maintains, benefit humankind through persuasion, and the fundamental means both utilize is "prosodic repetition or 'equivalence' including doubling both of phrases and of phrase lengths such as we find in parallelism and diphrasis" (11).

Much early lyric (seventh and sixth centuries) may be classified as "versified or sung oration, a variety of epideictic discourse" (Walker 2000, 155; see also Johnson 1982). The epideictic character of early lyric becomes apparent when we recognize that many of the extant lyrics seem to have been occasioned by ceremonies such as marriages and funerals or as praise songs or as hymns and prayers to particular divinities. Others were composed and circulated for use at private drinking parties (*symposia*), where they won applause when they expressed appropriately convivial and/or invective sentiments in fresh ways. In either case, sympotic songs presented views that were meant to engage the minds and emotions of the guests. As versified oratory for a literate readership, a later lyric might also address a larger audience: viewed simply with respect to their epideictic content, many of Horace's odes, presumably based on earlier Greek texts, read nowadays as elegant op-ed letters.

Drama

As staged drama was an imitation of human action, written drama could be viewed as an imitation of that imitation. As such, it was animated by dialogue, a form of discourse that is inherently here and now (the narratological "scene"). While episodic durations in monologic narrative are frequently condensed or expanded, direct dialogic discourse, even in writing, appears to unfold in real time. This requires of the reader far more attention, for no dialogic utterance is fully meaningful in itself but is responsive to some foregoing utterance. That is, since every reply is a re-plication, a verbal "folding back" upon an earlier statement, which was itself a "folding back" upon a previous utterance, no reply makes sense unless one understands what it is folded back upon. Conversation, be it actual, dramatized, or textualized, is thus both collaborative and competitive—in short: agonistic.[6]

Dramatic speech, when addressed to a tragic chorus and enunciating timeless, universal principles, is often identical to epideictic speech. When addressed to another character, it becomes a plot-contingent utterance. In either case, a dramatic speech has a personal context: the addresser and the addressee(s) are all participants in a moment of action (the particular enunciation of the speech), which has been preceded by other moments and now propels the personae to yet other actions and consequences. Unlike epideictic, which is a formal speech of indeterminate length, any lyric that borrows the discourse style of the dramatic speech cannot exceed the time it would take for the other party to interrupt with a rejoinder or in some way terminate the conversation. Our sense that a specific persona is the direct addressee of this speech colors our sense of the perlocutionary force of its words.

The eighteenth-century French literary theorist Charles Batteux (1713–1780) recognized dramatic speech as one of the roots of lyric. Patterned upon Greco-Roman tragedy, the plays of Corneille and Racine include rather long monologues addressed to onstage characters:

> As long as the action in drama and epic moves forward, the poetry is epic or dramatic; when the action stops and the poetry portrays nothing except the unique state of the soul, the pure feeling it is experiencing, the poetry in itself is lyric; to be set to song, it need only be given the suitable form. The monologues of Polyeucte, Camille, and Chimène [leading characters in three tragedies of Pierre Corneille] are lyrical pieces (*morceaux*).
>
> (Batteux 1824, 269)

Obedient as these playwrights were to the Aristotelian "unity of time" or their interpretation of it, their plots strictly adhere to "story time," but, into this chronological sequence, these long monologues, or *tirades*, inject "discourse time" both by representing the psychological time sense of the speaker and by constituting in themselves "slowdowns" and "pauses." Thus temporal variability has always been a feature of lyric both as song and as theatrical set piece.

As outer monologues, dramatic speeches, even when they are answered, are not so much parts of a conversation as they are a series of more or less confrontational exchanges, characteristic of an *agôn,* in this sense a verbal contest. Greek culture as a whole was highly competitive. In athletics, architecture, and performing arts, cities vied with cities, tribes with tribes, teams with teams, and individuals with individuals, a form of social play that resembles outright hostility while being peaceably rule governed. As Hesiod taught, there is bad strife that produces litigation, feuds, and war, but there is also good strife in which "potters disparage potters; carpenters, carpenters; beggars envy beggars; singers, singers" (*Works and Days* 25–26). The Greeks dearly loved song contests—Hesiod himself was once a winner of a song contest held at Calchis, and according to one legend, his opponent was no less than Homer himself.

For writers who still regarded themselves as devotees of the traditional Muses, the problem was how to produce texts that, when recited by a nonprofessional reader, in private or before a group of friends, could produce some of the *agalma,* the magical charm, of an actual oral performance. A century after Sophron and Xenarchus, the oral/agonistic style was brought to a new level of sophistication in the mimes of another Sicilian, Theocritus (ca. 325–267 BCE). His took the form of short vignettes, which he called idylls (*eidullia,* i.e., little images, or sketches). Like Sophron, whom he said to have imitated in several of them, he drew his material from conversations one might overhear between herders, reapers, fishermen, and housewives on holiday. But unlike his predecessor, Theocritus wrote his mimes in verse, dactylic hexameter, the meter of singers of epic, and in them he often thematized the act of singing. In *Idyll* 1, for example, singing provides a goatherd and a shepherd the opportunity to exchange gifts; in *Idyll* 2, singing is used to weave a magic spell; in *Idylls* 3 and 11, to serenade girls; in *Idyll* 7, to pass the time while traveling to a festival; and in *Idyll* 20 to bewail an unhappy love affair.

The song motif is especially foregrounded in *Idylls* 5, 6, 8, and 9, which represent song contests. In the typical pastoral song contest, two herdsmen meet, place wagers, e.g., a lamb, a carved wooden bowl, or a flute, and choose another person as a judge. The contest begins when one participant improvises several metrically equal lines, which the second contestant must match in theme while inserting pointedly different words. This kind of verbal dueling was known as *amoebean,* or alternating, song.

This agonistic give and take could be reproduced on a small scale simply by presenting two persons in rapid statement and counterstatement. When this occurred in metered verse, it was termed *stichomythia,* i.e., "talking in lines," a familiar device in Greek drama, where it was often used to portray sharp clashes of will. Except for only two lines, Theocritus's *Idyll* 27 is an entire series of one-liners between the seduction-minded cowherd Daphnis and an unnamed shepherdess. In English, one good example displaying amoebean substitutions appears in act 3 of Shakespeare's *Hamlet*:

> GERTRUD: Hamlet, thou hast thy father much offended.
> HAMLET: Mother, you have my father much offended.
> GERTRUD: Come, come, you answer with an idle tongue.
> HAMLET: Go, go, you question with a wicked tongue.

An even nimbler form of repartee is *antilabe,* a device in which two voices share the utterance of a single line. Though often associated with clever debating, it has other applications. In the balcony scene in *Romeo and Juliet,* for example, the two seem to demonstrate their attunement by breathlessly completing a single iambic pentameter line (see the three bracketed speeches in figure 6.2).

Theocritus's most extended use of antilabe is found in his fifteenth *Idyll,* the conversation of the two Sicilian housewives in Alexandria en route to the festival of Adonis. Here it portrays no clash of wills but rather the excitement of two women so eager to exchange news that each interrupts the other and completes her metrical lines. As this idyll demonstrates, we should not consign such dialogic exchanges exclusively to *ant*agonistic events. Call and response, antiphony, and refrains serve many other purposes. When, for example, the Muses give delight to the gods and goddesses on Olympus, they themselves sing amoebaean songs (*Iliad* 1:604).[7]

Inscription

A century or so after alphabetic script was introduced into Greece (ca. 800 BCE), public writing began to appear, first in temples, later in municipal sites. By the sixth century, they came to be used to commemorate persons and events, as I remarked upon in chapter 4. Though in the

Romeo: See, how she leans her cheek upon her hand!

O, that I were a glove upon that hand,

That I might touch that cheek!

Juliet: Ay me!

Romeo: She speaks:

O, speak again, bright angel! for thou art

As glorious to this night . . .

Figure 6.2 Antilabe in *Romeo and Juliet.*

archaic age very few people could read these carved letters, by the fifth century literacy had begun to establish itself among the privileged urban elites. In style and content, these inscriptions were brief texts in the style of epideictic monologues that served to reaffirm traditional principles of reverence for the dead, family solidarity, patriotism, cultic allegiance, and the praise of personal virtues. As epideictic speech, they normally avoided the topic of blame, for who would want to spend good money memorializing bad behavior?

A major literary development of the Hellenistic era was the adaptation of the inscription to the written page. The Libyan-born poet of Alexandria Callimachus (ca. 305–240 BCE) is credited with successfully forging this new genre and the Spanish-born Roman poet Martial (ca. 40–103 CE) with sharpening its cutting edge. Just as the dramatic model had constrained epideictic expansiveness in the lyric, this new literary genre, epigram, was to encourage conciseness. No longer dependent on oral mnemonics, epigram and epigram-influenced lyric could now dispense with repetitions, formulae, balanced phrases, and ring structure. Now repetition meant emphasis, and, in the place of comfortable, traditional formulae, the reader encountered compound nouns and newly coined epithets. Balanced phrases became antitheses, and ring structure, similarly condensed, became verbal chiasmus.

The Personal Letter

This too is a "made thing," its words affixed to a tangible object existing in space, but, when compared to words carved upon a brass or stone monument, a letter written upon a surface such as papyrus or parchment has a briefer time span. It can, however, be more easily collected, copied, published, and circulated. In that respect a letter is like any other literary document and, like a biological organism, survives through reproduction.

The letters of notable personages were indeed collected, published, and avidly read in the ancient world, but these were for the most part of a didactic nature. A personal letter is quite a different kind of document: it is a unique, not a multiple, missive—a private, not a public, message. Any text that a named addresser directs toward a named addressee, be it prose or verse, is normally recognizable as an intimate, personally motivated communication not intended to be shared with others, certainly not to be performed for the entertainment of strangers.

When Eva Stehle published *Gender and Performance in Ancient Greece* (1997), she suggested that some of "Sappho's poems may have been intended as gifts to particular former pupils of hers and that Sappho's poetry survived because it was designed to escape the tyranny of the performance culture and remain her 'voice,' that is, to circulate as text from the beginning" (323). Granted, this is an unprovable theory, but it is one that I do not find implausible. Even if Sappho created these poems as performance pieces, there is no reason to suppose she did not dictate them or write them down herself to provide singers with the original texts and no reason to suppose she did not send copies to the persons whose names appear in them.

Writing could obviously become a powerful medium when used to convey personal information withheld from others, secrets about oneself and about the personal lives of others. Of Antiphanes' third-century comedy *Sappho*, all we now have is a riddle proposed by that unlikely title character to a group of gentlemen, as quoted by Athenaeus in his *Deipnosophists*, or "Feast of the Learned": "There is one of feminine gender, who keeps her little ones tucked safely in her womb where, though voiceless, they raise a loud cry that passes over the raging sea and all the land to whatever person she wishes. Though anyone standing beside that person will hear nothing, a deaf man may well have the sensation of hearing." After someone makes a wrong guess, Sappho gives the answer: "The one of feminine gender is the letter (*epistolê*). The little ones she carries about within her are written characters (*grammata*). Though voiceless, they talk from far away to anyone they wish, but anyone else standing close by will hear nothing."

Why Antiphanes chose Sappho for this play we will never know. Did she have a reputation for letter writing? Was she a wit or a sphinxlike riddler? Though we now have little more than fragments of what had once been nine volumes of her collected poems, the scant evidence we do have tells us that in her poems she addressed members of her circle by name, including self-reference and self-address. While none of the fragments is what we might call a versified personal letter, many are personalized songs that may well have been sent to the named addressees—and to others—as gifts, as Eva Steele (1997) suggested.

Stehle's critics, ranging from the skeptical to the dismissive, have assumed that she meant that Sappho sent her verses, especially her love lyrics, as *letters*. Jeffrey Walker (2000, 233) asked, "Is it necessary to consider the poem [fr. 31, beginning 'He seems to me equal to the gods'] as a 'love letter' to an actual specific person . . . ?" For, "if the poem had really been meant as a 'letter' or gift for a single, specific person, who would then perform it to herself in solitude, it is difficult to understand how it would be disseminated and preserved." Gentili and Catenacci (2007, 85) read Stehle as asserting this poem to be "*una sorte di epistola d'amore,*" and Paola Ceccarelli (2013, 33) states flatly that Sappho's poetry lacks "markers of epistolarity," concluding that "Sappho's poems cannot be taken to imply anything in connection with letter writing." The problem with these critiques is that Stehle in no way identifies Sappho's poems as "letters." Though she asserts that they were meant to be independently freestanding externalizations of the poet's voice, the model she explicitly refers to, citing Jesper Svenbro twelve times, is *inscription*, not the personal letter.

Now, having cleared Eva Stehle of any suspicion of having claimed lyric epistolarity in respect to Sappho's poetry, I want to incriminate myself.

If by "epistolarity" we mean a versified imitation of the personal-letter form such as we find in Horace's *Epistles* and in Ovid's *Heroides* and *Epistulae ex Ponto,* his verse letters from exile, then we do not find epistolarity in the body of Greek and Latin lyric poetry. But our task is easier if we look for evidence of epistolary *influence* upon lyric poetry and establish in advance three criteria: the composition must be (1) outer monologue (2) directly addressed in the vocative case (3) to a nonpresent person. These three epistolary features characterize a multitude of Greek and Latin lyrics, not to mention later European lyrics. Of course, when a poet directly addresses an absent addressee by name, he or she is simply taking advantage of the *I–You* epistolary format to explore certain complexities of thought and feeling (cf. apostrophe). To acknowledge this fictive convention is merely to accept that *poiêsis* is grounded in verbal play.

Furthermore, it seems quite reasonable to me to suppose that in any culture in which spatially separated individuals can contact and converse with one another only through texts written and read in solitude and silence, this experience will impress itself powerfully in other literate genres similarly dedicated to intimate discourse.

The Dramatic Lyric

Our concept of poetry as distinct from other literary genres such as prose fiction and drama texts tends to be that of a disembodied, self-reflexive mind thinking about what it is powerless to do anything about—emotions of love and grief, the relentless passage of time, the inevitability of death, etc. These are universal themes that men and women in all times and places have sought to clarify by composing verbal artifacts, but the "I" that ponders them is always a cultural construct adjusted to the media and institutions available to register those thoughts.

If we have accepted what we call the "lyric poem" as the prototype of poetry, we also may assume that it is modifiable in outward details but immutable at its core. But, if we accept the propositions that genres are not timeless entities and that we derive our concept of the prototypical poem, the lyric, from philosophers and critics of the late eighteenth and early nineteenth centuries, we must try in other ways to account for the cultural evolution of poetry. One way is to project ourselves into different historical milieus. When we do so, we note that works classifiable as "poetry" exist now and have always existed in cultures other than our own but that they have certain peculiar characteristics, as indeed our works have when viewed from the points of view of other cultures. The principal advantage of studying cultures other than our own and learning to view verbal artifacts from their perspective is that, coming back to our own culture, we realize how wonderfully and beautifully strange our own achievements are.

With that in mind I want now to examine some of the features of Roman written poetry, the texts they called *carmina* (songs), even if they were never set to music. As we have seen, a significant aspect of the literate imagination, especially in preprint cultures, is the *auditory* imagination. Hence, the fashioning of written verbal artifacts into virtual orality devices, empowering readers to use the silent page to evoke the sounds of voices in lively conversational interchanges or, typical of lyric, in single-voiced speech. But before I examine several monologic lyrics of Horace, I will briefly discuss some examples of verse dialogue.

Perhaps as early as 42 BCE a tall, shy, bookish young man from a farm in northern Italy, Publius Vergilius Maro, then residing in Rome, began showing a collection of pastoral poems that over the next five years he would winnow down to ten and publish as the *Eclogues* ("Selections"). Most of these Latin adaptations of Theocritus's *Idylls* presented scenes of herdsmen, sometimes swapping songs, sometimes singing solos, sometimes competing in amoebaean contests. In the background of his Arcadian landscapes, however, readers could recognize contemporary events, such as the assassination of Julius Caesar, the ascendency of his adopted son, Octavian, later to be named "Augustus," and the seismic shifts of power leading to the crucial Battle of Philippi, followed by a series of aftershocks that would take the form of proscriptions and confiscations.

In 41 BCE, another young man, this one of short and sturdy stature, from the southern Italian countryside, Quintus Horatius Flaccus, drew Vergil's attention. Once they met, the two must have discovered they had much in common. Each was deeply read in Greek philosophy and poetry, especially the teachings of Epicurus, and each had lost his inheritance when, under the Second Triumvirate (of Antony, Lepidus, and Octavian), parcels of agricultural land were redistributed among soldiers returning from Philippi and the East. This was a policy of regional retaliation against municipalities that had sided with Brutus and his faction. (Horace, though, had other reasons to feel apprehensive. He had been studying in Athens then, and when Brutus and Cassius sought to regroup after the assassination of Caesar, he joined their army, became a commissioned officer, and saw battle at Philippi.) Whatever Vergil read of his new friend's work, short lyric poems in the manner of Archilochus, perhaps, plus some equally acerbic satires, he passed his copies on to Maecenas, a wealthy and influential adviser to Octavian. His earlier allegiances notwithstanding, Horace, like Vergil became a protégé of Maecenas, who in about 31 BCE presented him with a house and property in the Sabine Hills east of Rome, where he would live the life of a gentleman farmer and full-time poet.[8]

After the publication of the *Eclogues* in 37 BCE, Vergil would not again compose poetry in a dialogically dramatic format, though his unfinished epic, the *Aeneid*, like his Homeric models, is filled with speeches and counterspeeches. Horace, however, continued to be intrigued by the possibilities of representing voices in writing: (1) the monologic voice of one person, usually himself, addressing others, in a particular time and place; (2) the voices of others in dialogue with himself; (3) the voice of the epigrammatic wit; and (4) the monologic voice of a letter writer pretending the immediacy of face-to-face speech.

The first of Horace's two books of satires appeared ca. 35, the second in 30 BCE, but instead of calling them "satires" he evidently preferred the term *sermones*, i.e., talks, informal speeches, or simply conversations. Book 1, though predominantly epideictic monologues, also includes snatches of conversations with named and unnamed persons. These dialogic others typically represent traits and attitudes Horace finds laughable, vicious, or otherwise reprehensible. They appear suddenly, say their words, then disappear, having succeeded only in further exasperating the poet and sending him on to his next point.

In book 2, these voices dominate, and the satire comes to resemble literary mime. Now the poet plays the straight man, asking questions that propel his characters to reveal their absurdities and, sometimes, to expose his own. *Satires* 1, 5, 7, and 8 are direct dialogues without narrative frames or bridges.[9] *Satires* 2 and 6 are monologues that conclude with a monologue by another speaker. In 2, in which Horace appears as though delivering an impromptu *sermo* on the rewards of the simple life to a group of dinner guests, he tells them that this is the teaching of a certain peasant farmer, Ofellus, an unschooled, homespun philosopher he had come to know in boyhood. In response to a guest who questions the relevance of this discourse, Horace asks him to visualize this simple man. Ofellus's vocal presence then emerges in the last twenty lines of the poem, as Horace quotes him encouraging his sons as they work with him for the new owner of the land that once was his. This *rusticus* perhaps reminded the poet of own beloved father, had he been living in 42 when their family farm was confiscated in Venusia.

At about the same time that Horace published his second book of satires (30 BCE), he also published his seventeen-poem book of *Epodes*. For these, his models were the lyrics of Archilochus (circa 680–640 BCE), a soldier-poet known for his sharp wit and rancorous frankness. It was he that popularized the iambic meter, later used in both tragic and comic speeches, a rhythm that Aristotle said most resembled speech.[10] In the fifth century, iamb was already associated with lampoon and invective, the poetry of blame.

Archilochus had over time earned a reputation for shameless self-advertisement, but Aristotle viewed him in a different light. In the *Rhetoric* (1418b16) he cites him as using voices other than his own to help him make his points, a stylistic tactic, he says, that makes an author seem less self-absorbed, longwinded, or boorishly abusive. This suggests that Archilochus, like Homer, had also composed mimetic, i.e., dramatic, poetry, the sort of role-playing that Plato in his *Republic* would decry as representing oneself as "manifold" (*pollaplous*). We now have only fragments of his works, but Aristotle and, presumably, Horace would have had access to his collected works. The throng of voices that fill Horace's *Epodes*, as they do his second book of *Satires*, may indeed be an Archilochean touch that at least some of his contemporary readers would have recognized (West 1974, Lavigne 2005, Rotstein 2010).

Some of the features that typify a dramatic lyric are (1) *monologue* (a single person utters all the words), (2) *addressee(s)* (one or more persons, physically present, that are the speaker's intended audience but do not verbally respond), (3) *setting* (a given time and place are supplied), and (4) a *triggering circumstance* (some action or state of affairs that provokes the utterance). The first three features pertain to the physical speech event—someone speaks to someone at some time and in some place. But, in the absence of physical performers and scenic details, these features must be imagined. There is a benefit in this: when we read or recite a dramatic lyric, our attention is not distracted by the visual or vocal peculiarities of a particular dramatic performance.

The fourth feature of a dramatic lyric, the triggering event, has already happened. It is therefore neither heard nor seen nor imagined by the mind's ear or the mind's eye but is simply *known*. Just as every drama and every scene within it has a before and an after, every dramatic lyric has by definition an action or circumstance that precedes the utterance and future consequences that may flow from the utterance itself. Furthermore, like every drama, a dramatic lyric is problem centered. As Jeffrey Walker (2000) reminded us, this is the function of epideictic rhetoric: if something is good, it needs to be recognized and honored; if something is bad, it needs, through the utterance, to be corrected or otherwise dealt with.

Unlike a dramatic script, however, a dramatic lyric usually implies, rather than spells out, the state of affairs that provokes its utterance. For this purpose, implicit metaphor is a favorite device. The advantage of implicit metaphor is its generality. Readers may emotionally respond to it without possessing the author's personal life experiences. In *Epode* 13, for example, the causal circumstance is represented by the setting, the "savage storm" (*horrida tempestas*), an image of mindless, uncontrollable, inhuman ferocity.

EPODE 13

A savage storm has tightened down the sky and Jove
  descends in rain and snow. Now sea, now woods

roar with the North Wind out of Thrace. Let us, friends, snatch
  the chance this day affords. While we are young,

we have the right to clear the cloudy brow of age.
  Get me my jars of wine, my birth year's vintage.

Stop talk of other things. Perhaps some god will yet
  turn things around for us. For now, some drops

of Persian nard will help, plus Hermes' kithara,
  to lift ill-omened worries from our chests.

As that famed centaur to his ward once prophesied:
  "Unconquered mortal boy of goddess born,

Assaracus' domain awaits you, split apart
by cold Scamander and swift Simois.

From there, for you, the weaving Fates grant no return
nor shall your sea-blue mother bring you home.

Once there, lighten the weight of woe with wine and song,
that are for twisted trouble solace sweet."[11]

In this epode, someone, i.e., the poet, addresses several friends indoors while a rain- and snowstorm rages outside. The geographical setting is not precisely stated but is somewhere south of Thrace, the home of the North Wind, Boreas (Latin: Aquilo), a wild region with a cold rainy season from October through March. Alcaeus starts a poem (fragment 338) with almost identical references to the season and the descent of Zeus. This may, therefore, just be an homage to one of Horace's favorite poets, and "North Wind out of Thrace" may simply be a Greek commonplace, but it may also have had historical significance. Horace does not tell us here, but we know from other epodes, e.g., *Epodes* 7 and 16, that in the 30s, the decade of the Second Triumvirate, he was extremely worried that the Roman homeland would once again collapse into civil war. Of course, we cannot specify what the topics of their discussion were, but, since their talk is part of the dramatic context and the cause that provokes the speaker to interrupt, neither should we ignore it.[12] Fraenkel is unnecessarily rigid when he says: "My interpretations are, without exception, based on the conviction that Horace throughout his work, shows himself both determined and able to express everything that is relevant to the understanding and the appreciation of the poem, either by saying it in so many words or by implying it through unambiguous hints" (1957, 26). Horace sought to address an audience not only of skilled readers of poetic texts but also of contemporaries upon whom recent events had imprinted much that is now lost to us. His command to his friends to stop talking about other matters (*cetera*) invites them to collaborate with him in a *paralipsis*, a "leaving aside," the rhetorical device by which a speaker draws attention to something by pointedly refusing to discuss it. What accounts for the "ill-omened worries," the *dirae sollicitudines*, that oppress these young men? The political conflict that had torn the Republic apart, climaxing with the disastrous final Battle of Philippi with its consequences for Horace and his fellow soldiers seems a most likely topic of their conversation.

The graceful *exemplum* that concludes the poem also points toward Philippi. In it Horace assumes the voice of the wise centaur, Cheiron, teacher of the young Achilles, as he tells the lad he must travel to Troy (Assaracus was one of its early kings) but from that land will never return home. As everyone knew, Achilles had been reluctant to join Agamemnon's expedition yet knew that everlasting fame would be his if he did. This theme of going to war and never returning adds a note of fatality to this sympotic lyric. It also demonstrates the rhetorical effect achieved by an author who lets another person speak his words, the device that Aristotle attributed to Archilochus and that Horace had exploited in his second book of *Satires*. Here, as the poet's uninterrupted monologue blends into the voice of Cheiron, his immediate addressees partake for that brief moment in the glory of the doomed Achilles.

The Epistolary Lyric

One of the social consequences of literacy was the promotion of private entertainment. We can see this trend developing as early as the seventh century in the Hellenic world as private symposia and private clubs (*hetairiai*) appeared alongside, but separate from, the urban institutions that sponsored musical competitions, Homeric recitations, theatrical productions, and other festivals. Writers of lyric poetry found these groups congenial venues in which to try out their new and sometimes unconventional works. As already noted, closet drama in the form of written mimes emerged in the late fifth century and continued to be bought and circulated throughout the Hellenistic era.

Third- and second-century Roman poets such as Naevius, Ennius, Pacuvius, and Accius had resisted this tendency, preferring instead to emulate Greek epic and Greek tragedy. In the first century, however, a younger generation, which we now associate with Catullus and his circle, found more recent Greek models provided by the works of Theocritus and Callimachus much more congenial. This cohort of Roman poets, the *neoteroi*, found their predecessors' heroic attitudinizing no less tedious than were early twentieth-century modernists, such as Pound and Apollinaire, to find Tennyson and Lamartine. These new Roman poets regarded their art as a way to explore the personal—the vicissitudes of friendship and erotic love, loneliness and loss—and felt drawn to the philosophies of Epicureanism and Stoicism, which, despite their many differences, promised them ways to confront with equanimity a world no longer governable by reason and rhetoric.

For this new generation of poets, the personal letter provided an attractive model. Now as year by year Rome's dominion stretched further and further, northwest to Gaul and Spain and southeast to Greece and the Levant, Romans—poets among them—depended more and more on the letter to maintain their personal contacts. Here was a quintessentially private kind of writing, a medium that had always permitted individuals to overcome the distances that separated them from loved ones. This was as true for Catullus, briefly serving a thousand miles away in the province of Bithynia, as it was for Horace, playing the country mouse on his Sabine farm. It seemed only natural for first-century poets to mail their friends not only poems *in* letters but also poems *as* letters. Moreover, the ease with which the concept of "a letter" could be assimilated into the concept of "literature" was enhanced by the fact that those plural nouns *grammata* and *litterae* meant both "literature" and the individual "letter" one mailed to someone.

Before I examine some of the epistolary structures that underlie much of Roman lyric poetry, I should first briefly comment on several works that overtly imitate the structure of the personal letter. Horace's *Epistles*, a two-volume collection of twenty-three versified letters, written in dactylic hexameter during his later years and published in 20 and 13 BCE, include the addressee's name in the first several lines of each poem and end with a personal closure by way of valediction. Ovid's first published works, the *Heroides*, published some time between 25 and 16 BCE, are fictional letters written by women of mythology to their absent lovers, sometimes including their lovers' return letters. While they adhere to the epistolary form more strictly than Horace's, they lack his thematic and formal variation. Representing a psychological rather than a stylistic mixture, Ovid's epistolary poems explore the minds of lonely, sometimes abandoned women as they express the fluctuating emotions of yearning, nostalgia, guilt, and accusation. The first letter, from Penelope to Ulysses, reminds us that the separation of lovers, of husbands from wives and of fathers from sons, was a theme deeply woven into the Homeric epics and that Ovid was using the epistolary format to express those timeless, all-too-human longings. When he was sent off into exile him by Augustus in 9 BCE, he came to experience some of his heroines' emotions firsthand. His *Tristia* (Poems of sadness) were clearly letter-like, and his later *Epistulae ex Ponto* (Black Sea letters) were explicitly versified messages written to his wife and to prominent Romans who, he hoped, would intercede for him.

It is essential to keep in mind that a personal letter is meant to convey the oral presence of the writer-as-speaker. As I pointed out in chapter 4, this is an imaginative act: everyone knows that "hearing" the addresser's voice in the written words is social play, yet writer and reader persist in playing this game. Since writing a letter is like composing one part in a face-to-face conversation, epistolary lyrics, like dramatic lyrics, are virtual-orality devices. The only difference is that the dramatic lyric is *imagined* as a monologue transmitted from addresser to addressee in a single place and in an uninterrupted duration of time, whereas the epistolary lyric, though imagined as similarly unified in time and space, is also *understood* as a speech event interrupted in time and dislocated in space.

The epistolarity of Roman poetry, be it explicit or implied, may be examined at two levels of magnitude: the book and the individual poem. At the level of the book, an entire collection of poems is a written message conventionally understood as sent forth from an authorial addresser to one principal addressee. At the level of the individual poem, when the poet names an addressee who is clearly not present, that person must be imagined as situated at what might be called an epistolary distance normally traversed only by a written message. I will begin with an examination of the book and then go on to the individual poem.

As outlined in chapter 4, the essential structure of a Greek or Roman personal letter includes: (1) the names of the addresser and of the addressee in some variation of the formula "A greets B," (2) the message (news, advice, requests, encouragement, consolation, invitation, etc.), and finally (3) the valediction. The epistolary addressee of a book of poetry, the person named in the dedication, is usually announced early in the work and, in a book of lyrics, in the first poem. In a didactic work, e.g., Lucretius's *On the Nature of Things,* his address to Memmius is inserted early in the work. In Vergil's *Georgics,* an address to his patron, Maecenas, appears early in each of the four books. A dedicated book, no matter how many copies have been placed with booksellers, is said to be a unique gift sent to this patron of the arts, *litterae* that constitute both a mailed message and a literary composition.[13]

Just as a letter specifies the addresser as well as the addressee, first-century Roman poets, e.g., Catullus, Horace, Propertius, and Ovid, would occasionally refer to themselves by name and were often autobiographical. Even Vergil, despite his notable shyness and pastoral masks, speaks *in propria persona* in the *Georgics.* Authorial self-reference by name or by some other form of identification is what philologists have termed a *sphragis,* meaning in Greek the impression left by a signet ring on the wax used to seal a letter that indicated the addresser's identity and the intention that only the addressee should break the seal. The last eight lines of his fourth and final *Georgic* constitute a formal example of this feature:

> I have been singing here of the tending of fields and herds, / and of trees as well, while mighty Caesar / hurls lightning over the deep Euphrates, / gives a conqueror's laws to willing nations, and pursues his pathway to Olympus. / All the while, sweet Parthenope has been nourishing me, / Vergil, as I blossomed in the studies of quiet obscurity, / I who had once played shepherds' songs and, youthfully bold, / sang of you, Tityrus, there under the spreading beech tree.
>
> (559–566)[14]

When, in 1923, Richard Heinze published his monograph on the Horatian ode, he placed first among its defining features its address to a person present to the poet.[15] This dictum, a magisterial mixture of clarity and absolutism typical of nineteenth- and early twentieth-century German philology, has had its adherents and its critics. Among the latter, Kenneth Quinn ([1963] 2014) commented that while some were clearly dramatic odes, others qualified as lyrical epistles. Michèle Lowrie (2009, 83) added, contra Heinze, that in the Roman lyric "the addressee need not be present with the speaker to read a poetic letter and for its speech to be operative. Such would be the case if an actual invitation were sent in verse." As examples of the latter, we might name Catullus's C.13 and Horace's C.1.20.

Lyrical monologues addressed to named addressees, present or absent, invite us as readers, in the spirit of poetic play, to pretend that the addresser has turned away from us to speak to particular other persons. In the case of a dramatic lyric, we imagine addresser and addressee as copresent participants. In the case of the epistolary lyric, we imagine the addresser *imagining* the addressee as a present hearer. This metaimagining adds another layer of play to the way letter writers conceptualize the act of writing: readers mime the dramatic encounter because letter writing pretends to be direct oral address. As Jonathan Culler (2009, 892) maintains, lyric address, be it dramatic or epistolary, "does not function in a radically different way from apostrophic address." To illustrate the epistolary lyric, I have translated the seventh ode from Horace's first book of *Odes*, "*Laudabunt alii*":[16]

Odes 1.7

Others will praise illustrious Rhodes or Mytilene,
  Ephesus or sea-clipped Corinth,
Thebes famed for Bacchus, Delphi for Apollo famed,
  or Thessaly's fair Vale of Tempe.
Some only live to celebrate Athena's city
  perpetually in song and wreathe
their brows with sprigs of olive plucked from everywhere.
  Many will chant in Juno's honor
of Argos, land of horses, and of rich Mycenae.
  But as for me, I am not struck
so much by rugged Sparta or by lush Larisa
  as by deep-toned Albunea's home,
the falls of Anio, Tibur's sacred grove and orchards
  watered by streaming rivulets.
As South winds often wipe the dark sky clear of clouds
  and do not bring forth lingering storms,
so you are wise when you remember to set bounds
  to gloom and life's fatiguing toil
with smooth wine, Plancus, be you now encamped amid
  bright flashing pennants or at home
in shady Tibur. When his father banished him
  from Salamis, they say that Teucer
yet placed upon his brow, wine-wet, a poplar crown
  and thus addressed his downcast friends:
"Wherever fortune, kinder than a parent, takes us,
  there we shall we go. O friends and allies,

never despair with Teucer as your guide and seer,
  for surely has Apollo promised
another Salamis for us in some new land.
  Brave men, who often suffered worse
with me, now drive away your cares with wine—tomorrow
  we take again to the vast sea."

Since poems are made to be resaid or reread, we know when we revisit this text that Horace is not talking to us directly but to someone named Plancus. We also gather that Horace is not actually *talking* to him either. He does not even know where Plancus is. Perhaps he is in some Roman military camp. Perhaps he is home in his country villa in the Tiburtine Hills east of Rome. If this poem had been an actual letter, Horace would have had no way of knowing where to send it. This, then, is a virtual letter, an epistolary lyric that plays with the form of a letter. Historians tell us some facts about its addressee: Lucius Munatius Plancus had already had a long and diverse career. He had served as one of Caesar's chief lieutenants in Gaul, founded the city of Lyon, and was elected consul in 42 BCE. Plancus had supported Antony but switched sides to Octavian before the Battle of Actium (31 BCE), and in the year 27 it was he who proposed to the Senate that Octavian be granted the title "Augustus." Just how significant the biography of this or any other addressee is to the meaning of this or any other Roman poet's ode has been a matter of debate over the years. For contemporary readers, we should note, some of these addressees were more widely known than the poets themselves and, as I suggested in reference to *Epode* 13, pertinent historical details could have been read back into their poems. Whenever a contemporary public figure is addressed in a poem, the poet has to understand that whatever knowledge contemporary readers have of that personage may, when appropriate, enter into their experience of the poem, but, as with many of the lyrical addressees of this period, we no longer have historical knowledge detailed enough to evaluate this factor with any accuracy. Viewing Plancus from the window of this poem, however, we may assume that he had had a strenuous and troubled career and should now have been ready to come home and enjoy his wine under his shade trees at Tibur (Tivoli). As Michèle Lowrie (1997) observed, this poem questions the meanings of home and of *patria*.

The lyric begins with a graceful series of parallel clauses: others will praise this or that location. Here the places are all in Greece, as befits a poem modeled on the Greek lyric tradition. Over the course of its thirty-two lines the poet has inserted a series of shifts, some predictable, others abrupt, as though to convey a traveler's sense of displacement. As the initial topographical details build up, we understand that this set of false alternatives is a classical *priamel*, similar to Sappho's fragment 16 ("Some say that the most beautiful thing is . . ."). Horace begins with "Others will praise . . ." and we wait for the inevitable shift toward the poet's own preference. When it comes, it turns out to be the hilly landscape near the Anio Falls, where the Roman elite had built vacation homes twenty miles east-northeast of Rome, along the Via Tiburtina. Horace knew this landscape well. Some eight miles further upcountry in the Sabine Hills lay his own small estate.

At line 15 the poet transits from priamel to exhortation (*pareinesis*), pivoting on the metaphor of stormy weather = personal troubles and reintroducing the sympotic praise of wine as a mood-altering medication. Here, a little past the midpoint, we also have what corresponds to the epistolary salutation: Plancus, wherever you are now as I write this or when you receive this message, you should remember (*memento*) to set bounds to your gloomy feelings and incessant labor with a drink of smooth unmixed wine (*merum*), a stout, soldierly drink. Then he makes another shift and, as in *Epode* 13, it is toward a mythic *exemplum*.

An actual personal letter is likely to contain in-group gossip and allusions to mutually shared experiences. Since such references would exclude the general reader, Greek and Roman poets found that mythology––what Eric Havelock (1963) called the "tribal encyclopedia"—furnished culturally shared equivalents to personal references. The mythic moment he chose for his exemplum here would have been familiar to most Romans because it had been acted out in the popular Latin tragedy *Teucer*, by Pacuvius. According to this narrative, when Teucer left home, he had solemnly sworn to his father, Telamon, that after the war with Troy he would bring his half-brother, Ajax, safely home to Salamis. When he had to tell his father that Ajax had committed suicide, his father abruptly banished him. Fortunately for Teucer, it had been prophesied that he was to found another Salamis (he later did so in Cyprus). This is the message he tells his men as he invites them to fortify their spirits with wine. Here, as in *Epode* 13, Horace augments his own exhortation by assuming another persona.

As we see from this text, the epistolary lyric, like the dramatic, is firmly anchored in the present tense of the oral speech event. In a monologic text, the speaker may be interpreted as the poet, some version of his or her self, or someone with a completely separate identity. In any case, the lyric voice maintains a temporal and spatial presence in our imagination that distinguishes it from third-person narrative, such as epic, and first-person narrative, such as Wordsworthian recollection. By using the vocative case, the speaker is requesting the named person's attention in the here and now, which implies the imperative "listen." The frequency of the imperative mood reminds us how lyric poetry has absorbed the intentionality of epideictic oratory—not only to persuade others by argumentation but to urge a viewpoint and a course of action and do so by the shortest means available, a directive speech act.[17]

Voice and the Afterlife of Poets

If any thought of death is more terrifying than the annihilation of consciousness, it is the thought that one might continue somewhere to exist and somehow be aware of one's utter solitude, cut off forever from one's family and friends or from every other living thing. Of course, most of us would prefer an afterlife, one similar to our current life, only better. In ancient Greek culture, virtue and achievement conferred fame (*kleos*), which in turn assured the dead the mindfulness of the living, a connectedness that made tolerable their existence in the shadow world beneath the earth. Some of the dead retained some powers: the Greeks had their hero cults centered on particular burial places, and the Romans revered their ancestors as protecting spirits.

A common vision of postmortem existence is that the deceased somehow continue as the persons they were just prior to their death. For the aged and the terminally ill, this meant an afterlife of enfeeblement and loss of memory. Those prematurely killed by accident or violence were visualized as emotionally volatile—sorrowful, angry, and resentful—but otherwise mere bodiless shadows. The living, fearful of the restless dead, appeased them with ancestor and hero cults and, for extra security, might ritually burn their bodies or place blocks of stones upon their graves (Ogden 2009). The formula *requiescat in pace* is evidently rooted in "laying of ghosts" ceremonies.

Another curious feature of the ghostly dead is their mode of locomotion: they flutter about like birds or bats. We find this depiction in Egyptian images of the *ba*, the soul that flies forth from the mouth of the dying, a winged creature with the face of a human. As the Neo-Babylonian "Descent of Ishtar" tells us, the dead are dressed in winged garments. This avian metaphor implies that the animating spirit, the breath, once it escapes the body, is like an uncaged bird that will never again return. The God Hades is always depicted holding a scepter topped with a carved bird, and a bird emblem often appears in Greek funerary images. In one striking depiction on a vase, Charon, the boatman of the River Styx, ferries a flock of birds with human faces while Hermes, the conductor of souls (*psuchopompos*), leads them toward a female figure, perhaps the queen of the dead, Persephone (Vermeule 1979).

Though associated with the head and with the uniquely identifiable features of the face, after the body dies, the *psuchê*, or life spirit, normally becomes mute. It cannot speak because speech requires the *thumos*, another invisible element, which resides in the chest, the *phrenes*. According to the belief system we find in the Homeric epics, the *psuchê* after death cannot enter the security of the House of Hades until its former body is cremated and properly buried. Until then, it may haunt the living and appear as a speaking ghost, as in *Iliad* 23, when the ghost of Patroclus appears to Achilles and demands that his body be burned. Elpenor in *Odyssey* 11 appears to Odysseus and speaks without needing a taste of sacrificial blood. He still has the power of speech because he too has not been properly buried, and apparently for that same reason the *psuchai* of the Suitors (*Odyssey* 24) are able to tell their story. As Circe told Odysseus, in the land of the dead only the blind prophet, Tiresias, has been granted an unimpaired *phrenes* and a mind able to draw breath (*pepnusthai*) (10.490–495). In archaic Greek physiology, the conscious powers resided in the chest, within the *phrenes*, which had the ability to think in language, and in the *thumos*, which had the physical power (*menos*) to utter those thoughts in overt speech. Tiresias demonstrates his speech capacity even before he is given the blood to taste, and it is he that shows Odysseus how to use that blood to revive the speech of the otherwise strengthless heads (*amenêna karêna*) of the shades.[18]

Breath (Gk. *pneuma*, L. *spiritus*) is a circular rhythm: one breathes air out, or a portion of it, then breathes it back in. In the vital respiratory cycle, inspiration perfectly balances expiration. There is yet another mode of expiration that is noncyclic. This is the linear, breath-mediated, *thumos*-propelled process of utterance, a birdlike flight that is distinct from the expiring escape of the *psuchê*. This is perhaps implied in the Homeric formula "winged words" (*epea pteroenta*), a phrase that has perplexed Homerists for centuries. Was it meant to characterize especially swift, emotive speech, as most have claimed, or was it, as Parry (1937) concluded, simply a formula that an improvising singer inserted for metrical convenience in suitable contexts? Or was it an epithet descriptive of all oral speech acts, as Combellack (1950) argued? If the latter interpretation is valid and, in every speech event, it is words that are sent flying forth from one person's mouth to another person's ears, then the wingedness of uttered speech and the wingedness of the expiring soul can both be understood in terms of the physiology of respiration.

For poets, writing was the means of acquiring a name—of being remembered by future generations, thereby guaranteeing a kind of afterlife. An early expression of this thought is Sappho's four-line text known as fragment 55. In it the poet addresses a person traditionally identified as an "ignorant woman" but who may in fact have been a girl who chose to leave Sappho's choral school, dedicated to the Muses' art: "When you die you shall lie there, and no memory of you will linger in later time, for you have no share in the roses of Pieria. Unnoticed in Hades' House as well, you will pass to and fro among the obscure dead, having now flown from our midst." The implication of these verses is that having a "share in the roses of Pieria," i.e., participating in the rites of the sacred Muses, including perhaps the wearing of chaplets of roses, gave one the hope of a privileged afterlife instead of the common fate of mortals, which, as Homer described in the eleventh book of the *Odyssey*, was to forget and be forgotten. She had "flown from our midst" and, after death, would keep flying about, but now among the equally mindless shades in the cavernous House of Hades (Hardie 2005, 20). The notion that poet-singers, thanks to the intercession of the sacred Muses, could continue to perform there among the dead was a happy thought that poet-singers like Sappho were quite happy to think.

Poets came to regard their works as somehow affecting the condition of their afterlife. The early Roman poet Ennius (c. 239–169 BCE) wrote his epitaph in this characteristically alliterative elegiac couplet:

Nemo me decoret dacrumis nec funera fletu
Faxit. Cur? Volito vivu' per ora virûm.

Let no man honor me with his tears or bury me in lamentation.
Why should he? I flutter alive through the mouths of men.[19]

The idea represented in the phrase "volito vivu' per ora virûm" struck a resonant chord for later Roman writers who interpreted it as an afterlife to which writers could aspire. Vergil in his *Georgic* 3 (8–9) echoes his predecessor's words:

Temptanda via est, qua me quoque possim
tollere humo victorque virûm volitare per ora.

I must try to find a path on which I too might
lift myself from the ground and victoriously flutter through the
mouths of men.[20]

With book publishing, poetry writing became a recognized profession, though, in the absence of copyright laws, a poorly remunerated one in need of patrons for support. Poets' choice of theme and genre gradually entered their texts as a metapoetic theme, a sort of Judgment of Paris—no, not that genre or that, but *this*. Vergil in the above quote is promising himself an Ennian immortality, but if only he can find himself a suitable *via*, a theme or genre that did not simply rehash an already published (*iam vulgata*) story.

Other poets placed what we might now recognize as self-advertising blurbs at crucial places in their publications, often at the end of an edition, where it functioned rather like an authorial colophon. Horace concluded his third and, as he then thought, his final book of *Odes* with the metapoetic statement that he had herein "completed a personal monument more durable than bronze statuary, loftier than the royal pyramids of Egypt." By implication, his three-book *monumentum* is his tomb, his words the inscription that will make him forever memorable. But not just memorable—immortal, for "I shall not wholly die and much of me will escape the public registry of deaths (*non omnis moriar multaque pars mei / vitabit Libitinam*). Year after year, he predicts, he will grow larger in the praise of all those who will read his *Odes* and remember that it was he "who first transposed the singing of Sappho and Alcaeus into Italian sounds."[21] Ovid, when he concluded the first book of his erotic poems, the *Amores*, claimed for himself a similar destiny. He tells the reader that, having dropped out of the worldly race for power and prestige, he was no longer competing for a merely mortal prize. "My quest is to achieve lasting fame such that I may be sung forever throughout the entire world [*Mihi fama perennis / quaeritur, in toto semper ut orbe canar*]." Poems are exempt from death, so even after I am cremated, I shall live, and much of me (*parsque mei multa*) will survive (*Amores* 1.15, 7–8, 42).

Here Ovid seems to allude to Horace's earlier poem, but, literary allusion aside, what did the two poets mean by that "muchness," that "considerable part of me," as it is sometimes rendered? Probably what they thought Ennius had meant when he said he would live on, fluttering through the mouths of men, a phrase often translated as "fluttering upon the lips of men." This is, I submit, the movement of the lips as the mouth utters written words, a fluttering quite unlike the mouth of a singer producing sustained notes. Like the poets who followed him, Ennius was writing for the voice and sought his afterlife among living readers who would continue to utter his words.

Roman poems, though called "songs" (*carmina*), were not published as song sheets. Rhythmical, often exquisitely so, they were not more tonally varied, when recited, than were oratorical declamations. They could be set to music, but, as Heinze (1923) maintained, they were primarily intended to be read by solitary readers in a low murmur or recited before an audience. Writing was indeed a visual instrument of memory, but, like any well-crafted instrument in the hands of a skilled worker, it was not the focalized object of the reader's attention. The mouth, not the eyes, was considered the main organ of reading, which for another thousand years would be experienced primarily as a motor, rather than a perceptual, process. One would not *see* writing so much as one would *say* writing.

Ovid makes this clear in the nine-line statement he used to conclude his *Metamorphoses* (15.871–879), another homage to Horace's closing statement to book 3 of his *Odes*.

> And now I have completed a work (*Iamque opus exegi*) that neither the wrath of Jove nor fire / nor sword nor the devouring passage of time shall have the power to destroy. / The day that claims the right to this mere body of mine /—let it arrive whenever it chooses to terminate the span of this uncertain life. / But in the better part of me (*parte tamen meliore*) I shall be carried / forever beyond the high stars and my name will be indelible. / Wherever Roman power extends

> over the tamed lands, / *I shall be read in the mouth of the people* (*ore legar populi*) and through all the ages to come, / if the poets' presentiments (*vatum praesagia*) have any truth to them, by fame shall I live.

These *vates* are no mere writers of verses but inspired prophets, a title that Roman poets had come to assume when envisioning future ages. Like Tiresias, the Theban prophet who retained the power of speech even in death, these poets also live on as voices. But these voices do not now come to the living from beyond the grave. Instead they exist as rows of written characters contained within verbal artifacts that, as virtual-orality devices, replay their makers' voices whenever they are unrolled and read.

The common Latin verb for "to publish" was *edere*, literally, "to give out." There was something both oral and irrevocable about this verb: to die was *edere animam*, i.e., "to give up the ghost," and *edere clamorem* meant to "give out a cry." In his *Ars poetica* (*Epistles* 2.5), Horace advises the Piso brothers to revise their compositions well, for "you will be allowed to delete what ever you have not published—once sent, the voice does not know how to return [*delere licebit / quod non edideris—nescit vox missa reverti*]" (389–390).[22]

The metaliterary convention of addressing one's own book inside one's own book provides interesting evidence that writers viewed their books as voicelike emanations of themselves in the form of children or servants who were sent off to fend for themselves. We have already encountered this trope in Plato's *Phaedrus* (see chapter 5), when Socrates describes a written text as a child subject to abuse by readers and unable to protect itself in the absence of its father, the writer. For Roman poets, the author's address to his book often took the form of acquiescing to the manuscript's petition to leave home and earn success in Rome but warning it about the kind of reception it might get.

In Horace's first book of *Epistles* his final piece is not itself an epistle at all but rather a speech to the book itself: he has finally yielded to this boy's desire to be freed but warns him that once he leaves, he cannot come back, and that when his youthful beauty fades he will, at best, be used as a Latin textbook. Then, after this literary manumission, he gives his final injunction, dictating a sort of Roman wiki page: on some sunny day, just tell people about me, how I started poor and worked my way up, that I am short, prematurely gray, like to be out in the sun, am quick-tempered but quickly appeased and, oh yes, I'm now forty-four years old.

Similar in form but not in tone is the first poem in Ovid's four-book *Tristia*. This work, composed in the poet's exile on the northwestern coast of the Black Sea, was sent to Rome with the hope that the emperor Augustus would relent and let the poet return.[23] Like Horace, Ovid tells his book what to expect but envies the book's freedom to reenter the city from which its author is excluded:

> You! Go ahead in place of me, you who are permitted—gaze on Rome.
> Oh that the gods would make it possible now for me to be my book!
>
> *Tu tamen i pro me, tu, cui licet, aspice Romam.*
> *Di facerent, possem nunc meus esse liber!*
>
> (57–58)

The concept of writing is obviously meaningless without the concept of reading. Not only are the two practices correlative, but in literates the two skills are combined. We may safely assume that the writers who aspired to an afterlife *ore populi* had no doubt themselves experienced the way their reading and rereading of certain texts seemed to conjure up the persons that had created them. In this sense, to read was to perform a rite of raising the powerful dead, a *nekuia*, except that these dead had not wholly died, had never flown off to the House of Hades, but still inhabited their monumental verbal avatars.

Georges Poulet described this uncanny presence in his "Phenomenology of Reading" (1969, 157–159):

> When I am absorbed in reading, a second self takes over, a self which thinks and feels for me. Withdrawn in some recess of myself, do I then silently witness this dispossession? . . . This *I* who thinks in me when I read a book, is the *I* of the one who writes the book. . . . Thus a book is not only a book, it is the means by which an author actually preserves his ideas, his feelings, his modes of dreaming and living. It is his means of saving his identity from death. . . . And so I ought not to hesitate to recognize that so long as it is animated by this vital inbreathing inspired by the act of reading, a work of literature becomes (at the expense of the reader whose own life it suspends) a sort of human being, that it is a mind conscious of itself and constituting itself in me as the subject of its own objects.

Summary

Writing began as a way to enumerate objects and store records of debts, laws, and other social obligations. A later use was to store cultural information deemed important for the young to learn. Yet later, writing became a means of imagining performances, such as narratives, dramas, and songs, in the absence of performers. In this phase of literate culture, represented in Greco-Roman society from the fifth through the first century BCE, writing was designed primarily to evoke the sounds of speech, the essential meanings of words enunciated with the tones and intentions of given speakers, whose visual presence might also be imagined. Written artifacts were thus produced as vocal recordings to be mentally performed by readers through the simulation of verbal audition and verbal articulation and secondarily as representations of persons, objects, spaces, and actions. As has been long recognized, ancient readers preferred to recite prose and verse aloud, even when alone, though, when necessary, most of them could evidently do so silently. Their format of choice, *scriptio continua,* may have reflected their strong audio-motor bias in reading.

Over time, literacy created what we now know as "literature" in both prose and verse formats. As a mnemonic system superior to long-term memory, writing made prose possible and, with it, history and novelistic fiction, two literary genres that have influenced each other through the centuries. Lyric poetry also began to absorb new features: it could still provide the words for a song, but writing now allowed it to exist apart from the performance venue. Once placed, in the form of a bookroll, in a private library, it began to function as a script for the solitary performance of reading.

In the first century BCE, Roman poets became fascinated with archaic Greek song, both elegiac and strophic, and sought to Latinize it. Though the Roman lyric was not composed to be sung, as was the Greek, it was composed to be *voiced.* As such, it was shaped by two oral genres, *epideictic oratory* (the persuasive voice of wisdom, exhortation, and praise) and *drama* (the personally engaged voice of an onstage character). It was also shaped by two written genres that mimicked oral address, *inscription* (the impersonal voice of a monument, a voice from the dead, or from the mind of a cool, epigrammatic observer) and the *personal letter* (the voice of one who shares thoughts and feelings with an absent friend through the only means available, namely, writing and the literate imagination). Epitaphic inscription, typically written in hexameters or elegiac couplets, had aimed for the musical memorability of lyric—and now lyric borrowed from inscription its power of epigrammatic brevity. Letter writing, for its part, had aspired to the candor and formal elegance of lyric expression—now lyric came to assume the silence, solitude, and intimate voice of the letter writer.

Of these four models, the most directly influential were drama and the personal letter. The lyrical speaker within the text would typically be represented as addressing a person or persons either physically present through dramatic monologue or absent through epistolary monologue. In any actual address, whether face to face or via a written message, the speaker would likely refer to privately held information. But if poets wished to produce poems that would be understood, as well as "heard," they needed to situate their statements within a widely available cultural context. For Romans, as it was for Greeks, this meant referring to mythology through brief allusions or more extended exempla. Presenting two of Horace's lyrics, one dramatic, the other epistolary, I showed how the poet concluded each by artfully blending his voice with that of a figure of Greek myth.

The monologic voice that sounds in classical lyric poetry is quite different from the voice that has typified the modern lyric—the meditative, musing inner voice of the soliloquizing thinker. Roman lyric poems nevertheless continued the trend of archaic Greek lyrics toward self-reference and the construction of a personal authorial identity independent of family, class, and state. This identification of poet with poem led to the curious concept that reading the words of a long-dead poet, imagining his or her voice and its rhythmic breathings, could somehow bring that person back to life, a happy thought that seemed to confer a measure of earthly immortality on these artists. Poetry, as written for the voice, may also be implicated in the images of the happy *un*earthly afterlife that major religions promise their believers. No matter how intensely religionists pore over and revere their holy books in this life, paradise is a purely oral state devoid of books and resonant with song.

Seven

Writing and the Reading Mind

In the last chapter, I discussed writing as a means of reenacting speech events such as dialogues and monologues, either for others or for oneself. If for oneself, this performance could be a silent simulation of speech using the reader's articulatory and auditory imagination. In this concluding chapter I will sketch out some of the unique ways in which writing also began to represent aspects of the mind that only writing could reveal. We have no reason to assume that the new model of the mind that literate culture introduced was the inevitable outcome of evolutionary forces. For Europe, it was simply what sociocultural institutions, over twenty-five centuries of historical change, constructed and bequeathed. Thanks to its adaptive plasticity, the human brain will undoubtedly generate new versions of itself and new paradigms of reality in the future—as long as we do not prematurely cut short this experiment in speciation.

I begin this chapter with a wide-angled review of some of the fundamental evolutionary adaptations that underlie our biocultural profile, specifically our primate capacity to process parallel and serial information, our human skills in executing these two processes at the same time, and our capacity through language, oral and written, to take optimal advantage of these two processes. Then, drawing on literary examples, I explore how early preprint literacy served to restructure our conception of a reflective, autobiographical self and how specific features such as the metered line and visual imagery contributed to the formation of the literate mind.

Dyadic Stylistics

Every successful innovation improves the operation of a system by foregrounding one or several of its mechanisms while reserving its other mechanisms for background tasks. If the innovation is one that affects our capacity to represent the world, i.e., perceive our environment, categorize its contents, manipulate concepts, and communicate our thoughts to others, it will tend to function as a filter in the process of selecting and clarifying certain aspects of the world while obscuring others. In human evolution every new form of communication, from gesture to protolanguage to syntactical speech, had to have altered the way our ancestors knew their world, enhancing some cognitive skills at the expense of others. Yet at no point along the way did the universe of discourse supplant the universe of sights, sounds, smells, tastes, and touches. We obviously continued to enjoy moving our bodies through space, falling in love, and refreshing ourselves with nourishment and sleep, long after we learned to put these experiences into words.

In the foregoing chapters I have been focusing on the cognitive changes that followed from the introduction of writing to oral cultures. But before I continue to explore those effects in this final chapter, I want to turn our attention, Janus-like, from that point at which oral culture incorporated literate culture back to the much earlier point at which nonlinguistic culture incorporated oral culture and ask how language, when it first emerged, innovated our being-in-the-world. By asking how language itself altered our noetic *umwelt* I do not mean to present those sets of features that differentiate natural languages but those that unite them all. So, for the purpose of this inquiry I am as much a linguistic universalist as a linguistic relativist. My universals are not, however, grammatical deep structures lying beneath the different surface structures of spoken languages, nor are they word-order universals; they are instead design features that characterize all languages as communicative codes.

Human consciousness, as I have argued here (chapters 3 and 4) and elsewhere (Collins 2013), is essentially *dyadic*. By this I mean that we have evolved the knack of zooming in on fine details while simultaneously scanning the larger perceptual fields surrounding them. We are able to perceive a figure *and* at the same time peripherally take in its ground; in other words, we can examine a part *while* not neglecting the whole. Though this pattern seems most important to us in visual cognition, it also is apparent within and across other perceptual systems. We can, for example, focus on details of environmental sounds while broadly noting changes in wind velocity and temperature, center our attention on nuances of flavor while observing more generally the color or texture of foods, and pick out precise melodic variations in a musical performance while also being aware of its rhythms. This dyadic pattern, this unity of unequal elements, seems, like the Golden Ratio and tonal harmony, hardwired in the human brain.

Is this multitasking facility unique to humans? Other animals can certainly multitask, but we are unique in our capacity to reflect on our multitasking. Why would our species evolve this two-in-one cognitive capacity? In *Paleopoetics* (2013, 56), I speculated that it might have been an adaptive response to the relative rapidity of human evolution over the past 2.5 million years:

> The dyadic pattern might best be regarded as an extremely flexible template, an evolutionary strategy that allowed early humans to ease their way into early Paleolithic culture without losing the mother wit of their primate past, to evolve finer and finer micrometric muscular controls and focus on more and more intricate sequential routines, while at the same time monitoring their environment and parallel-processing its multimodal streams of information.

The principle central to dyadic cognition is the coordination of parallel and serial processing. When an animal shifts its attention to parallel mode, it receives input from multiple sources and automatically sorts them in terms of importance. But when it needs to narrow its attention, it blocks out virtually all but certain kinds of input. When, for example, a squirrel searches for acorns or a chicken searches for insects, the animal's attention proceeds from one point on the ground to another in a simple succession of fixations with no set order. But when an animal forms a plan, succession morphs into a new kind of seriality, *sequence*. Here again, we are not the only animals capable of this procedure. Most birds, for example, when building their nests, begin with a matrix of strong twigs, then a weave of thinner twigs, and end by lining the concave structure with softer materials such as plant fluff and feathers.

Our early ancestors, no doubt, also sought found objects to execute plans, but when they eventually saw the advantage of modifying objects for specific tasks, they began to fashion specialized tools. To do so required that they learn set sequences of steps. Sequential order and improved manual dexterity must also have reinforced their ability to exchange meaningful strings of manual gestures. They found, however, that vocal sounds were more efficient in certain activities, such as instructing younger members of the band in a manual skill, because the teacher's hands could now remain fully engaged in demonstrating that skill.

As tool use and spoken language coevolved, the two conferred selective advantages, but humans still needed to maintain their multisensory, parallel cognitive processes in good working condition. When they conversed with one another using sequences of symbolic signs, they also needed to attend to the meanings of holistic indices, such as posture, gaze, facial expression, and vocal tone.

Walter Ong's typology of oral language in *Orality and Literacy: The Technologizing of the Word* ([1982] 2013) was meant to contrast primary oral to later literate cultures, but his list may also provide us a useful way to explore the initial emergence of language from prelanguage and the ongoing implications of that innovation. The first entry on his list of oral characteristics, "additive rather than subordinate," refers to the utterance of phrases and sentences linked only by conjunctions such as "and" or "but" or by adverbs such as "then." A simple sequence of events so organized typifies the *paratactic* style of Homeric epic and the early chapters of Genesis.

If oral verbal artifacts remained purely additive/paratactic, they would have the narrow linearity of inventories and would tax not only the memory of listeners but also their comprehension. In different ways, the eight traits that Ong next enumerates serve to mitigate the severe seriality of language by activating the cognitive mechanisms associated with parallel processing. Consider Ong's second characteristic of oral language, "aggregated rather than analytic":

> This characteristic is closely tied to reliance on formulas to implement memory. The elements of orally based thought and expression tend to be not so much simple integers as clusters of integers, such as parallel terms or phrases or clauses, antithetical terms or phrases

> or clauses, epithets. Oral folk prefer, especially in formal discourse, not the soldier, but the brave soldier; not the princess, but the beautiful princess; not the oak, but the sturdy oak.
>
> ([1982] 2013, 38)

These culturally familiar word clusters slow down the forward motion of unpredictable, discrete syllables long enough for hearers of an oral discourse to grasp its recently heard units in short-term (working) memory and anticipate the next phrases. His third characteristic, "redundant or 'copious,'" also addresses the needs of short-term memory but does so through repeating ("backlooping" to) previously mentioned items.[1] His fourth characteristic, "conservative or traditionalist," refers to the retelling of entire verbal artifacts, e.g., narratives, songs, and religious dramas. Just as internal repetition supports short-term memory, reperformance supports long-term memory.

In his comments on redundancy in primary oral cultures, Ong asserts that "sparsely linear or analytic thought and speech is an artificial creation, structured by the technology of writing" ([1982] 2013, 39). By "sparse linearity" he evidently means listlike discourse devoid of redundancy, but, if we consider language in terms of production and reception, we must recognize that linearity is an essential property of this communicative code. Every medium of language, be it gestural, oral, or written, exercises the brain's capacity to organize extended strings of information. All of Ong's oral characteristics except the first, the "additive," are stylistic strategies aimed at rebalancing a system excessively weighted toward linearity. To communicate with others, we must rely on this extremely serial—indeed sequential—cognitive process, but to engage our whole minds in any speech event we also need to draw upon that other pole of the dyad, parallel process. We need to call up those prelinguistic skills of wide-angled perception and do so by reconfiguring this one-dimensional linear medium to form networks capable of conjuring up two-, three-, and even four-dimensional representations.[2]

In table 7.1, the features listed under code, medium, and style are not *all* the features a thoroughgoing analysis would include. Hockett ([1960] 1982), for example, listed under "code" thirteen separate "design features." Here I have tried merely to indicate how this topic might be arranged to illuminate issues of style.

Spoken and written media use arbitrary symbolic signs to access the contents of semantic memory and thereby represent our perceptual

TABLE 7.1
Code, medium, style

Code	Medium	Style
Language • linearity • displacement of reference	**Oral** • phonic-auditory • speaker and hearer copresent in time and space	• paratactic (additive) • formulaic (aggregative) • repetitive (redundant) • re-performed (traditional)
	Literate • graphic-visual • writer and reader separated in time and space • private	• paratactic • hypotactic • rereadable • republished

experience of the world. The peculiar nature of these representational media tends to take experience that is normally perceived as parallel input and convert it into a serial format. As a linear flow of discrete phonemes or graphemes, discourse is briefly perceived before it vanishes and must be continuously serially processed in short-term memory (Baddeley 2000). In its oral medium, discourse is absolutely evanescent, its phonic-auditory pulses irretrievable. In its literate medium, it is temporarily evanescent: the eyes fixate briefly on the letters of a word and then pass on, but the text remains and may be returned to. In short, in oral discourse the hearer of speech is stationary while the words keep moving on; in reading the words are stationary while the reader's eyes keep moving on.

In both media, language can refer to currently present objects deictically and imperatively, e.g., the exclamation "Look at that hawk!" or a flashing road sign that says "Eat here!" But this is not what makes language special—prelanguage could presumably accomplish these sorts of tasks through indexical gestures. What proved truly innovative was its capacity to represent nonpresent persons, places, things, states, and actions, referents that, displaced in space and time, could only be thought or imagined. In the oral medium this gap separating the speech event from the event spoken about could be bridged when a narrator imitated the speech or affect of a character or played the role of a witness reliving a past episode. This performative style Plato found especially troubling. In the literate medium, the reader must wholly reconstruct the nonpresent world referred to in the text. This inner performance is, however, facilitated by the fact that no living, breathing narrator is visually present to distract the mind. Even though the literate medium recruits the visual channel, the perception of letters does not interfere with the visual imagination because these arbitrary shapes have no iconic values. (I will examine how writing enhances mental imaging later in this chapter.)

As code and medium constitute the fixed features of linguistic communication, style supplies its toolkit of variations. What I am proposing is that the fundamental function of style is to counterbalance the sequential linearity of language by inducing verbal artifacts to exhibit parallel features. Since I have already discussed Ong's first four stylistic features of orality (in the top-right box of my table), I will now comment on the four features of literate style I have listed below them, beginning with parataxis, which, for reasons I will now explain, should not be relegated to orality alone.

"Parataxis is first of all a state of mind" (Notopoulos 1949, 11). If so, what does it represent? John Burt (2012, 650) suggests that "paratactic style lends itself naturally to the rush and chaos of life as it is lived in the immediate first-person." Life exhibits "rush and chaos" when many things are happening at once, as, for example, on the battlefield of Troy. But how can a style of one-after-the-other phrases, clauses, and sentences represent this state of affairs? I think we might best regard parataxis as a stylistic equivalent to the way the brain has of responding to a set of events occurring in parallel by reducing them to minimal details, thereby protecting itself from sensory overload. Thus parataxis can signify the "rush and chaos" of experience not by representing that actual situation but by representing our defensive response to it. The fact that paratactic sentences are typical of oral discourse may be because a narrow linearity is the way we reduce the confusion in our minds as we improvise speech. Similarly, this may explain why inner speech is also paratactic (Vygotsky [1934] 1986).

I have included parataxis here as a stylistic feature of the literate medium because writers can avail themselves of this linear, discontinuous style of narration whenever they wish to convey this paratactic "state of mind." As Wulf Oesterreicher (1997) pointed out, this orality-based style is not restricted to the oral medium but can be transferred to writing as "conceptional orality." My references in the last chapter to Roman lyric poetry as attempting to recreate oral *energeia* through dramatic and epistolary monologue underscores Oesterreicher's argument.

Hypotaxis, i.e., grammatical subordination, also has stylistic implications particularly pertinent to written discourse. To contrast this with parataxis, in what follows I have summarized and converted Horace's *Epode* 13 (see chapter 6) first into a set of strictly paratactic (coordinate) sentences, then into a single hypotactic (subordinate) sentence with recursive embedding.

PARATAXIS

> It is storming outside. Let us make the best of a bad situation. We are young and have the right to spend this day in wine and song. The centaur said to Achilles: "you are doomed, but you may drink wine and sing and thus lighten your troubles on the shores of Troy."

HYPOTAXIS

> Since it is storming outside, *let us make the best of a bad situation*, for, being young, we have the right to spend this day in wine and song, just as the centaur told Achilles he should do, once, being doomed, he landed on the shores of Troy.

Parataxis, as a coordination of grammatically equal clauses and sentences, may be visualized as movement along a horizontal line. Hypotaxis, as the arrangement of one or more subordinate clauses in relation to a single main clause (italicized in the example above) may be visualized as a vertical hierarchy. As parataxis strings a series of units together one after the other, hypotaxis places them in a structure that, like a living body, exists in all its parallel parts at the same time.[3]

My third literate style feature, rereadability, corresponds to those last three oral features, formulaic, repetitive, and reperformed (Ong's terms are in parentheses in the top box), that serve to counterbalance the extreme linearity of an oral performance by helping audiences follow the rapid, sometimes discontinuous progression of a narrative. Written texts could include formulas and repetitions, but there was no necessity to do so. Over the centuries, first prose writers, then poets, introduced hypotactic structures both on the level of the sentence and on that of the entire text. But the simple fact that readers, when having direct access to texts, could reread a poem or a paragraph meant that even the most discontinuous and digressive texts, as well as the most complexly multi-voiced texts, could be fully integrated in all their parts. In this way, literature developed stylistic resources capable of activating the brain's serial and parallel mechanisms and restoring to it an intensely satisfying measure of dyadic equilibrium.

Reading Inner Writing

The dyadic pattern that we observed in parallel and serial information processing, specifically in the processing of language and of verbal artifacts, may also be observed more generally in biocultural evolution. Those older adaptations that, as *innovanda,* need to be innovated remain even after they have become altered and the relation between the older and the newer skill is dyadically patterned. A transition such as that between orality and literacy should not be considered in dualistic or dichotomous terms. Egbert Bakker (2005, 45) makes that point clear:

> Oral composition presupposes an oral style as opposed to a literate style, but, as we already saw, the very notion of "oral style" betrays the perspectival bias of the literate classicist. "Oral style" presupposes "literate style" in the same way that "archaic style" presupposes "classical style," or "paratactic style" presupposes "hypotactic style," and so on: the oral nature of this style is of our own making, and defined with respect to literate style in the black-and-white contrast that I have already argued against.

A literate culture assumes some of the functions hitherto performed by the oral culture within which it now coexists. Both linguistic media share in the maintenance and production of meaning through a process of mutual accommodation. As Oesterreicher demonstrated, one form of accommodation was the stylistic imitation on the part of writers of the oral medium, which he called "conceptional orality." What I will now suggest is that oral culture also made concessions, and these took the form of "conceptional literacy."

How our memory works, that is, how the brain is thought to receive, store, and access information, has always been approached by analogy to whatever seems at the time the most sophisticated form of informatics technology. Nowadays computer circuitry is a popular model, and before that it was the telephone switchboard, before that the telegraph, and even before that the machinery of pulleys and springs. Early literate cultures, taken with the new technology of writing, pictured memory as a writable surface. The biblical phrase was "tablet of the heart" (Proverbs 3.3, 7.3; Jeremiah 17.1)—one ponders Torah by reading inner writing. As we have seen, the archaic Greeks thought of memory as stored within the chest in an organ called the *phrên* (pl. *phrenes*), so, after writing was reintroduced (ca. 800 BCE), this organ became the inner writing surface. The phrase "to write on the *phrên*" came to mean to preserve and rehearse a speech or text, and as verbatim recall became an educational goal, the Greeks trained themselves in the ability to practice what one might justifiably call *phrenography*.

To understand how this analogy worked, we need to review what writing surfaces were available to the ancients. The earliest Greek reference to writing appears in the sixth book of the *Iliad* (line 169), when a warrior in a battlefield conversation tells a story that features a letter inscribed on a "folded board" (*pinax ptuktos*). Since the Homeric epics portray men and women who do not communicate through writing, this single reference is a curious exception. As scholars have interpreted it, a *pinax* was made of two hinged pieces of wood inscribed with letters either in ink, on wax, or carved directly on the wood. When it was folded over, it was closed with warm wax upon which a carved image was impressed, a distinctive seal that could not be secretly tampered with before it reached its intended recipient. This was an established form of messaging during the ascendancy of the Neo-Assyrian Empire (911–609 BCE) and seems to have then been imported by the Greeks in the eighth century, along with the Phoenician alphabet (Bellamy 1989).

While the word *pinax* continued in later centuries to be used to designate a formal list, e.g., of laws, dynasties, authors, and book titles, for everyday note taking and letter writing the Greeks used what they called a *deltos*, a smaller but similarly hinged wooden tablet about the size of a twenty-first-century electronic tablet. Having one or several (usually two) thin boards thinly coated with blackened wax, a *deltos* (L. *tabula* or *tabella*) could be easily carried about, opened, and written upon with a metal instrument (Gk. *graphis*, L. *stilus* or *ferrum*). This was sharply pointed at one end and flattened wide at the other, the latter used to smudge the wax and thus erase some, or all, of the text.[4]

In the fifth century, expressions of conceptional literacy began to appear. Pindar, as early as 474 BCE, had declared: "Read [*anagnote*] the name of the Olympic victor where it is written (*gegraptai*) in my *phrên*" (*Olympian* 10.1–3). Jesper Svenbro (1976) observed that dramatic and choral poets would be especially inclined to use this analogy because their texts were meant to be learned verbatim by hired performers: "the interior of the actor is a space for writing" (200). Aeschylus seemed fond of this analogy. Consider Elektra's words to Orestes in the *Choephoroi*: " 'listen and write this in your *phrenes*' (450)—precisely what a dramatist might have said to an actor in the rehearsal" (Svenbro 1976, 201). In Aeschylus's *Prometheus Bound* (788–789), the tormented Titan commiserates with the Zeus-betrayed girl, Io: "I will tell you, Io, the wanderings of your turbulent course: inscribe them on the mindful tablets [*mnêmosin deltois*] of your *phrenes*." In *The Suppliants* King Danaus tells his daughters to preserve his advice to them well "entabletted" (*deltoumenas*) (178–179), and in the *Eumenides* (273–275) Aeschylus writes that "Hades' *phrên* is a careful transcriber [*deltographos*] of the accounts of human lives."

Plato, as we have seen, regarded writing as, at best, a means of making memoranda. For him, memory was important not for what could be inscribed on it in this life but for what could be retrieved from it as recollections of a former existence in the transcendent world of divine ideas. In his reference to the inner writing the image he plays with is not writing itself but the wax used to seal a letter or a *deltos*. In the *Theaetetus*, shortly after Socrates declares that thought is a dialogue of the soul with itself (190a), he directly presents an analogy: memory is like a square of wax within the soul, a gift of Mnemosyne, mother of the Muses. "Whenever we wish to remember anything we see, hear, or think, we subject this wax to sensations and thoughts and imprint them on it just as we form impressions from signet rings" (191c9–191d8). If thought is a dialogue of the soul with itself, and what Plato is referring to is rationally controlled inner *speech*, not inner *writing*, then the wax impression stands for the intensity of the perceptual input, not for its meaning. When later, adding pun to metaphor, Socrates says that the wax (*kêros*) of memory is located in the heart (*kêr*) of the soul (194c6–10), he does seem to acknowledge, albeit fancifully, the traditional belief that auditory memories of speech somehow lodge in the chest. We still speak of rote learning as "learning by heart." The Latin verb *recordari* (to remember) meant to retrieve something from one's heart.

What I have been examining as analogy may also be formulated as a conceptual metaphor, *memory is inner writing.* In this Lakovian metaphor, the properties mapped onto memory from the source domain of writing are not the lines and squiggles of graphemes but their fixed sequentiality, their permanence *as though* they were inscribed text. This conceptual metaphor thus implies rote memorization, the sort of teaching characterized by catechizing, an action that literally means making learners *echo* someone else's words by having them repeat them over and over (Collins 1996).

As a primary metaphor, *memory is inner writing* entails an important secondary metaphor, namely, *memory retrieval is inner reading.* To be somewhat more precise, the memory system that is here likened to readable writing is not episodic memory with all its rich, multimodal sensory details but rather speech-mediated semantic memory. It is the once-heard words of a speaker that are written on one's *phrên.* These words are imagined not as though they were available as a visually legible transcription taken down from dictation but as an auditorily imagined replay of a speaker's vocalized words. Phrenography turns out, after all, to be *phon*ography.

The Romans imported their psychological concepts from Greek philosophy and medicine. The vital principle, *anima,* like the Greek *psuchê,* was located in the head, and the *animus,* like the Greek *phrên,* was located in the chest.[5] But, like its Greek counterpart, the *animus* was also regarded as an inner writing surface. We find idiomatic usage as early as 166 BCE, when in his *Andria* the playwright Terence has a character say: "I, mindful [*memor*]? Those words . . . that were told to me [*illa dicta*] . . . are written [*scripta*] in my *animus*" (2.282–283). Cicero repeats this trope when in *De oratore* he advises attorneys to develop a verbatim memory for opponents' public statements, which should not simply be poured into one's ear but be inscribed in one's *animus* (2.355). In effect, the *animus* was an impressionable storage soul, the site of verbal voice memory.

Most Latin verbs associated with the *memory retrieval is inner reading* metaphor derive from the practice of reading papyrus scrolls, not tablets. In order to visualize the *animus* as a writable/readable text we should first consider the nature and use of the scroll, or bookroll, the dominant textual format up until the last centuries of the Roman Empire. Wilhelm Schubart (1921) summarized the ways in which reliefs and statues depict the reader:

> [He] sits and holds with both hands the open scroll, which lies on his knees. He does not have the scroll extended to its entire length: instead, the beginning and the end are each rolled up, the latter securely grasped by the right hand, the former by the left. In the middle, between these two rolled portions, lies one small open surface, the portion of immediate reading that is one column of script or, if they are narrow, as many as four. In this way a scroll of several meters' length could be held as a small object no bigger than a [modern] book. As one proceeds to read a scroll, one's left hand draws the read column and rolls it up while the right hand loosens its grip on the cylinder it holds and lets a new column slide toward the left.
>
> (110, my translation)

When a Roman used a scroll, reading it always meant rolling it, hence the noun *volumen* (volume), from the common Latin verb *volvere* (to roll). So when Cicero says that "Cato's books ought to be read," he chooses to say that his books ought to be rolled: "*volvendi . . . sunt libri . . . Catonis*" (*Brutus* 8.7.298). When a scroll is rolled forward (the text passing to the left), the proper verb for this action is *evolvere*, to unroll. That is, the rows of letters were optically read from left to right, but the rectangular blocks of text, the *paginae*, were manually propelled from right to left—which, by the way, is exactly how the modern spine-bound book is perused. When it is rolled backward (the read text passing to the right), the verb is *revolvere*, an action implying either a rereading of an earlier passage or a complete rewinding of the papyrus onto the rod, which, apparently on account of its central position, was termed the book's navel (Gk. *omphalos*, L. *umbilicus*). If the *animus* includes as one of its faculties the memory of speech, and if this memory is by analogy likened to a scroll, the recollection of verbally mediated information will be portrayed as an act of reading that scroll. But where we might expect "reading in the *animus*" to be expressed as "*legere in animo*," we find that this process of reflection is now consistently likened to the work of hands moving a scroll in the manner described by Schubart.[6]

The commonest instance of this analogy occurs with the word *volvere*. When a person is represented in a text as thinking and this verb is used, it is customarily translated in English as "to ponder" or, when paired with *animus*, as "to turn over in the mind." Such translations are reasonable because we at our point in literate history no longer analogize thinking as reading or, using our modern spine-bound book as analogue, to speak of thinking as "flipping over in one's mind" or "thumbing through one's thoughts," but the Romans evidently did so. To translate *volvere* as "to ponder" is therefore to blur the analogical nuances of this verb. When, for example, Tacitus describes a person as *"futura volvens,"* literally "rolling future events," he implies the act of reading ahead in the scroll (*Annales* 1.64). When Statius speaks of a person who "rolls the advice the sage of Gargettus [Epicurus] gives," he describes the act of turning an internal papyrus inscribed with admonitions (*Silvae* 2.2). Turning this metaphorical scroll is often an act of recollecting the cultural past: when Silius portrays Hannibal thinking of his noble lineage, he has the Carthaginian general "rolling the ancient chronicles of his ancestors" (*volvens veterum memorata antiqua parentum*) (*Punica* 13.35).

This metaphorical scroll reading as a writing-influenced mode of inner speech also characterizes that thought process when it is repetitive, even obsessive: one "rolls" thoughts over and over—and over—when one engages in rumination and worry (see chapter 3). Reading and rereading any oral discourse transcribed upon one's *animus* takes time. Unlike a mental image that might be called up quickly and scanned, or as a spine-bound book might be opened and its pages flipped to arrive at some sought-for passage, a scroll-like memory must be wound and rewound, its grievances and worries serially retrieved. Hence the verb *volvere,* when it refers to the mental rehearsal of a set of thoughts, often appears in the form of the present participle, as *volvens,* and in the imperfect tense, as *volvebat* (he or she kept turning over . . .).

Latin has yet another way to signify an action that continues over time. This is the frequentative aspect. Should one want to describe the action of repeatedly rolling something, one might use the verb *volutare.* An early example, coming from Plautus's comedy *The Captives,* conveys the obsessive, sometimes frantic, effects of anxiety: "The more I keep rolling over [*voluto*] this business in my chest [*pectore*], the more my distress [*aegritudo*] builds up in my mind [*in animo*]" (4.2.1–2). When the historian Titus Livy tells us that he has often wondered how the Roman state would have fared in a war with Alexander of Macedon, he phrases it this way: "During these periods of thought, I often silently kept rolling [*volutavi*] my *animus*" (9.17.2). Here his acquired knowledge of history is imprinted directly on his *animus,* where it can be accessed and read silently, i.e., subvocally.[7]

So far I have explored this conceptual metaphor as it appears in the Latin verbs for "roll" derived from the *volv-* root: *volvere, evolvere, revolvere,* and *volutare.* There is yet another related verb, *vertere,* to turn. This too is an action associated with scroll reading. Rolling is what readers do to the text, but turning those left and right rolls is *how* they do it. When a "turning" verb is used, it is commonly in the frequentative, a form expressing repeated action. For *vertere* the frequentative is *versare,* or *vorsare,* a rapid, repetitive turning that I will translate "versate." (The *OED* actually allows for "to versate," meaning "to turn about.") Plautus has one of his characters declare, "I am versating many items of business at once in my heart" (*multas res simitu in meo corde vorso*) (*Trinummus* 2.1.1). Livy writes: "they were to versate in their *animi* within themselves each particular item" (*versarent in animis secum unamquemque*) (*Ab Urbe Condita,* 3.34).[8]

When Horace advises the Piso brothers, two aspiring playwrights, he literally tells them to "keep turning the Greek literary models by nightly and daily hand" (*vos exemplaria Graeca / nocturna manu, versate diurna*) (*Ars poetica* 268–269). What could the brothers understand by this expression? Certainly not the flipping of pages in a spine-bound book. They and their contemporaries could only have understood that they should constantly turn their rolled-up copies of the Greek; that is, they should read and reread these texts. Horace's phrasing suggests the figurative "turning" of the mind, but, by including the reference to the "hand" (*manu*), he is reasserting the literal meaning of this conceptual metaphor as the turning of a scroll in the physical act of reading.

Lyric and Soliloquy

Horace was advising that these two brothers immerse themselves in Greek dramatic texts and, as he himself had done by studying the great lyric poets of the seventh and sixth centuries, absorb their various styles before developing their own. This is not the message we read in post-Renaissance poetics, such as in the first poem in Sir Philip Sidney's sonnet sequence *Astrophil and Stella.* Wishing to address his love in verse, he first "sought fit words to paint the blackest face of woe, / Studying inventions fine, her wits to entertain: / Often turning others' leaves." He could not invent, i.e., write out of a fluent imagination: "biting my truant pen, beating myself for spite— / 'Fool,' said my Muse to me, 'look in thy heart and write.'" As indicated earlier, this copying of *exemplaria*, or "turning others' leaves," would have been stored in semantic memory, perhaps also in procedural memory. When the Muse told him to look in his heart, Sidney found, besides general knowledge of style and substance, the affect- and image-rich personal contents of episodic memory.

Looking into one's heart and seeing something other than what one had dutifully transcribed within it would not have seemed to a Roman poet a very promising agenda. Using episodic memory to inspire one's writing had always defined the lyric poet. But to be usable this material needed stylistic processing: recruiting inner speech to mediate episodic material directly, i.e., to soliloquize, was rarely practiced by Greek or Roman poets.

In the first century BCE, Roman poetry blossomed in a wide array of forms and themes besides dramatic and epistolary lyrics. Some poems were written for social occasions, e.g., Catullus's epithalamium for Julia and Manlius (*C.* 61); others might be publicly presented, e.g., Horace's patriotic ode *Carmen Saeculare*, commissioned by Augustus and performed by a chorus of boys and girls. A number of Horace's lyrical monologues, including his "Roman Odes" (*Odes* 3.1–6), were epideictic exhortations on issues of contemporary concern. In a similar vein, Vergil's *Aeneid* was a twelve-book epic celebrating the mythic origins of Rome while reflecting his hopeful and anxious responses to recent Roman history. Ovid's collection of myths, the *Metamorphoses*, sometimes classified as epic, displayed a much more Alexandrian lightness of touch. Among the newer genres was the *epyllion*, the "little epic," a short narrative on mythological themes, also in the Alexandrian style, such as Catullus's *On the Marriage of Peleus and Thetis* (*C.* 64). Some were didactic poems, such as Lucretius's *On the Nature of Things* and Vergil's *Georgics*, as was Ovid's mock-didactic *Ars amatoria*, advice for men on how to seduce women and for women on how to seduce men, and his *Remedia amoris*, on how to prevent, or treat, a broken heart.

Besides these genres and subgenres addressing issues of general concern there was, however, a growing body of poems that reflected a turning away from the public sphere and a focus on particular erotic relationships, liaisons that were marked by a flux of emotions from hope and desire to anger, remorse, and desperation. In accordance with the general outwardness of classical writing, there was something epical about the erotic, what with its verbal dueling, its thrust and parry, flight and pursuit, exclusion and penetration, victory and surrender. This was a popular metaphor: "I have lived as one desirable to the girls / and have soldiered [*militavi*] not without glory" (Horace, *Odes* 3.26), implying he now had the scars to show for it, and Ovid declared, "Every lover soldiers [*militat*] and Cupid is his camp commander" (*Amores* 1.9.1). Unlike epic conflict, however, virtually all humans, male and female, from adolescence on could participate in this agon or remember having done so. Here was a powerful set of experiences that almost everyone had in common, yet at the same time the most universal aspect of every erotic relationship seemed, paradoxically, to be its radical difference from everyone else's erotic relationship.

One need not endorse Bruno Snell's theory of cognitive evolution to recognize in archaic Greek lyric a defining of autobiographical selfhood in erotic terms. Later elegiac poets, by choosing this theme, put themselves forward as the protagonists of their own life stories. We cannot minimize the voyeuristic, gossipy allure these poems had for readers, imagining the amours of Catullus and Lesbia, Gallus and Lycoris, Horace and Canidia, Propertius and Cynthia, Tibullus and Delia, Ovid and Corinna, and Sulpicia and Cerinthus.[9]

Probably the most famous, certainly the most concise, expression of this theme is Catullus's two-line epigram (C. 85):

> *Odi et amo. Quare id faciam, fortasse requiris*
> *Nescio sed fieri sentio et excrucior.*

> I hate and I love. How do I manage to do that, perhaps you are asking?
> I don't know, but I feel it's happening and am nailed to my cross.

The Greeks and Romans admired epigrams fashioned into elegiac couplets and were familiar with witty paradoxes, but this epigram is presented not as a rhetorical figure but as a simple fact. They were also familiar with emotional conflicts, but these were usually expressed as alternating, not simultaneous, emotions. How does Catullus do this? He doesn't *do* it—he feels it *being done* (*fieri*, the passive infinitive), and he is now the passive recipient of torture (*excrucior*, first-person passive present tense). In Latin the verbs *odi* (I hate) and *amo* (I love) usually have objects, but in this couplet they are used intransitively: the site of this excruciating conflict is wholly within the speaker and is disconnected from the third-person agent, presumably Lesbia, who has provoked these emotions yet is not here blamed for them. Nevertheless, despite its insightful inward gaze, this is not a soliloquy. Catullus is not alone. For here, too, dramatic lyric is the default mode, and, brief as it is, this poem is a dramatic monologue addressed to some "you" who "perhaps" wishes an explanation. Even when Roman lyric appears, as here, to be a soliloquy, it is framed as a dramatic speech. When he addresses himself as "*Miser Catulle*" (Wretched Catullus) (*C.* 8), he does so as though there were two Catulli, one the adviser, the other the unfortunate wretch.

I referred earlier to Kenneth Quinn's classification of Horace's odes along a spectrum from the dramatic to the epistolary. He went on to add that there are "odes of course between the two extremes, neither clearly epistolary nor clearly dramatic, whose status is occasionally hard to determine." These he referred to as "dramatic monologues," among which there are a few odes in which "it seems to me the monologue is a soliloquy: though the addressee is named, it is not supposed to hear what is said. In an ode which is a soliloquy, communication is of course non-existent. These odes are, in fact, a kind of adaptation of the sustained aside, just that sort of inward meditation that Heinze said could not exist in ancient poetry" ([1963] 2014, 87). There are, I grant, degrees of soliloquial speech, that is, of solitary utterance. Some instances represent speakers addressing absent addressees in asides that we may call apostrophes to persons imagined or, in extreme states of mind, hallucinated, but these may be distinguished from soliloquies that are not only not meant to communicate with anyone but are there simply to reveal one's own thoughts and feelings.

When in 1833 John Stuart Mill penned his influential essay "What Is Poetry?" he defined "poetry" as that latter, inward mode of soliloquy, contrasting it with "eloquence," i.e., oratorical or dramatic speech:

> Poetry and eloquence are both alike the expression or utterance of feeling: but, if we may be excused the antithesis, we should say that eloquence is *heard*; poetry is *over*heard. Eloquence supposes an audience. The peculiarity of poetry appears to us to lie in the poet's utter unconsciousness of a listener. Poetry is feeling confessing itself to itself in moments of solitude, and embodying itself in symbols which are the nearest possible representations of the feeling in

> the exact shape in which it exists in the poet's mind. Eloquence is feeling pouring itself out to other minds, courting their sympathy, or endeavoring to influence their belief, or move them to passion or to action.
>
> (1859, 71)

Succeeding generations would regard this formulation as eminently reasonable even though it pushed to the margins those two popular nineteenth-century verse genres, narrative and dramatic monologue. Most nineteenth- and twentieth-century critics agreed that lyric, which Mill deemed the prototypical poetic genre, was in its purest form the emotionally toned inner thoughts of a solitary thinker, identifiable as the poet.

If Greek or Roman poets and readers had heard the Muse advise them to look into their hearts and write, they might have felt, well, bemused. But if they had heard Mill's definition, they would have regarded it as singularly odd. For them, eloquence, i.e., outwardly voiced or auditorily imagined speech, was the only proper way to frame a poem. For example, Horace's second epode seems a soliloquy in praise of rural peacefulness far from the emotional stresses of city life and its cycles of getting and spending. After sixty-six earnest lines, the last four lines come as a jolt: "When Alfius, the money lender, now on the point of turning countryman, had said this, he called in all his loans on the Ides, intending to invest that money again on the Kalends." With that sly coda, the poet steps in and reframes this first-person reverie as an embedded monologue, the reported speech of a third person, identified by name, a man who would never realize this dream because all along he never really wanted it.

Classical poets consistently represented thoughts as overt, not covert, speech. When they wanted to embed the monologue of a solitary speaker, they had recourse to the convention of apostrophe, the device by which a speaker turns away from auditors, real or potential, in order to address an absent or nonconscious person, thing, or personified idea. After introducing his speaker, Corydon, Vergil in *Eclogue* 2 vanishes and lets the shepherd address the absent Alexis. While Corydon's words, some of them addressed to himself, are indeed overheard, they are very much framed in what Mill would recognize as eloquence, futile though it be. As for addresses to things and ideas, Horace presents himself as alone when addressing his Sabine fountain of Bandusia (*Odes* 3.13) and the Roman state, personified as a ship on a stormy sea (*Odes* 1.14), as though both fountain and ship (of state) were conscious addressees. Anyone who writes a personal letter or applies the epistolary model to a lyric poem may also be said to apostrophize his or her addressee, e.g., Horace to Plancus (*Odes* 1.7). The one Horatian ode that some scholars consider a true soliloquy, in that it represents the thoughts of a solitary person not directed as apostrophe toward some imaginary other, is the speech of Neobule, a girl who complains that her gender is unfairly constrained (*Odes* 3.12). If we interpret this ode as wholly in her voice, then it is she that addresses herself in lines 2–3, using the same sort of self-address as Catullus's "*Miser Catulle*," which seems to draw it back from pure soliloquy to direct address. But if, as others interpret it, the speaker is Horace, sympathizing with the girl's plight (Neobule was the name of Archilochus's much-maligned girlfriend), this ode becomes a standard dramatic monologue addressed to her as though present to him.[10]

As a means of problem solving, humans have been talking to themselves no doubt from the beginning of spoken language, and before that they may at times have even gestured to themselves. But no word for it existed in any known European language until Augustine of Hippo coined one in the late fourth century CE (Stock 2010). Oddly enough, however, "soliloquy" is introduced in a dramatic dialogue between "Reason" and Augustine. When Augustine tells her he is ashamed to admit his intellectual lapses, Reason, like Sidney's Muse, loses her patience for a moment: it is ridiculous to be embarrassed by one's own reason. "Because we are speaking to ourselves alone," she tells him she has chosen to entitle this dialogue *Soliloquies*, "a name which is, to be sure, a new one and perhaps an awkward one, but one that is quite suitable to indicate its purpose" (*Soliloquies* 2.14). Though in this two-volume work, which focuses on the power of the mind to know itself, Reason and Augustine are two aspects of the same person, the form of discourse is neither a monologue nor solitary speech; it is dramatic dialogue. Even though it is an internal debate, these meditations retain the form of a Platonic or Ciceronian philosophical dialogue. With Augustine, having come close to observing the emergence of true soliloquial self-revelation, we need to turn from modes of address to modes of reference.

When Mill wrote "Poetry is feeling confessing itself to itself in moments of solitude, and embodying itself in symbols which are the nearest possible representations of the feeling in the exact shape in which it exists in the poet's mind," those "symbols" were not what semioticians call "symbolic signs," those conventional designations such as words, numerals, and graphic letters. Mill's symbols are images that function primarily in metaphor and to a lesser extent in metonymy. Because a metaphor is based on a relation of partial similarity between two items, it relies on iconic thinking, and because metonymy is based on a relationship of cause and effect or of contiguity in space or time, it relies on indexical thinking. So, when Mill speaks of "symbol" as the special embodiment of poetic feeling, he actually refers to the semiotic functions of the *pre*symbolic mind, the older prelinguistic brain.

Here again, I think we have a feature associated with soliloquial monologue that reveals a difference between the outwardly projected poetry that Greeks and Romans practiced and the poetry of meditation and self-examination that later emerged in the Christian West and the Moslem East. Classical poetry had relied primarily on mythic *exempla* and employed metaphor and metonymy sparingly, regarding them as vivid forms of ornamentation rather than as structural elements. Introspective soliloquy, however, opens up an inner, ahistorical universe of feelings and intimations and can no longer draw its referents solely from the sunlit world of cities and countryside and of men and women in real or mythicized interaction. It must therefore find in language the imagistic means to substantiate that shadowy inner world, thus lending its mental entities a local habitation and a name.

What Meter Does

When we see a structure in a biologically evolved organism, we presume it has a correspondent function. In the culturally evolved world of intentional entities, that same presumption also makes sense, especially when a structure seems to persist for centuries or millennia. Metrical verse seems just about as close to qualifying as a cultural universal as do toolmaking, cooking, and weaving. But, while we know what the latter three are "for," what is metered speech "for"?

Every linguistic community has its traditional metrical system, but not only does each system have different features, but every generation or so it seems to require a retuning—sometimes a major overhaul. Why do metrical structures change over time? Why for example did Saturnian verse in Latin and other Italic dialects vanish in the third century BCE, to be replaced by dactylic hexameter? Why was Ennius's alliterative style derided in the first century BCE? Why were Greek lyric and elegiac meters imported from seventh-century Greece and third-century Alexandria and naturalized as Italic measures? Later in post-Roman Europe, why were accentual meters prized for centuries until in the thirteenth century Italy, Spain, and France adopted end-rhymed syllabic? Why in fourteenth-century England did Old and Middle English alliterative/accentual forms give way to accentual-syllabic meter? Why did a prosodic feud between the advocates of blank verse and rhymed couplets simmer from the seventeenth to the nineteenth centuries? And why in the twentieth century did the latter give way to a variety of phrase-centered, unrhymed "free verse," accentuals, and syllabics?

Why, after all this time and effort, haven't we arrived at a standard metrical design, one as standard as, say, the claw hammer, the shovel, or the loom? We have no need to reinvent the wheel, we tell ourselves, so, even within the same language tradition, why is there this perennial impulse to reinvent the poetic line?

Undoubtedly, languages change, and altered phonology has to have been a factor. The cosmopolitan prestige of foreign prosodic systems has sometimes played a role. But I suspect that something else was driving these innovations. We innovate, after all, when the way we provide for some natural need no longer works properly. When, for whatever reason, the tools and techniques that we have been using to secure an adequate food supply or protect ourselves from environmental dangers fail us, we change them. But before we attempt to use this pragmatic principle to understand the curious mutability of *specific* prosodic systems, we should try to establish what poetic prosody *generally* seems to do.

The most commonly recognized feature of what we have come to call "poetry" is its metrical structure, i.e., the measured count of specific rhythmical units.[11] In an oral culture meter would be heard in recurrent rhythmic patterns of similar lengths. Whether chanted or sung, this metrical structure would allow hearers to anticipate the occurrence of prominent syllables and thereby invite them to participate in the rhythmic unfolding of the performance. In a literate culture, these serial patterns are usually formatted in lines that continue in one direction, in most languages from left to right, until they complete a metrical count, at which point the writing and the reader's eyes turn back to the left margin, a turn, or *versus*, that characterizes this sort of verbal artifact.[12]

An oral performance of song, such as Homer described it, was an energetic bodily action involving words and dancing. In a rural community, groups of men and women might sing while dancing. In royal palaces a singer might perform accompanying himself on a lyre while others danced. In the Greek tradition the rhythm of the words and of the dance seem to have been synchronized, and for that reason a poetic line was said to consist of "feet," each representing a step, which in turn consisted of the placement (*thesis*) of one foot on the ground or dancing floor and the raising (*arsis*) of the other foot. In the singer's uttered words, the *arsis* was represented by the short, rapid syllable(s), and the *thesis* was represented by the long syllable. The *thesis* would have been relatively longer for the dancers, too, since the planting of the foot would be followed by the shift of body weight to that foot before the other foot could be raised to execute the next step.

One of the first consequences of adapting song to the written medium was its distancing from dance. While it preserved the alternation of long and short syllables, the combination of which was still called a "foot" (Gk. *pous*, L. *pes*), the meaning of *arsis* and *thesis* became reversed, a terminological shift reflected in Latin usage. Since the voice, not the legs, was now the main focus of the performance, the raising, or *arsis*, became the raising of the voice while uttering the prominent syllable, and the placement, or *thesis*, became the lowering of the voice for the nonprominent syllable(s). This vocal raising and lowering may also suggest the influence of accent, or stress, on classical prosody.[13]

Every language determines the nature of each metrical unit within the line by assigning prominence to a semantically significant feature of that language. In Greek and Latin that feature is syllable length, or quantity. In Chinese and other tonal languages, it is pitch. In English and other Germanic languages, it is stress, often conveyed by amplitude. In the Romance languages, in which syllables are relatively equal in prominence, the significant feature has tended to be the syllable count of a line, its completion marked by a rhyming syllable. In actual practice, however, these phonological features do not strictly predetermine the verse. Metrists have pointed out, for example, that spoken Greek, and certainly spoken Latin, also employed lexical stress (or accent) and likely wove that feature into their verses alongside syllable length, or quantity (Kirby-Smith 1999). Stress and tonal patterns also appear in Romance poetries and, in English and other stress-dominant languages, quantity and pitch are also important concomitant features.

As the previous paragraph indicates, the level of granularity necessary to do justice to the poetic prosody of any language tradition is daunting and becomes even more so when one reflects on how poetic practice in each tradition changes over time. This makes writing a treatise on meter like trying to paint a sunset in progress. I would not, therefore, attempt it here even if I had the space to do so. We need instead to refocus our efforts on identifying what prosodic characteristics all poetic traditions share and on finding some of the reasons why measured discourse has proven so effective in verbal artifacts over the centuries both in oral and literate settings.

We naturally associate accelerated heartbeat and breathing with an increase of volume and speed in information processing. During a performance of dance or music, rapidly repeated movements and sounds can therefore induce in audience members a heightened state of consciousness. Ululation, drum beats, whirling, stamping: all these stylized sounds and movements use as their formula *rapid repetition = intensity* and serve to represent the expanded input of the working memory confronting an arousing, perhaps overwhelming, perception event. In such situations, rhythmic cries in various pitches, percussive sounds, and repeated movements serve to block distractive thoughts, i.e., self-selected directions of attention and cross-references to episodic-memory fragments that, as I suggested in chapter 3, include those anxious, intrusive thoughts that seem to flow just below the surface of consciousness.

At a subtler level of excitation, during the recitation or silent reading of a verbal artifact, phonological repetition serves to extend working memory into a lengthened, "intermediate-term" duration and at the same time prepares us to contemplate mental images. Visual arrays in a predominantly nonverbal performance, e.g., a military ceremony, a historical reenactment, or a sports pageant, would be meaningful to onlookers as iconic representations of a collective episodic memory. Actors and viewers would understand that these images do not represent their personally stored memories though they manifest *like* their own recollected episodic memories. When a spectacle portrays past events, it represents them in that liminal, intermediate space between the real and the unreal, the play-framed fictive zone. The mimetic performance, with or without spoken components, thus builds upon and exploits two modes of episodic cognition: one is an intense and complex experience of a rhythmically animated ongoing present; the other is a recollection of a past experience visually displayed. These two modalities, the auditory and the visual, also correspond to those two kinds of memory, the short-term, or working, memory and the long-term declarative memory, both the episodic and the semantic.

The complementary aspects of these two modalities align closely with Nietzsche's distinction in *The Birth of Tragedy* between the Dionysian and Apollonian forces that he found combined both in Greek tragedy and modern music drama. At the Dionysian pole, our sense of separate individuality with its unique autobiographical past yields to a shared reality, insistently present through a rhythmic repetition of sounds and movements. At the Apollonian pole, we step back and assess the world as a visual re-presentation of past events.

Here Merlin Donald's first two hominizing stages, the episodic and the mimetic, come together as dyadic opposites. The Dionysian pole represents the immediate experience of event perceptions, our awareness of an extended present as constituting a specific episode, a "going with the flow" that is further enlarged in a mimetic culture by certain communal "scripts," e.g., hunting, foraging, toolmaking, and food sharing, during which the perceiving self is submerged in social action. The Apollonian pole represents the new, individuated self that had coevolved with episodic memory. When these two episodic modes, that of the ongoing present and that of the retrospected past, appear together in the mimetic performance, each draws power from the other.[14]

When language became a medium for mimetic play (Donald's "mythic stage"), it became possible to incorporate these two complementary modes even in the absence of dancers, drummers, actors, and spectacle. The solo voice of the storyteller, singer, or rhapsode could serially produce prosodic patterning while at the same time conjuring up settings, persons, and actions that must be imagined as coexisting in parallel. The prosodic repetitions would weave the ground upon which visualizable figures were depicted. Audiences would remain broadly aware of the repetitive pulses in the background while their attention was focused on the images the words suggested. Table 7.2 indicates these pairings. This doubleness, which Nietzsche called the union of the Dionysian with the Apollonian,

TABLE 7.2
Correlation of prosodic form with semantic content

	Prosodic form (*how* the text is performed)	**Semantic content (*what* the text represents)**
Modality	auditory (metered rhythm)	visual (mental imagery)
Memory systems	short term (working)	long term (episodic and semantic)
Processing style	serial	parallel
Nietzschean principle	Dionysian	Apollonian
Attentional style	broad awareness	narrow attention

was possible only because linguistic prosody and linguistic semantics are noninterferent functions of the brain. We can do them both at the same time because our auditory and visual systems, while linked, are separately wired.

It is often assumed that metered prosody in a preliterate society served as a means of storing language in long-term memory. However, as I suggested in chapter 2, there are other features that exert much more mnemonic strength, e.g., the temporal sequence of events, the places in which they are set, and the person(s) central to the action. Moreover, as mentioned in chapter 3, when intrinsically interesting oral narratives are enhanced with anticipatable metrical beats, the brain's reward system is activated, and, at the same time, its worry-prone inner speech is inhibited, a two-for-one hedonic payoff that increases the chances that these compositions will be reperformed and their words even more firmly impressed upon the public memory.[15]

Orally transmitted narratives are normally not verbatim reproductions of previous performances, though their order of events, places, and persons, as well as much of their phrasing, will be preserved. When, however, particular bits of information must be accessed verbatim, e.g., the number of days in the months of the year and certain grammatical paradigms, meter and rhyme may be useful, but it is unlikely that the verbal content of narratives or lyrics was ever preserved simply by remembering what particular words go with what particular metrical unit. Metrical patterns, such as dactylic hexameter and iambic pentameter, the heroic couplet and the ballad quatrain, function like containers into which one can store any number of different verbal messages, from epics to pastorals, from satires to reminiscences to political rants. It is the prosodic *structure* of poetic genres, not their verbal *content*, that is preserved in long-term memory, specifically in procedural memory.

When writing eventually became accepted alongside oral performance as a valid medium for poetry, its prosodic structure continued to play an essential role even as its effects became less and less sounded out and more and more processed internally by the auditory imagination. That prosody is not, and probably never was, an efficient means of storing language in long-term memory is clear from the fact that metered poetry continued to be composed centuries after poems were preserved in the medium of writing. The true memory function that meter has always served is short-term, or working, memory, and it serves that function by imposing constraints, such as the number of metrically significant syllables per line, thereby increasing the possibility of anticipatable elements (Fabb 2015).

If, as I proposed in chapter 3, rhythm operates below the level of the conscious mind, where it activates pleasurable hormonal releases while inhibiting inner speech, it can only accomplish this when it works on the preconscious level, i.e., when we are only broadly aware of it, when we hear it, as it were, out of the corner of our ear. Rhythm in the form of poetic meter cannot do its work when it becomes an overtly conscious feature, when it is taught as just another technical device, like metaphor or chiasmus, just another *gradus* up the path to Parnassus, another trick for polished young gentlemen to master. Whenever prosodic form becomes imitative or is otherwise enlisted as semantic content, it loses its power. Now instead of operating behind the scenes, it is forced on stage and made to pantomime the message of the words.

> 'Tis not enough no harshness gives offence,
> The sound must seem an echo to the sense.
> Soft is the strain when Zephyr gently blows,
> And the smooth stream in smoother numbers flows;
> But when loud surges lash the sounding shore,
> The hoarse, rough verse should like the torrent roar.
> When Ajax strives some rock's vast weight to throw,
> The line too labours, and the words move slow;
> Not so, when swift Camilla scours the plain,
> Flies o'er th' unbending corn, and skims along the main.
>
> ("Essay on Criticism II," 364–373)

It is now three centuries since Alexander Pope wrote this dazzling passage, three centuries for poets to scramble in search of subtler prosodies, searching for ways to undo its demystification.

Finely metered rhythm inhibits one's private inner distractions, thereby making it easier to process the words on the page and use them to visualize the specific images that the words suggest. William Butler Yeats (1903, 247) described this working relationship between rhythm and imagery as follows:

> The purpose of rhythm, it has always seemed to me, is to prolong the moment of contemplation, the moment when we are both asleep and awake, which is the one moment of creation, by hushing us with an alluring monotony, while it holds us waking by variety, to keep us in that state of perhaps real trance, in which the mind liberated from the pressure of the will is unfolded in symbols.[16]

This "purpose of rhythm" is not simply to lull the mind with a predictable meter or to induce a state of hyperalertness. Anticipation and surprise, co-occurring dyadically, help the reader maintain a play frame, which in this case is marked by the indexical sign of *fiction* and the iconic sign of *truth*. Rhythm permits the mind to lay aside its ego-based will and become, as he says, "unfolded in symbols." By "symbols," of course, Yeats, like Mill, does not mean symbolic signs, those arbitrary signifiers such as words, but rather word-cued icons, mental images of complex significance. In an actual mimetic spectacle, such imagery might be displayed in masks, costumes, banners, and processions of dancers, as well as by sacred architecture. In the silent, solitary contemplation of written words, these images emerge from the mind itself as responses to the words. Note: the "mind is . . . *un*folded" in them, not, as in an outward pageant, *en*folded in them. By this choice of words, Yeats implies that in response to verbal suggestions the mind opens up like a flower's bud to reveal the images that embody its stored and hidden meanings.

Seeing What We Read

When stories are told, it is the voice that is foregrounded, for it is the voice that conveys the voices of the characters, often in dialogic exchanges, and describes the setting of these exchanges. During such oral performances, visual perception strongly constrains the mental imagery that the narrative can evoke. Most content words, e.g., nouns, verbs, and modifiers, as distinct from function words, e.g., conjunctions, prepositions, and pronouns, evoke visual representations, but, when they are orally cued, these mental images must compete in the brain with visual perceptions of the face and gestures of the speaker, his or her clothing and posture, as well as the entire setting of the performance. Undoubtedly, one of the most profound effects of writing was the creation of a new audience, readers who could now, in the physical absence of the speaker, auditorily imagine the voices and visually imagine the world evoked by the silent written page.

How our minds convert words into mental images, especially visual images, has intrigued me for many years. A quarter-century ago, when I was preparing the book *Poetics of the Mind's Eye* (1991), I thought I might find excellent examples of visual imaging in Shakespeare's plays, but when I searched, I soon discovered that carefully constructed imaginary vistas rarely appeared. The main reason for this, I concluded, is that a playwright intends that the audience focus on the visual scene unfolding before their eyes, not shift their visual attention to mental imagery. So, when a speaker constructs an imaginary display, it is the reactions of the onstage addressees, also visible and audible to the audience, that engage the audience's immediate perceptions.[17]

Orally performed verbal artifacts, be they narratives, lyric songs, or dramas, whenever they introduce visual images, tend to insert them as brief pauses in the action or the argument. Speakers are always the visual center of attention, and verbal images are what they selectively deploy to make their points: each image momentarily attracts attention as a facet of the speaker's spellbinding art flashes and then is gone. As insertions, placed side by side with the action they interrupt, Homeric similes serve as paratactic elements that clarify the moment that provokes them. The same is true of epithets and extended descriptive elaborations. Texts that transfer the oral-performance style to the written page, e.g., the Theocritean idyll, the Vergilian epic, and the Horatian lyric, also tend to treat descriptive passages as paratactic inserts, though at this stage in literary culture these set pieces now begin to assume more important thematic functions.

The visual elements of a traditional oral or a virtual oral (literate) discourse will be organized paratactically. There are, however, two different ways in which such imagery is projected, depending on how the hearer as viewer is situated. I will call these Paratactic 1 (P1) and Paratactic 2 (P2). P1 corresponds to what the Greeks meant by *diêgêsis*, the narrator's art of leading an audience into a succession of imaginary spaces among imaginary persons. As the narrator guides them, the audience members seem to move from one eventful setting to another, accumulating a series of mental images the significance of which they must decide upon before these parts can cohere as a narrative. This is the paratactic style generally associated with Homeric epics. It may also correspond to what Aristotle called the "strung-together style" (*lexis eiromenê*). Since P1 assumes the imagined movement of the viewers as the speaker leads them through the landscape of the narrative, each viewer may visualize this scene and its individual images from multiple perspectives, especially when visual references are repeated. This perspective shifting can render an array of objects oddly cubistic.

In P2, the speaker invites the audience to imagine a person in a fixed location while separate images appear to him or her in a particular sequence. Though Homeric imagery is predominantly P1, we also find this second effect achieved in several episodes, e.g., in *Iliad* 3, during the *Teichoskopia*, as Helen points out the Greeks from the ramparts, and in *Odyssey* 11, the *Nekuia*, or conjuration of the dead. In the latter episode, Odysseus and his men perform a ritual on a particularly remote shore of River Ocean: at first there is rush of ghosts thirsty for the blood of the sacrifice, but Odysseus blocks them, and soon an order of sorts emerges, a procession of the dead, each with information to divulge. These appear also as separate images, but, as items in a single action, they seem arranged in sequential order, not simply in succession, as in P1. This is the order of images shown in a public ritual, such as an initiation ceremony in which a prescripted series of figures appears before the initiand(s). We also find this sequence of visitants in narratives centered on an immobilized figure, e.g., Prometheus chained to the cliff, Job on the dunghill, Daphnis dying of love in Theocritus's first *Idyll*, Gallus less seriously afflicted in Vergil's tenth *Eclogue*, and Aeneas with his father reviewing the procession of future Romans in *Aeneid* 6.752–886.[18]

Bruno Snell (1953) was the first to my knowledge to link both narrative and pictorial representation as ways of seeing. Thomas Webster (1958) followed up by using this grammatical distinction to connect Greek geometric art with Homeric narrative and noting a common "disregard of spatial relationships. The parts are given their known value even when this conflicts with their seen relationship to the whole. The result is a *paratactic aggregate* and not a *hypotactic* system, and in detail this love of parataxis is what distinguishes Homeric sentence construction from the fully developed periodic style of fifth century prose" (207, emphasis added). In 1975, Paul Feyerabend, borrowing Webster's phrase, proposed a pre-Socratic cosmic ontology as a "paratactic aggregate," arguing that a world viewed from a single perspective, unified as though it was a three-dimensional projection, was not any "realer" than a world viewed as an assemblage of individual wholes and parts—and a single perspective may in fact prove illusory.

The extent to which a pictorial style either reflects or influences the way the brains of a given people at a given point in its history process visual information is bewilderingly difficult to assess. But that painters and writers in a particular culture might share a particular visual style, the one in depiction, the other in description, is easier to establish as well as to understand. Hence the paratactic grammatical style of traditional oral narrative corresponds to a paratactic visual style that we can clearly recognize as the dominant tradition of classical painting.

The ancients were quite aware that poetry and painting were *somehow* related. Simonides had said that painting is silent poetry and poetry is painting that speaks. When Horace in perhaps his last piece of writing, the epistle traditionally named the *Ars poetica*, declared that a poem is like a painting (*ut pictura poesis*), he went on to explain that, like a painting, one verbal artifact "may grab you [*te capiat*] more if you stand closer to it, another when you stand further away; this one loves obscurity, that one wants to be viewed in the light and does not dread the sharp gaze of the critic; this one pleased but once, that one will please when repeated ten times over" (361–365).

A poem, though, is not in all respects like a painting, nor is a painting necessarily like a poem. Every verbal artifact, spoken or written, is a one-dimensional sequence of words. Though every pictorial surface is a two-dimensional plane, the contents of which may be "read" in any direction, painters have ways to attract and direct the gaze of the viewer in a temporal order: foregrounded, brighter, larger, and more centrally located figures are seen sooner than partly occluded, dimmer, smaller, or more peripheral elements. This "hierarchical time" (Small 1999) suggests a sequence of fixations that may evoke a narrative, but only if the viewer has recognized the figures and knows the story. In chapter 4, I cited Longus's introduction to his prose novella, *Daphnis and Chloe*, where he tells of coming upon a beautiful but baffling mural depicting a set of individual images that apparently represent a cast of characters involved in a series of episodes. Wishing to connect these images into a single narrative, he sought out a local person to interpret them. One reason for Longus's perplexity lies in a stylistic peculiarity that the classical-art historian Jocelyn Small (1999) has identified: "Most Greek pictures are strikingly without any sense of place. . . . In most cases, all action takes place against a solid undifferentiated background. The effect is one of timelessness" (567). For if, as Aristotle claimed, "time was not duration but movement though space . . . the only way to notice that movement has indeed occurred in a pictorial representation is to change the setting" (570).

Discourse, the medium of verbal artifacts, seems rather more suited to the representation of time, albeit a narrow, linear version of its passage. The literary subgenre *ekphrasis* demonstrates that when human figures in a depiction are converted into words, the result cannot remain purely a spatial description but readily morphs into narration. (The Greeks seemed to acknowledge this by defining *diêgêsis* as a descriptive passage or a narrative.) Homer's extended ekphrasis of the shield of Achilles renders this complex artifact intelligible in large part because he narrates the depictions using knowledge that we certainly would not have had as ordinary viewers of the shield. How could we have known that the judgment scene (*Iliad* 18.497–500) involved one man accused of murder by a kinsman of the deceased or that, while one of them promised to pay up, the other wanted another outcome? In the *Aeneid* (1.453–493), Vergil goes a bit "meta" when he narrates the actions of Aeneas narrating the actions depicted in separate images on a mural that had been placed in the Carthaginian temple of Juno. That is, instead of interpreting the painting himself, the poet has his hero perform a dramatic ekphrasis for the benefit of his faithful companion Achates. It is as a uniquely privileged exegete, one who even finds himself depicted in one section, that Aeneas narrativizes his interpretations.

I will now conclude with some brief observations on how Horace, as lyric poet, deployed visual imagery. We have already seen how, in *Epode* 13 and *Ode* 1.7, he artfully pivoted from natural settings to epideictic exhortations to dramatically inserted mythic exempla, but in the following ode (1.9) his pivotings have perplexed many and dismayed some. Though it has remained one of his most cherished poems, it has also proven one of his most problematic.

You see how dazzling, deep in its high snow,
Soracte stands and how the struggling woods
 cannot sustain their burden now
 and streams are clogged in jagged ice.

Thaw all this cold by piling on the hearth
an extra load of logs and, better yet,
 bring out the jug, O Thaliarch,
 of Sabine vintage four years old.

As for the rest, consign that to the gods
who have no sooner calmed the battling winds
 wild on the sea than cypresses
 and ancient ash trees cease to quake.

Please, no more asking what tomorrow brings,
but claim as profit every day that luck
 will grant to you and, being young,
 spurn neither sweet love nor the dance.

While peevish white-haired age is still far off,
now park and plazas and the whisperings
 softly as night descends, the hour
 appointed—these are yours to claim,

as is the girl's glad laugh that now betrays
her hiding place, yours too the trinkets stripped,
 in lieu of fuller payment, from
 her arms or finger scarce resisting.

The ode begins: "You see" the snow-covered top of Soracte, a mountain some twenty-five miles north of Rome. This is not a question or a command but a simple observation that initiates a dramatic lyric the addressee of which is either the person designated as "master of revels," the Thaliarch, or some other unnamed young man. A series of images follows: the "you" also sees woods loaded with snow and streams clogged with ice. Then the poet calls for wood to be placed on the hearth, so we know the speaker is now referring to a cold interior space that needs a thawing—*dissolve frigus*—and at his command the visual images of snow and frozen streams that began the poem also dissolve into spreading warmth and flowing wine. As in *Epode* 13, he asks for talking to stop, in this case talking about what tomorrow may bring—if it simply brings itself, that is good enough for him. The exhortation is simple enough: stop worrying, have a cup of wine—no, several cups. Since bad times await you in the future anyway, stop wasting your time anticipating them. Life isn't so bad if you accept it on its own terms. Just remember there is a proportion to everything—a *modus in rebus*.

Then, in the final two strophes, Horace seems to shift the scene: suddenly it is a warm early evening in Rome, young people whispering together in the dark, a girl's laugh, the playful theft of an armband or ring, and with that the poem is over. These rapid changes have left a number of readers professedly bewildered. Some commentators have felt it crucial to establish one single, specific location for the poem: a country home, perhaps Horace's Sabine villa, a house in Rome, or even some location closer to the sea where cypresses and ash trees grow. Others have tried to deny the temporal shift by suggesting, for example, that Soracte could still be snowcapped even as spring brought mild nights to Roman streets. These commentators seemingly require Horace to adhere to the Aristotelian dramatic unities. I think they are quite right up to a point. After all, the entire ode, all 105 words, is a dramatic monologue uttered to some young man physically present to the speaker. Accordingly, all the references to locations, weather, and seasons are supposed to make sense to that person as he drinks wine with Horace before a warm fireplace in winter. It would be dramatically absurd to think this addressee would be magically transported in time and space or that the speaker is experiencing some sort of *Quo-me-Bacche-rapis* hallucination.

An alternate solution has been to read the last two strophes metaphorically. The word "now" (*nunc*) emphatically asserted in lines 18 and 21 would elaborate line 17 and mean "now" while you are still young, as Orelli (1843, 62) suggested. The settings and seasons would then represent states of mind, a favorite Alexandrian trope, as Pasquali (1920) pointed out, adding, however, that the rapid shift in argument is a Horatian innovation (81–82). I have followed this line of reasoning by translating that one verb that governs all the nouns from line 18 to 24, *repetantur,* in terms of ownership or entitlement. If, when you are young, every new day should be jotted down as profit, then, among other pleasures, the wealth that you should rightfully claim as yours includes warm nights and amorous play.[19]

Despite confessing his fondness for this poem, Eduard Fraenkel (1957, 177) judged it seriously flawed: "Its heterogeneous elements have not merged into a harmonious unit. Line 18 *nunc et campus et areae* and what follows suggest a season wholly different from the severe winter at the beginning. The incongruity cannot be removed by any device of apologetic interpretation." The temporal leap was what most troubled Fraenkel, but he did not seem to perceive that all the separate images in the poem constitute a space that is also heterogeneous. The initial image, that of a dazzling white mountain standing in snow, is seen as a huge, single image. The next image, that of the woods staggering under their load of snow, is a separate image viewed from a much closer vantage point, and the streams clogged with ice is yet another image. (Cf. the various viewing distances that, as Horace explains, make *ut pictura poesis.*) We may agree that they all could be observed on a cold winter day, but not as parts of a single visual field.[20] These are as much paratactic images as are the later, temporally disjunct images of the playing fields (the Campus Martius), the city squares, the night, the trinket, the girl (albeit hidden), the arm, and the finger. As our mental vision moves from the monumental to the miniature, from white, cold day to dark, warm night, the poet projects for us a montage of images that do not coexist as details viewed from a single vantage point. In pictorial terms, that organization of space would require linear perspective, the sort of optically integrated design that would not become the norm until Bruneleschi and Alberti made it so in fifteenth-century Italy. Some Roman wall paintings may "exhibit linear perspective but not necessarily throughout the whole painting. . . . As a rule, they tend to combine various perspectives in a manner that often jars our modern eyes with its seeming inconsistencies" (Small 2009, 155n2, 151).

The visual imagination is a volatile faculty and can sometimes act independently of the verbal cues that activate it. We see this in the paratactic structure that the Soracte ode exemplifies, as images demonstrate a degree of autonomy, allowing them to interact with one another. In that respect, they behave similarly to images in metaphors and antitheses. Metaphor sources, or vehicles, are not really *there* in the made-up world of the poem but *here* in the word-stocked mind of the reader. Likewise, when antitheses are represented as sensory images, e.g., cold/warm, frozen/flowing, old/young, bright/dark, each in turn becomes a figure embedded in the ground of its opposite, both opposites hovering in the mind as dyadic complements. Once again, as we generally find in his lyric *ars poetica*, Horace, the consummate metrist, has used his intricate strophes to allow him—and us—to think in images or, as Yeats would put it, "unfold the mind in symbols."

In an image-rich poem such as this that manages to feature metaphor-inflected antitheses, we find ourselves, each time we read it, including more or less of each opposite when imagining a particular antithetical component. This creates an elusive variety in our experience of the poem. Donald Vessey put it well when he wrote: "the sweep of the six stanzas is broad and any reductive approach demeans it. The text is both encompassed and lost in each reading and rereading, and it always surpasses them" (1985, 38).

Summary

The underlying theme of *Neopoetics* has been the adjustments on the part of an oral culture to the introduction and increasing influence of literate institutions. Each previous chapter has touched on one or another aspect of this centuries-long process of accommodation. This final chapter began with a section meant to draw together a number of these conceptual threads and reweave them into a new design, one that aligns the two grammatical arrangements, parataxis and hypotaxis, with the two ways the brain sorts incoming data, viz. serial and parallel information processing. As I then attempted to demonstrate, these two aligned pairs, (1) the parataxis + serial process and (2) the hypotaxis + parallel process, together constitute a *dyad*, i.e., two functions that our brain can execute at the same time as a single cognitive experience. The paratactic style associated with oral-performance culture and the hypotactic style associated with literate culture are not decoupled or dichotomous. Aligned as they are with the brain's two information-processing modes, they represent complementary opposites that lie along an inclusive, dyadic continuum.

As the written page became the new performance space for verbal artifacts, the sensory centers of the brain needed to learn new and more finely tuned ways to simulate auditory and visual experience. This was no easy feat to accomplish. As I pointed out in chapter 6, Greek and Roman literary genres consistently imitated oral performance, a phenomenon that has been termed "conceptual orality." In the second section of this final chapter, I suggested that orality, as it came to terms with literacy, exhibited a correlative reaction, which might be called "conceptional literacy." It began to reconceive its essential memory system, long-term semantic memory, as inner writing. As it did so, oral-performance style became less and less improvisatory and increasingly dependent on verbatim memory.

The mutual accommodations made by oral institutions such as storytelling, singing, and stagecraft, on the one hand, to book culture, on the other, produced a unique body of texts that in many respects seems quite alien to the post-Renaissance norms we have come to accept as "literary." For us, prose fiction and poetry have become ways to explore inner thoughts and emotions with all their complexities and contradictions, but for Greek and Roman readers, immersed as they were in outward vocal communication, literature offered a variety of virtual-orality devices that one could purchase, bring home, and listen to, either by reciting these texts or imagining their sounds in private, relatively silent readings. The concept of a literary text as a transcript of introspective contemplation, self-examination, and soliloquy was foreign to archaic and classical cultures.

They fully understood, however, that verbal artifacts have uncanny powers to move one deeply and to alter one's outlook toward the world. In chapter 3, I examined some of the claims made on behalf of *mousikê*, the Muses' combined art of narrative, song, and dancing. The ancients understood that metered rhythm, one feature that survived the transfer of sung performance to the written page, was somehow responsible for the magical effects of poetry. The recent findings of neuroscience that I had earlier applied to music I now applied to poetic meter, concluding that meter produces two effects: it quiets the reader's inner speech while it disposes the mind to contemplate the images that the text evokes.

In my closing pages I have reexamined visuality, a topic I discussed earlier (in chapters 1, 2, and 4). Here, in focusing on the styles of visual imagery in classical poetry, I reviewed the work of scholars who have applied the term "parataxis" to pictorial and literary imagery. Using once more an ode of Horace, I analyzed his images as representing multiple perspectives, a visual effect observable in ancient paintings. This way of representing a visual array is quite different from ours, which privileges a single perspective and an integrated field of objects. Just as oral culture visually imagines a paratactic world, literate culture endeavors to imagine a hypotactic world, one in which images are hierarchically arranged, share the same three-dimensional homogeneous space, and are viewed from a single perspective.

The innovatory developments that have led to the emergence of the literate imagination also created the emergent roles, first of the author, then of the reader. But, as the preceding chapters indicate, it has been difficult to sequester one from the other, for, to some degree, every writer is also a reader, and every reader is a writer. Each role must include the other in its definition. Anyone able to understand the one without the other would probably also be able to hear the sound of one hand clapping. My approach to this particular pair of complements, the writer and the reader, has been dyadic, in that I have focused closely on each in turn while keeping broadly aware of the other. As this book now draws to a close, our two principal players, writer and reader, take their curtain calls. They stand for a moment, hand in hand, acknowledging the applause, then one, then the other, strides to the footlights and bows. We in the audience *look* at the performer who bows, gestures hand to heart, and blows a kiss to the box seats, but we also *see* the other who smiles and claps in the background.

Epilogue

Poetics and the Making of the Modern Self

It is more than mere passage of time that separates the poets of Rome's Golden Age from those that John Stuart Mill had in mind when, in 1833, he contrasted poetry with eloquence. Both these discourses, he said, are "alike the expression or uttering forth of feeling. But . . . eloquence is *heard*; poetry is *over*heard. Eloquence supposes an audience; the peculiarity of poetry appears to us to lie in the poet's utter unconsciousness of a listener. Poetry is feeling confessing itself to itself, in moments of solitude. . . . All poetry is of the nature of soliloquy" (Mill 1965, 109–110). Classical poets experienced the human feelings we still recognize as our own, but they conveyed them, as we have seen, with a public outwardness and an intent to persuade, whereas the typical post-Renaissance poet revealed the thoughts and feelings of an increasingly private, problematic "self."

How may we account for this change? One factor had been the spread of a new theogony, the story of a god who appeared on earth in the incarnated form of a Jewish reformer, Jesus of Nazareth. Before his execution at the hands of the Roman occupation authorities of Judea, he told his followers that he not only was the son of the one god of the universe but that they could, too, become the adopted sons and daughters of this god if they accepted his messiahship and faithfully followed his precepts. Eventually they would rule the earth with him when he returned to judge the living and the dead and establish his thousand-year theocratic kingdom. This meant that those then living would never die and that those who had already died would be resurrected into everlasting life.

At first regarded as just another exotic import, Christianity by the turn of the fourth century had become firmly rooted in the Greek- and Latin-speaking cities lining the Mediterranean. Instead of a multitude of superhuman beings, this religion worshiped one "almighty father, maker of heaven and earth, and of all things visible and invisible" (Nicene Creed, 325 CE). This "maker," for which the Greek theologians had chosen the title *poiêtês,* was the single solitary "poet" that had created both the world that could be perceived by the senses and the equally real world of unseen powers—souls, spirits, demons, and angels. This reclusive, celibate creator, himself invisible, was present throughout the visible world. Since Christians believed there was no hiding from his gaze and, ultimately, from his justice, they seemed trustworthy stewards of private and governmental resources, a consideration that no doubt contributed to the toleration of this religion (313 CE) and, finally, to its advancement as the single state religion (380 CE).

Over a century and a half earlier, as Emperor Hadrian lay dying, he addressed a short poem of farewell to his *animula,* his little soul, the life principle that governs bodily actions, as it departed from him. While he, as well as most of his countrymen, believed he *had* a soul, Christians were taught they *were* souls. This shift, which created a new "self," also created for it two possible final destinations: no longer that subterranean cavern of flitting shadows, it would either be a place of torment or of bliss. Where it would abide for all eternity would depend on how well, in the eyes of its celestial judge, it had lived out its brief span of earthly probation.

Central, therefore, to the practice of this new religion was self-examination. "Be ye perfect," Jesus had exhorted his disciples, "as your Father in Heaven is perfect" (Matt. 5.48). Inspired by this heroic injunction and expecting the immediate return of Jesus, the Christ, as victorious ruler of the world, men and women became anchorites, i.e., persons who had withdrawn from secular society, establishing monastic settlements first in the northwestern Egyptian desert in the late third century and then soon afterward in Italy and Gaul. Theirs was a life of rigorous *askêsis,* a psychophysical regimen analogous to the training of soldiers and athletes. It was in such ascetic communities of monks and nuns that a new inner world began to open up, a psychic world permeable to legions of spirits, both good and evil. This was the psychic space that Augustine explored in his *Confessions*, a book that would introduce Europe to the topography of this inner realm.

From the early Middle Ages through the Renaissance, European thinkers would debate the nature of the immortal self. What were the properties of the soul (Gk. *psuchê*, L. *anima*) that made us a "little lower than the angels" (Psalms 8.5) and a little higher than the beasts? What was it about *being* an immortal soul that made one morally responsible enough to deserve one of those two extremes, eternal torment or eternal bliss? And if the brain was the seat of the soul, how did it subserve the actions of the rational soul; more specifically, how did it weigh options in the process of making morally responsible decisions?

The Christian Middle Ages formulated a theory of the brain incorporating the findings of the third-century BCE Alexandrian anatomists Herophilus and Erisistratus and the work of the Rome-based physician Galen (129–200 CE). This theory claimed that consciousness is the result of an airlike substance, *pneuma* (L. *spiritus*), that channels sense impressions from the front to the back of the of the skull through a series of cavities, called "ventricles," within which thoughts are processed, judgments evaluated, images displayed, and memories stored.

Right reason required that the soul first sense the world correctly, despite its natural tendency to misperceive or be deluded by demonic agents. Perceiving, however, was no simple impression of objects upon the brain, for the brain received five different kinds of *sensa*—from the organs of sight, hearing, smell, taste, and touch. In order to be recognized, some objects required us to integrate two or more senses and do so within a ventricular module they called the "common sense." This account of sense representation was further complicated by the multifunctionality of mental imagery, which suggested the notion of "internal senses." The latter did not correspond to the five external senses but were sensory displays, predominantly visual, that accompanied each principal cognitive faculty. In a world that the Nicene Creed divided into *visibilia* and *invisibilia*, it was the psyche's imagination that could make the invisible visible. Imagination had extraordinary powers, not least of which was that of arousing emotions (Gk. *pathêmata*, L. *passiones*). Many devotional exercises relied on it to construct scenes from the Gospels and from saints' lives and through these images to move souls to emotions of awe and thanksgiving, but, by the same measure, demonic spirits could exploit its powers to generate illusions and phantasms.

The word "passion," as a standard medical term, referred to a disability, e.g., an illness of the body caused by an imbalance of the humors or a state of mind agitated by unresolvable thoughts and feelings. Such conditions rendered one helpless (cf. the noun "patient" and the adjective "passive"). To strengthen and protect the soul from passional suffering: this was the purpose of every virtue. The four self-corrective natural virtues, drawn from the moral teachings of classical antiquity, were the cardinal virtues of *prudence*, countering impulsiveness; *justice*, countering oppressive force; *temperance*, countering appetitiveness; and *fortitude*, countering timidity. To these, Christianity added three more, the supernatural, or theological, virtues of *Faith*, *Hope*, and *Charity*. These were wholly new and not at all the natural attributes of fallen humanity. Not even the sages of antiquity, Jewish as well as pagan, could deploy them. To these worthies, what constituted Faith would seem a creed completely incredible, Hope an absurd expectation, and Charity, defined as the love not only of one's neighbor but also of one's enemy, a symptom of mental derangement. It was therefore only the gift of divine grace that could produce in a Christian that total psychic transmutation likened to a rebirth experience. These strengths, infused into the soul at baptism, reinforced the cardinal virtues while shielding the vulnerable soul from a host of other dysfunctional emotions.

Listed first, Faith was based on an acceptance of scriptural revelation and the moral and doctrinal teachings of the Church. The first fact that revelation proclaimed is that God made man good but that man willfully contaminated himself and his descendants with a weakness for evil. This "Fall," as narrated in Genesis and the agon of flesh versus spirit that the Apostle Paul describes (Rom. 7.1525; Gal. 5.13–24), seems like a way of addressing a genetic flaw that recent philosophers of mind have analyzed in terms of two cognitive strategies, one impulse driven, the other reflective and analytic (see the "dual-inheritance" theory, discussed in chapter 1). According to this theory, the instinctive responses of the prehuman primate brain continue to conflict with the learned, measured responses of the modern *Homo sapiens* brain. The virtue of faith supplements natural reason by relieving the emotional stress associated with moral uncertainty and with doubts regarding the righteousness of divine providence. Faith assures us that "nothing happens without a reason." It also endorses the truth and efficacy of the two other theological virtues.

Hope directly counters despair, a condition expressed in a number of distressful emotions, such as loneliness, fear of abandonment or confinement, and grief occasioned by the death of loved ones. It counters these by assuring Christians that they belong to the "Communion of Saints." Also visualized as the Mystical Body of Christ, this is the fellowship of the saved in heaven (the Church Triumphant), the eventually saved in purgatory (the Church Suffering), and the believers still on earth (the Church Militant). Hope also relieves the fear that one deserves to be expelled from this communion, abandoned on account of personal guilt. A dread of abandonment is further set to rest by the ritual of confession, a sacrament that entails contrition for one's sins, forgiveness through powers vested in the Church, and retributive acts of penance. Hope finally is invoked to counter one's fears of bodily abandonment in death by representing to the dying the peace of heaven, together with the joyful prospect of reunion with the souls of one's loved ones.

Charity is love (Gk. *agapê*, L. *caritas*). This grace-bestowed strength, which Paul in his listing of these virtues ranked the highest (1 Cor. 13.13), blunts our natural instincts to distrust and repel outsiders and to retaliate against those who have offended us. Its deepest expression, altruism, can even lead individuals to sacrifice their lives for others. The only others one could—and indeed *must*—hate were Satan, his demonic legions, the souls of the damned in hell, and anyone who has made common cause with these, such as witches and sorcerers. Charity thus relieves the mind of anger, nagging resentments, and envy. It also moderates the erotic drive by reminding Christians that the union of man and woman for the purpose of progeny is in partnership with God and that sexual love, *amor*, is a species of higher love, *caritas*. This doctrine, having first evolved during the twelfth century into the cultural phenomenon of courtly love, was later supplemented by Platonic concepts found in the *Symposium* and the *Phaedrus* and yet later established the conventions of "romantic love." Christian culture thus succeeded in blending Hope and Charity into the attenuated emotion of erotic yearning, or a hoping for love.

For all but a small restive minority, this proved a quite satisfying system, a universe of discourses governed by the "Queen of Sciences," theology, a world in which, as Scholastic philosophers claimed, the dictates of reason and the truths of religion were in perfect harmony. But this was also a realm in which some 90 percent of the population was illiterate, civil authority intolerant of dissent, heretics publicly burned alive, and other religions, e.g., Judaism and Islam, demonized. If communities anywhere on earth were preached the Word of God yet refused conversion, the Pope could declare them enemies of God and exhort Christian armies to destroy them, as in the multiple campaigns waged during the four-centuries-long Crusades (1096–1487), a series of disasters only briefly interrupted midway by the Black Death (1346–1353). The final Crusade, the suppression of the heterodox Christian sect the Waldensians, foreshadowed the Wars of Religion that broke out a generation later. This nine-decade conflict (1562–1651) coincided with the European invasion of the continents of the Western Hemisphere, the genocide of native populations, and the beginnings of the African slave trade.

During this period of violent conflict and conquest, a new weapon emerged in the form of the printing press. As one might suppose, it took extreme ingenuity to align Christians' actions with Christian principles, but polemicists on all sides seemed up for the task. The Faith upon which the beautiful structures of Hope and Charity had been erected could only be further undermined by such attempts to use it to justify persecution, genocide, and chattel slavery.

From the seventeenth century on, the availability of relatively inexpensive books led to a gradual increase in literacy and a wider exchange of ideas. With the general revulsion toward the faith-based bigotry of the past came a fresh appreciation of reason as a guide to conduct. Most readers still acknowledged Christian values and honored the virtues of Hope and Charity, but, as a cultural foundation, the traditional Faith was weakened as new criteria for truth were put forward. As natural sciences revived, philosophers no longer felt obliged to bend their conclusions to accommodate religious dogma.

Hope and Charity, as the church taught, were supernatural gifts when aligned to scriptural Faith, but as natural responses they may be understood differently. Hope, whether or not it envisions an otherworldly afterlife, projects the mind into a future time as it aspires to personal survival. Charity, or love, projects the mind into the space around it and, as a prosocial trait, focuses on group survival while ensuring the survival of the species through sexual reproduction.

These two evolutionary traits accomplish their purposes, but not without considerable psychic cost. Faith's narrative of salvation had sought to divert the mind from its discontents, but when this narrative became less plausible, the emotional expense became all too evident. For most, life had never been easy the under the rule of Faith, but, if one imagined life to be a book of problems, one could simply flip to the appendix at the back to find catechetical solutions to every one of them.

Through the eighteenth and nineteenth centuries, however, writings began to appear that represented the human condition without those consolatory answers. Their subject matter increasingly became responses to that fundamental question, "What's the matter?" and took, in effect, a medical approach to a psychosocial malaise. In one way or another, this approach involved a careful observation of symptoms and a review of past circumstances, leading to a diagnosis of causes. The prescribed treatment and prognosis would then extend the curative process into an uncertain future. What did seem certain, though, was that faith healing would no longer produce successful outcomes. Of course, literary therapy did not cure the ailment either, but it could lay bare and illuminate the problems and, equally important, reassure readers that they were not alone. Thus the Republic of Letters offered itself in place of the Communion of Saints.

Both prose fiction and poetry thematized those perennial problems that both science and religion now seemed powerless to address: anxiety, loneliness, grief, the inevitability of death, unsatisfied longings, violence, betrayal, ennui, impotence in the face of injustice. In prose fiction these issues were viewed from the perspectives of different characters; in poetry they were viewed from a single perspective, usually the author's. Despite its singularity, however, the lyrical soliloquy, as Mill defined it, was as multifaceted as the self that pondered the problem at hand. It displayed its complexity by its application of various cognitive modes, e.g., memory (episodic details and semantic knowledge of facts and concepts), immediate perception (in various sense modalities), simulated perception (unbidden and constructed mental imagery), expectation (prediction, planning), and judgment (evaluation of the practical or ethical consequences of an action). Any of these modes could become more complex by simply embedding other modes within it. For example, the episodic memory of a past action might include a judgment on its moral worth. As this example illustrates, different cognitive modes, be they embedded or successive, can clash with one another to produce emotional reactions. Thus a judgment on a remembered action may produce any number of affects—a twinge of remorse, a bittersweet nostalgia, or a glow of satisfaction. Since most modern lyrics are no more than a page in length, we may easily reread one several times at a sitting, allowing the hypotactic complexity of the state of mind it represents to unfold itself gradually.

To a growing literate public, post-Renaissance writing promulgated this complexly examinable self, but it had also had a hand in creating it. What revealed these heretofore unrealized resources of the human mind was silent literacy, the capacity of a solitary reader to use the words of a solitary writer to enact a single, momentarily shared soliloquy. The cultural evolution of the modern self cannot be chronicled by reference to technological innovations or the rise and fall of nations. It is affected by historical forces, certainly, but to trace the genealogy of the self itself, the entity we literates refer to now by the pronoun "I," we must find it in the intimate communion of those two solitaries, writer and reader, at the fragile surface of the printed page.

Appendix

Three Horatian Texts

Epode 13

Horrida tempestas caelum contraxit et imbres
  nivesque deducunt Iovem; nunc mare, nunc siluae
Threicio Aquilone sonant. Rapiamus, amici,
  occasionem de die, dumque virent genua
et decet, obducta solvatur fronte senectus.
  Tu vina Torquato move consule pressa meo.
Cetera mitte loqui. Deus haec fortasse benigna
  reducet in sedem vice. Nunc et Achaemenio
perfundi nardo iuvat et fide Cyllenea
  levare diris pectora sollicitudinibus,
nobilis ut grandi cecinit Centaurus alumno:
  "Invicte, mortalis dea nate puer Thetide,
te manet Assaraci tellus, quam frigida parvi
  findunt Scamandri flumina lubricus et Simois,
unde tibi reditum certo subtemine Parcae
  rupere, nec mater domum caerula te revehet.
illic omne malum vino cantuque levato,
  deformis aegrimoniae dulcibus adloquiis."

Ode 1.7

Laudabunt alii claram Rhodon aut Mytilenen
  aut Epheson bimarisve Corinthi
moenia vel Baccho Thebas vel Apolline Delphos
  insignis aut Thessala Tempe.
Sunt quibus unum opus est intactae Palladis urbem
  carmine perpetuo celebrare et
undique decerptam fronti praeponere olivam.
  Plurimus in Iunonis honorem
aptum dicet equis Argos ditisque Mycenas.
  Me nec tam patiens Lacedaemon
nec tam Larisae percussit campus opimae
  quam domus Albuneae resonantis
et praeceps Anio ac Tiburni lucus et uda
  mobilibus pomaria rivis.
Albus ut obscuro deterget nubila caelo
  saepe Notus neque parturit imbris
perpetuos, sic tu sapiens finire memento
  tristitiam vitaeque labores
molli, Plance, mero, seu te fulgentia signis
  castra tenent seu densa tenebit
Tiburis umbra tui. Teucer Salamina patremque
  cum fugeret, tamen uda Lyaeo
tempora populea fertur vinxisse corona
  sic tristis adfatus amicos:
"Quo nos cumque feret melior fortuna parente,
  ibimus, O socii comitesque,
nil desperandum Teucro duce et auspice Teucro,
  certus enim promisit Apollo
ambiguam tellure nova Salamina futuram.
  O fortes peioraque passi
mecum saepe viri, nunc vino pellite curas;
  cras ingens iterabimus aequor."

Ode 1.9

Vides ut alta stet nive candidum
Soracte, nec iam sustineant onus
    silvae laborantes, geluque
        flumina constiterint acuto.

Dissolve frigus, ligna super foco
large reponens, atque benignius
    deprome quadrimum Sabina,
        O Thaliarche, merum diota.

Permitte divis cetera, qui simul
stravere ventos aequore fervido
    deproeliantis, nec cupressi
        nec veteres agitantur orni.

Quid sit futurum cras fuge quaerere, et
quem fors dierum cumque dabit lucro
    appone, nec dulcis amores
        sperne puer neque tu choreas,

donec virenti canities abest
morosa. Nunc et campus et areae
    lenesque sub noctem susurri
        composita repetantur hora,

nunc et latentis proditor intimo
gratus puellae risus ab angulo
    pignusque dereptum lacertis
        aut digito male pertinaci.

Notes

1. Innovating Ourselves

1. The German humanist Christoph Cellarius (1638–1707) is credited with popularizing this tripartite periodization. The title of his major work, translated from Latin to English, reads: *Universal History Divided Into an Ancient, Medieval, and New Period* (1683).
2. Line 5 states that the "great series of ages is born anew" (*magnus ab integro saeclorum nascitur ordo*), and the rest of the text makes it clear that the ancient pattern will now be reproduced. Vergil did not, however, mean anything like the concept of "eternal return" that Nietzsche toyed with. The motto that is found on the back of every U.S. one-dollar bill has "*seclorum*" rather than "*saeclorum*" or "*saeculorum*" ("of ages") because Charles Thomson, designer of the Great Seal and good Latinist, wanted to make it resound like the last half of a dactylic hexameter line and so used a standard variant of the otherwise unmetrical plural possessive of *saeculum* to do so.
3. "An individual has a theory of mind if he imputes mental states to himself and others. A system of inferences of this kind is properly viewed as a theory because such states are not directly observable, and the system can be used to make predictions about the behavior of others" (Premack and Woodruff 1978, 515). They argued that chimps do possess this cognitive skill, a claim that has been generally confirmed by other investigators, though to what extent nonhuman primates use this skill is still a debated question. See Call and Tomasello (2008) and Rosati, Santos, and Hare (2010).
4. Being multimodal sets of generic features, they are more properly "perceptual schemas," (Jeannerod et al. 1995), but I will use the term "image schemas" because

visual perception is our predominant link to our habitat. See also Mandler (2004) and Barsalou (2008, 2009).

5. In the late 1990s a particular gene, *FOXP2*, was implicated in speech production. A family in England had been found with a hereditary language disorder and, when their DNA was analyzed, it was revealed that their *FOXP2* had become defective. Estimating that it first appeared in its current form in anatomically modern humans about 100,000 years ago, some conjectured that this was the genetic key to Chomsky's hypothetical language-acquisition device (LAD). Despite the caution of their peers, some scientists began referring to it as the "language gene," even the "grammar gene." Whatever its function—and most authorities now assume it is associated with the coordination of mouth and tongue muscles—the discovery announced in 2007 that the Neanderthals possessed the identical gene sent linguistic speculators back to the drawing board: if *FOXP2* makes language possible, it would have theoretically done so for the last common ancestor of humans and Neanderthals before their mutual divergence 300,000 to 400,000 years ago.
6. For other reactions to *The Symbolic Species*, see Hudson (2001) and Sonesson (2006).
7. The coincidence of play behavior and early-childhood language acquisition suggests a cognitive/developmental link. See McCune-Nicolich (1981).
8. In his article "Some Universals of Grammar with Particular Reference to the Order of Meaningful Elements" (1960), reprinted in *On Language* (1990, 40–70), Greenberg did not, of course, claim that all languages have the same sequence of syntactical units. He argued instead that if a given language tended to use a particular order of, say, subject, verb, and object, it would tend to use another order when arranging nouns and adjectives. For example, according to his fifth universal, "if a language has dominant SOV [subject–object–verb] order and the genitive follows the governing noun, then the adjective likewise follows the noun." These correlations he called "implicational universals." Despite their differences, he maintained, all languages employ rule-governed sequences.
9. In speech-act theory, "locutionary" refers to the form and content of a meaningful utterance. "Illocutionary" refers to the motivation of a speaker as revealed in a given utterance. Searle's five kinds of speech acts are defined in terms of their illocutionary force—what they are intended to do. "Perlocutionary" refers to the effect, intentional or not, that the utterance has on hearers, e.g., they feel insulted, encouraged, persuaded, etc. By "responses" I mean both their physical and their subsequent verbal reactions as the erstwhile *You* role shifts into an *I*.
10. I'm aware of the awkwardness of these terms and, having been converted into nouns, their ungrammaticality. In self-defense I would say that these terms retain a familiar concreteness that a more analytical phraseology would lose. See also Collins (1996; 2007, xii).
11. For example: "(I say) I do not feel lonely" (expressive); "(I say) Do not touch that apple" (directive); "(I say) The hunter did not bring back any meat" (assertive). Henk Haverkate (1990, 89) claimed that declaratives cannot be ironic, thus im-

plying that they cannot be construed in the negative, because they use "performative formulas, to which no sincerity condition applies."

12. For an interesting application of Bateson's play theory to ancient Greek comedy, see Kidd (2014).
13. I borrowed this scenario from Robert Frost's poem "The Wood-Pile," in which the poet represents in his sequence of cognitive events the precise way the visual brain parses reality through oculomotor shifts.
14. In addition to visual perspective taking, we should note kinetic and emotional mirroring. Kinetic mirroring occurs when our mirror-neuron system prompts us to mime the actions of others. Emotional mirroring occurs when we react to affective signs, e.g., smiles, grimaces, weeping, looks of terror, etc. These kinetic and emotional reactions are for the most part preattentive.

2. Narrative Memory

1. Exceptions to this characterization are traumatic and "flashbulb" memories. Traumatic memories are those that record moments of extreme danger, when all one's senses are mobilized and the visual field is perceived in "preternatural" detail. (But see Stetson, Fiesta, and Eagleman [2007], whose experiments led them to conclude that perceptual acuity is not an online experience but rather a function of retrospection.) Flashbulb memories record the minute circumstances associated with the moment some surprising and consequential public occurrence was announced, e.g., an assassination or major catastrophe.
2. Cf. William James's fringe of forgotten time (James [1890] 1950, 258–262). As a detached portion of one's autobiographical continuum, an episode is nevertheless not alone. We understand its meaning, or feel we understand it, because all the other potentially retrievable episodes that constitute our selfhood provide its implicit context.
3. As a retrospective tactic, the othering of the self may be a way of achieving distance from some stressful episode in one's past, a "screen memory," as Freud would put it: that person is a third person and not really the *I* who now looks back at that episode. Third-person experiences, i.e., events in which one seems to observe oneself as others might observe one, we may link to the "perspective taking" that Brian MacWhinney (2001) said was an advantage enhanced by grammatical language. But this retrieval experience seems not at all that advantageous, since it is associated with moments of extreme self-consciousness, social phobic reactions, and, in terms of gender, female objectification (Rice and Rubin 2009). For such reasons it may shed light on the doppelgänger phenomenon. Nigro and Neisser (1983) call the first-person perspective "field memory" and the third-person perspective "observer memory." I prefer, however, Rice and Rubin's simpler, clearer terms "first-person" and "third-person" perspectives.
4. If, as I am suggesting, we regard episodic memory as the basis of narrative, we may have a clue to that enigmatic passage in Aristotle's *Poetics*. Pity is an emotion

we feel for others, fear an emotion we feel for ourselves. Since we are not actually implicated in the misfortunes portrayed in the drama or narrative, we can only feel fear by entering into the perspective of the endangered hero. The latter shift replicates the way the self is objectified in distressful episodic recall but without the residual regret or shame. The shift of perspective that comedic narratives promote is that between the emotions of gloating and embarrassment, gloating from the safe perspective of the superior first-person viewer and embarrassment when this viewer identifies with the comedic third person.

5. This one-dimensional linear concept Bergson critiqued as an unwarranted spatialization of time. It was not merely a different *kind* of time but a logical misconstrual of time, the reality of which for him was "pure duration," the unquantifiable ongoing present. It is what his contemporary Mallarmé (1967) celebrated as "Le vierge, le vivace et le bel aujourd'hui."
6. This clever insight has been attributed to a number of sources from Einstein to Woody Allen. The earliest source that I've found is a Bergson essay first published in Sweden in 1930, "Le possible et le réel," reprinted in a collection of his essays, *La pensée et le mouvant* (1934, 58): "À quoi sert le temps ? (Je parle du temps réel, concret, et non pas de ce temps abstrait qui n'est qu'une quatrième dimension de l'espace.) Tel fut jadis le point de depart de mes réflexions. Il y a quelque cinquante ans, j'étais fort attaché à la philosophie de Spencer. Je m'aperçus, un beau jour, que le temps n'y servait à rien, qu'il ne faisait rien. Or ce qui ne fait rien n'est rien. Pourtant, me disais-je, le temps est quelque chose. Donc il agit. Que peut-il bien faire ? Le simple bon sens répondait : *le temps est ce qui empêche que tout soit donné tout d'un coup*" (italics added).
7. This distinction corresponds to what the Russian formalists called *fabula* and *syuzhet*. It also reflects the influence of Emile Benveniste ([1966] 1973), who maintained that oral communication was marked by the copresence of the first and second persons in the act of speech (*discours*), whereas writing introduces the notion of an impersonal time, story (*histoire*). Unfortunately, the two terms have translated poorly into English, in which "discourse" suggests a formal lecture and "story" an informal telling. Though I have retained the narratological terms as they are now used, I wish that Chatman (1978) had translated *discours* as "utterance" or "tale" and *histoire* simply as "history."
8. Archaeological evidence indicates that in Neolithic and Bronze Age settlements fearsome animals, e.g., snakes, lions, bulls, and eagles, were often the outward form of gods. It is also interesting to note that in later iconography the dragon, which was represented as a winged serpent with horns like a bull and jaws like a lion, became the principal image of dread.
9. Etiology, broadly defined, is a discourse on origins. Etiological myths seek to explain how things got to be the way they are, and, as such, they are not limited to cosmogonic or creation myths.
10. Indo-European etymology of words meaning "show" and "tell" reveals how closely associated the two actions once seemed. The Greek verb *deiknumi* (show) and the Latin *dicere* (tell) have the same IE root *deik-* (Pokorny [1959] 1989, 188–189), as

do the English *teach* (OE *tæcan*) and the German *zeigen* (show). (I do not find support for including the Latin *docere* [show] as deriving from this root, but false cognates do tend to attract by semantic affinity.) The Greek verb *phrazô*, which came to mean "tell," meant "show" in Homeric Greek. Its IE root *phrad-* may have meant "tell through pointing out" (Mourelatos 1965). In Classical Greek the common word for "guide" was *phrastêr.*

11. Giants and dragons are common figures in ancient narratives. Most persons until recent centuries, it is safe to say, believed that tales of such beings were based on existing indexical evidence. One way to explain why otherwise rational persons from Herodotus and Aristotle to Saxo Germanicus and Cotton Mather believed they existed is to point out that huge bones had been found that seemed to belong to no known race of humans, beasts, or birds. In classical Greece two famous relics were the gigantic bones said to be those of Orestes and Theseus and the huge shoulder bone of Pelops. Assuming they actually existed, they were most likely those of prehistoric animals, e.g., mammoths. Vergil alludes to this belief when, referring to the site of the 42 BCE Battle of Philippi, he predicts that the time will come when some plowman will unearth rusted weapons and empty helmets and will marvel at the massive bones (*grandia . . . ossa*) exhumed from those graves (*Georgics* 1.493–497). As for dragons, the finding of saurian skeletons from the Jurassic period would have easily been so interpreted (Mayor 2011).
12. The early books of the Bible also tell a number of ruin-inspired narratives. The ruins of a great ziggurat, in Babylon, possibly destroyed by the conquering Hittites at about 1530 BCE (Albright 1939, 1957) led the writers of Genesis to compose the legend of the Tower of Babel as a cautionary tale of human arrogance and an etiological myth of linguistic diversity. The writers of the book of Joshua read into the scattered stones of Jericho and of Ai (which meant literally "ruin") evidence of their nation's history of god-favored invincibility.
13. For some quite perceptive approaches to the relation of language and language origins to wayfinding and landscape mapping, see Adornetti (2014a), Ferretti et al. (2013), and Meschiari (2009a, 2009b).
14. Cf. this excerpt from Barack Obama's second inaugural address: "We, the people, declare today that the most evident of truths—that all of us are created equal—is the star that guides us still; just as it guided our forebears through Seneca Falls, and Selma, and Stonewall; just as it guided all those men and women, sung and unsung, who left footprints along this great Mall, to hear a preacher say that we cannot walk alone; to hear a King proclaim that our individual freedom is inextricably bound to the freedom of every soul on Earth." Seneca Falls, New York, was the site of the women's rights convention of 1848; Selma, Alabama, was the starting point of the civil rights march to Montgomery led by Martin Luther King Jr. in 1965; Stonewall alludes to the tavern in Greenwich Village, New York City, where the gay patrons made a violent stand against police harassment in 1969; and "this Mall" refers to site of this speech and to the speech Martin Luther King Jr. made there in 1963.

15. We find this structure in shorter form in folktales and folk ballads. For a more extensive discussion of the adventure mnemonic see Collins (2013, 193–197).
16. That word "character" is misapplied to a person in an oral-narrative tradition. In this usage, the word itself belongs to a recent literate and more introspective age, the late seventeenth century. It was then that a number of writers became fascinated with the small book, *Characters*, by Theophrastus, a disciple of Aristotle. We do not know if Theophrastus gave his book that title, but we know that in the earliest listings it is referred to as *Ethical Characters*. We do know, however, that in Greek literature the word "character" did not mean a particular person in a piece of fiction but rather a distinctive mark, an ineradicable inscription engraved on a surface. A person is not a character—a person has a character or, rather, a set of characters, i.e., behavioral traits. We also know that in the book of that name Theophrastus never applies the word "character" to any of the thirty personality types he describes.

3. The Dancing, Singing Daughters of Memory

1. That suffix, *-sunê*, was simply a handy way the Greeks had to take an adjective, e.g., *mnêmôn*, and convert it into an abstract noun. For example, *dikaios*, "just," became *dikaiosunê*, "justice," and *sôphrôn*, a compound of *sôs* (healthy) and *phrên*, became *sôphrosunê*, healthy-mindedness, the Hellenic ideal of moderation and discretion.
2. Cf. the updated saintly responsibilities: St. Bernardine of Sienna (fifteenth c.), patron of advertising executives; St. Bona of Pisa (twelfth c.), patron of flight attendants; St. Drogo (twelfth c.), patron of coffeehouse owners; St. Apollonia of Alexandria (third c.), patron of dental technicians; St. Eligius of Noyon (seventh c.), patron of electricians and taxi drivers; and St. Clare of Assisi (thirteenth c.), patron of television writers.
3. By mentioning the Eleutherian Hills, he may be alluding to the cult of Dionysus Eleuthereus, Dionysus the Liberator, and suggesting that the delightful forgetfulness the Muses provide is, like the effect of wine, a temporary liberation from anxiety.
4. Leavitt and Christenfeld (2011) tested the responses of readers to texts with surprise endings, (1) some without foreknowledge of the ending, (2) some with that ending revealed at the beginning as though it were part of the text, and (3) some with an editorial introduction stating how the plot ends. What they discovered—a surprise ending for them—was that the third condition enhanced the experience "by actually increasing tension. [For example] knowing the ending of *Oedipus Rex* may heighten the pleasurable tension caused by the disparity in knowledge between the omniscient reader and the character marching to his doom. This notion is consistent with the assertion that stories can be reread with no diminution of suspense. . . . Although our results suggest that people are wasting their time avoiding spoilers, our data do not suggest that authors err by keeping things hidden" (1,153).

5. One nineteenth-century philosopher, Ludwig Noiré, proposed that language evolved from what he called "synergastic" vocalizations, a theory that the arch-anti-Darwinian linguist Friedrich Max Müller (1868) reframed and parodied as the "yo-he-ho" theory. Though it is unlikely that grammatical language was used before 200,000 years ago, it is quite possible that archaic *Homo sapiens* could have uttered a range of distinct phonemes that in some groups emerged as protolanguage (Wray 2002, Mithen 2006).
6. The dyadic patterns we find in language and music may be regarded as relatively recent adaptations of the much older, simpler coordinations. The inability to "walk and chew gum at the same time" might have applied—except for the gum reference—to early bipedal primates. In terms of developmental stages, modern human children also need to master multiple motor tasks. On a winter morning a six-year-old may need to stop walking in order to use her hands to button her coat. I haven't tested this on a population of six-year-olds; this, I confess, is an anecdote drawn from my experience once walking my daughter to her school bus stop. She didn't then appreciate my pointing out her inability but now, after forty years, will probably forgive my impertinence.
7. At this point I should say a word concerning Steven Brown's (2000) evolutionary theory of language and music. I agree that music is not the origin of language, nor is language the origin of music. I also agree with Brown's linking of music to prelinguistic emotive vocalization. Unlike Brown, however, I do not posit an ancestral state in which emotional expression and referentiality were somehow undifferentiated in a protolanguage he calls "musilanguage."
8. This diagram is a simplified representation of a much more complex series of evolutionary innovations both biological and cultural. Obviously, instrumental music is not pure melody without rhythmic structure, nor is dance necessarily devoid of melodic or instrumental accompaniment. If my primary concern had been tracing the evolution of melodic or rhythmic structure in instrumental music and dance, I would have had to construct different diagrams. My objective is instead to tease apart the various strands of sound and sight that are woven together in the performance of song.
9. The semioticians of music that I have so far encountered have tended to pitch their theorizing at a fine-grained level of detail, often drawing on divergent theories of sign, e.g., those of Greimas, Saussure, Barthes, Chomsky, and Peirce. The Peircean analysis that I apply here, admittedly elemental, is, I think, adequate to my purpose.
10. To my knowledge, Nicholas Cook (2014) was the first to notice this resemblance. Pinker's provocative assessment of music as "auditory cheesecake" ([1997] 2009, 534) rather than an evolutionary adaptation, has incited—and I use that word advisedly—much research into the cognitive and affective neuroscience of music in the new century. Fitch (2006, 199–200) responded to Pinker's claim by pointing out that the fossil evidence suggests that human anatomy could have produced musical sounds before the evolutionary split between Neanderthals and *Homo sapiens sapiens*; that bone flutes estimated to be over 40,000 years old

have been unearthed; and that, "unlike cheesecake," music and dance are behaviors enjoyed in all human communities and are as universal as language. Moreover, since music making, being a loud activity, would have attracted predators and was energy expensive to produce, it therefore had to have had offsetting benefits for our Paleolithic ancestors.

11. Jordania (2006) further speculates that one of the selective advantages of bipedalism was that an erect animal appears larger to four-legged predators. Apes when they engage in aggressive display stand erect, but this posture is only occasional. Our weaker, gracile ancestors had to develop other strategies to scavenge meat on the open savannas. "Rhythmically well-organized loud noise" (304), vocally projected, would also have been used to scare off large cats and may have been preadaptive to human speech.
12. Brain scans indicate that when viewers observe rhythmic processions or dance performances, the parts of the brain in both hemispheres that they would use themselves to execute these movements become activated. Interestingly, when persons expert in these same dance movements view them, their left motor cortex becomes more activated than their right, suggesting that they regard these actions as more possible and object directed (Liew et al. 2013). This is consistent with the evidence that expert musicians, unlike nonexperts, process musical sound predominantly in the left hemisphere.
13. In several important respects, Vygotsky's description of inner speech resembles Alison Wray's formulaic utterances (Wray 1998, Wray and Grace 2007, see also earlier in this chapter). As a protolinguistic vocabulary, the latter were single multisyllabic phrases that functioned as the directives within small social units. They were language for insiders—esoteric language. Formulaic commands in *our* language now take the form of agglutinative words constituting predicates, e.g., "wipeyourfeet," "comesitdownandeat," and "gotobednow." These casual speech idioms also carry the meanings associated over time by the addressees with the addresser, a further aspect of in-group semantics. Inner speech is therefore morphologically similar to formulaic speech but is intended for an even smaller audience—oneself alone.
14. No one who has seen Wilder Penfield's schematic drawing of the primary motor cortex as an upside-down homunculus will forget its huge hand, bulging eye, pouting lips, and protruding tongue. The area now found to represent the larynx lies at the very base of that vertical strip, just below the area that represents the movements of the tongue. Named in honor of the Italian anatomist Luigi Rolando, it is called the Rolandic operculum, or "little lid," and is slightly tucked under each temporal lobe.
15. Reviewers did not take kindly to what David had claimed was a radical and revolutionary approach to Homeric scholarship. Perhaps he overstated his case, but their unwillingness to entertain evidence drawn from anthropological parallels and the personal experience of reading the sounds of Homer's verses reveals the shortcomings of a discipline wary of venturing beyond textual proofs and philological dogma. How, one wonders, would they have reviewed Jane Harrison, Milman

Parry, T. B. L. Webster, Eric Havelock, or E. R. Dodds? Yeats's poem "The Scholars" is always worth rereading in this regard.

16. Inner speech may, in fact, be suppressed by any technique that deactivates the muscles used in overt speech. Thus, deep regularized breathing, holding one's tongue between one's teeth, or pressing its tip against one's hard palate will inhibit the inner speech impulse. Even chewing gum or tightly pressing one's lips together will have a certain beneficial effect. We can always tell when a conversation has turned unproductive if, as one person speaks, the other clamps his mouth tightly shut.
17. The poems I cite are Shelley's "Music When Soft Voices Die" and Keats's "Ode to a Nightingale."

4. Visual Instruments of Memory

1. This passage appears in his one extant book, *A Description of Greece* (10.24.6). Written ca. 150 CE, it is entitled a *periêgêsis*, a leading-around (cf. *diêgêsis*, a "leading-through"), so it should be translated as "A Tour Guide to Greece." The stone in question is apparently not the "navel stone," which was believed to mark the absolute central point on the surface of the earth. This was displayed in the Temple of Apollo (10.16.3). There were, and still are, plentiful stones in the soil around Delphi and Parnassus, more than enough for the Greek Noah, Deucalion, and his wife Pyrrha to throw over their shoulders, thereby magically creating a new race of humans after the Great Flood. Homer referred to "rocky Pytho" in *Iliad* 2.519.
2. If there is any inference to be drawn from this chronological fable, it would be that the Olympian pantheon came into being with the rise of Mycenaean power in late Bronze Age Greece and planted its cultic roots deep enough to survive the destruction of that culture circa 1150 BCE.
3. Etiological myths often promote ethnocentric biases. The heir apparent to the kingdom of Deucalion was Hellen, ancestor of the Hellenes. Cf. the favorite son of Noah, Shem, the ancestor of the Semites.
4. As fanciful as the Heraclid myth was, the founding myths of Thebes and Athens were arguably more bizarre. Thebans and Athenians based their land rights on the ultimate aboriginal myth, autochthony, the notion that their ancestors had grown from seeds sown directly in the soil. Isocrates in his *Fourth Oration*, the *Panegyric* (380 BCE), restates the traditional claims of Athens. Plato in book 3 of his *Republic* fashions his own "noble lie" of autochthony. See Collins (2007, 247–248n51). See also Detienne and Jones (2001).
5. The earliest writing, pictographic, ideographic, and syllabographic, would naturally separate each sign from the other, a practice that Western alphabetic script continued at the level of the phoneme-signifying letter without separating word from word. Word separation, as a consistent chirographic practice, began with Irish monks in the seventh century CE but did not fully catch on elsewhere till the tenth century (Saenger 1996).

6. The inscription beneath the pyramid on the Great Seal of the United States of America (on the back of the one dollar bill) proclaims a "*novus ordo seclorum*," a "new order of ages." It also says "*annuit cœptis*," "He (God) has given his approval [literally, 'nodded assent'] to our beginnings," a recognition that *ordo* meant the arrangement of a row, in this case the starting of a sequence of temporal units. In Vergil's so-called messianic poem, *Eclogue* 4, from which the *novus ordo seclorum* is derived, the poet prophesies a glorious future, a vision that closes with another textile allusion, this time to the spinning goddesses, the three Parcae, or Fates: " 'Talia saecla,' suis dixerunt, 'currite' fusis / concordes stabili fatorum numine Parcae' " (" 'Ages like these, run on,' to their spindles the Parcae have said in accord with the unshakable dictate [nod] of the fates") (46–47).
7. The earliest Greek references to writing as weaving come from the poems of Pindar and Bachylides (Scheid and Svenbro 1996, 111–119). The professional chanters of epic were known as *rhapsodes*, i.e., song stitchers, suggesting that each performance was a set of artfully sequenced segments cut from some immense tapestry. Cicero referred to the weaving words into his writings (*Epistulae ad familiares* 5.12, 9.21), and Quintilian used the noun *textum* (weave) to refer to a writer's style (*Institutio oratoria* 9.4.13, 17). Since a "line" derives from the Latin word *linea*, a flaxen thread used in weaving linen, a phrase such as "this text follows a linear order" reminds us how the craft of weaving is implicated (another textile word!) with reading and writing.
8. Perceiving individual facial differences is a very complex sort of visual scanning. Persons living in racially homogeneous societies can recognize among their kin and neighbors a great variety of physiognomic differences, but when they find themselves visiting other similarly homogeneous societies, they have difficulty telling persons apart: they all "look alike." Scripts are very much like faces in that respect. If we are unfamiliar with a given script, writing is a mystifying pattern of designs. Our eyes cannot easily pick out the small variations of shape. Consider, for example, the extra detail that in English capitals changes a P to an R, a C to a G, a P to a B, an O to a Q, an F to an E, and an I to a J or an L, distinctions that were even more difficult to perceive when handwritten and especially in cursive script.
9. Contrary to a common interpretation, Odysseus does not visit the underworld. Instead, he travels to a special place and performs a *nekuia*, a rite that calls up the dead and allows the living to interrogate them. This is distinct from a descent into the underworld, a *katabasis*, such as Vergil's Aeneas and Dante undertake.
10. The Statue of Liberty in New York Harbor preserves this relation of iconic and symbolic signs, the words of Emma Lazarus's poem represented as the statue's words: " 'Keep, ancient lands, your storied pomp!' cries she / With silent lips. 'Give me your tired, your poor, / Your huddled masses yearning to breathe free, / The wretched refuse of your teeming shore. / Send these, the homeless, tempest-tost to me, / I lift my lamp beside the golden door!' " The gate to Dante's

Hell "speaks" also: "Per me si va ne la città dolente, / per me si va ne l'etterno dolore . . . Lasciate ogni speranza, voi ch'intrate."

11. Graffiti is an ancient kind of unofficial public writing that is cheaply produced (scratched or drawn with paint or charcoal). Though it is often defamatory or crudely humorous, it is as concerned with attracting readers as any monumental inscription. Here, too, proper names are memorialized. Though it is usually in the assertive mode, the spontaneous circumstances of its composition often prompt expressives.
12. The prosody of this tiny masterpiece underscores this emotional sequence. The first line, written in dactylic hexameter, begins with five long syllables (a variation allowed by the meter), has its caesura in its traditional place (following the first long syllable of the third foot), then ends the line with the normative three dactyls and one trochee (or spondee). It ends with *têide*, "here," before launching into the pentameter with its two dactyls (the second foot a substituted spondee), the third-foot spondee split by its obligatory caesura, finally concluding with the—also obligatory—two anapests. The form of the elegiac couplet, though later put to many other uses, was a meter said to derive from funeral laments. Simonides seems to have drawn on this old tradition in shaping this couplet: the energy and freedom of the epic hexameter in the first line, a continuation of that thrust in the first two feet of the pentameter, the climax and stasis of the central spondee, concluding resignedly with a peripeteia, i.e., the reversed dactyls in the two anapests that form the ending.
13. Besides the transliterated Greek, added in brackets, I have slightly modified Svenbro's English translation for the sake of clarity and readability.
14. Artemon, the editor of Aristotle's letters (now lost), said, "a letter ought to be written in the same manner as a dialogue . . . as one of two sides of that dialogue" (Pseudo-Demetrius of Phalernum, *On Style*).
15. I have reduced this graph to its simplest form, so I have not tried to account for an ongoing correspondence. If I had, I would have divided *B* and *b* into a number of sectors representing return letters to which the current letter might refer. Such a graph would be required if one were to analyze most letters in an epistolary novel.
16. This pretense of presentness has now been masked by the "Dear ———" convention. As for the valediction, it was only one or two centuries ago that even family members would close a letter with such reminders of the ever-present status of their relationship with elaborate phrases such as "I am and remain yours in truth and hope" or "Believe me to be your ever respectful son ———."
17. Not all languages have these specific tenses. For example, while the future perfect is common in English and Latin, it is relatively rare in Classical Greek.
18. Of course, if the intention of the letter writer were to sever the relationship, the *F* interval of reception would not be linked with *D*. If any reference to the addresser's present-tense condition were made, it would stress her current separate status, as: "As you read this, understand that I'm far away and happily through with you."
19. For an excellent discussion of epistolary time, see Altman (1982, 117–142). See also Rosenmeyer (2001, 74–80).

5. Poets' Play and Plato's Poetics

1. The "god" (*theos*) in each instance is unspecified. Since it is masculine, it cannot be a Muse. The songs that Phemius sings, provided for him by this "*theos*," were *oimai*, variously translated as "paths," "melodies," and "lays" (see Ford 1992, Nagy 1996). I translated it "passages" because we still refer to verbal sequences as though we were moving through them. Elizabeth Minchin (2008, 28n84) speculates that the archaic Greeks were using a metaphor similar to that of the Australian Aboriginal pathways known as "songlines." On the question of Phemios's sources, see Penelope Murray's (2005) interpretation of "dual motivation." I will comment further on *thumos* and *phrenes* in chapter 7.
2. Whether Aristotle meant by "reading" a silent or a spoken process, his reference here to *enargeia* can only refer to a simulation of the oral performance through auditory and visual imagination. For his treatment of *enargeia*, see *Rhetoric* 1411b and *Poetics* 1455a22–26. See also Collins (1991b, 123–129).
3. When an animal selects an object and engages with it as though testing its properties or seeking novel stimuli, this behavior is referred to as *exploratory* play and is considered a subset of object play. For humans, exploratory play ranges from creating sand castles and solving crossword puzzles to formulating scientific hypotheses.
4. No species has developed the knack of playful mimicry to the same degree as humans, who regularly combine locomotor, object, and social play with imitative play especially in the form of role-playing. By "imitative play" I do not mean a young animal's imitations of an adult skill, the learning process by which it practices a parent's motor routines, such as swimming, browsing, tracking prey, vocalization, etc. I mean "playing" *someone*, temporarily assuming another's identity, a skill human children excel in but that chimps, for example, find extremely challenging.
5. Note that these English "pl-" words, *simple, complex, explicit, implicit, explicate*, and *implicate*, all of Latin origin, share the root meaning "to fold or to braid," a textile metaphor often associated with carefully chosen speech.
6. This process, known as "spreading activation," is a valuable concept in that it permits us to view both language and cognition as associative structures that stimulate attention outwardly from some activated center. In language this is relevant to semantic difference and category formation. In cognition, it is relevant to visual perception from retinal input to early visual organization in the primary visual cortex and to subsequent fine tuning in the extrastriate cortex. It is also useful in understanding associative memory networks. See Collins and Loftus (1975) and Deane (1992).
7. Not all epithets our metaphoric. Some focus on one feature of a noun: "swift-footed" (Achilles), "wind-footed" (Iris), "fair-haired" (Demeter), "great-hearted" (Odysseus), and "gray-eyed" (Athena). These epithets function as metonyms in that they isolate one detail, e.g., feet, hair, heart, and eyes, and invite us to contemplate the entire person beginning with that specific feature.

8. Besides Hesiod, Solon had said that "singers tell many lies" (fragment 29W), and Pindar had judged lying to be virtually inextricable from Homer's masterful art (*Nemean Ode* 7.20–27).
9. Those references to dogs and poverty in the *Republic* bring to mind that early disciple and associate of Socrates, Antisthenes, founder of the School of Cynics, who is mentioned as present at the death of the master. As for the quotes, no one has yet found them in the work of any known poet. Might one not speculate that these came from the works of Meletus and that these verses might have been part of that poet's indignant counterattack on this impudent gadfly of a philosopher?
10. Gorgias's reference to "plaything" (*paignion*) appears in the conclusion to his *Encomium of Helen*, a speech that sets out to prove that Helen was wholly blameless for the catastrophic Trojan War. This *jeu d'esprit* bears about the same resemblance to the Homeric epic as does Offenbach's *La belle Hélène*. His nimble statement on tragedy is listed as fragment 23 in Diehl and Kranz's collection of pre-Socratic fragments.
11. How serious Plato is in this mythmaking is questionable. If completely serious, he implies his city will be inhabited by a sheeplike flock of dolts. Philosopher-kings may have been exempt from the rule of *alêtheia* as long they thought their lies would prove beneficial to their subjects, but in no way were poets and sophists granted a similar waiver. These entertainers were, after all, motivated only by their desire for celebrity and for the personal wealth that flowed from it, so, as we say now, if you could see their lips moving, you could be sure they were lying. If, on the other hand, he is indulging in an Aristophanic flight of fancy, then this myth exists merely to amuse Glaucon (and the reader) or to satirize Athenian chauvinism.
12. Havelock (1963, 108–109) identifies the practical, as well as the ideological, function of poets in an oral society: a king's decrees were versified by poets in formats that made these dicta memorable for the people.
13. The scepter that the Muses gave to Hesiod symbolized the importance of his and other poets' role as authoritative speakers whose words must be heard and heeded. This staff was traditionally held by a speaker at a public assembly, often passed to him by a herald and from him to other speakers in turn. It was also used by rhapsodes during epic recitations (Stoddard 2003; Collins 1996, 29–35). Homer's kings kept their own scepters for addressing assemblies. I should note in passing that Hesiod's "kings" were not the blend of Bronze Age potentates and Iron Age warlords that Homer celebrated but rather tribal chieftains respected as judges and guardians of religious traditions (cf. the biblical Samuel).
14. It is interesting to note that Plato reapplied the phrase he had used in dismissing Ion, "*pantodapos gignê*" ("you become all sorts of things"): in the *Republic* the itinerant poet is clever enough "*pantodapon gignesthai*" ("to become all sorts of things").
15. This phrase is inscribed on a statuette found on the Athenian acropolis dating from the end of the sixth century: "To whoever asks me I reply with the same answer" (Svenbro 1993, 29).
16. On these issues, see Clay (2000).

17. Aristotle's version of this triad, as outlined in the *Poetics*, begins by assuming that, in effect, everything said by storytellers or poets is role-playing (*mimêsis*). Any first-person narrative is a copy or representation of a series of events, and, if only told in words, its persons, settings, and actions are reproduced as mental images by audience members. If acted out on stage, this copy is perceived both auditorily and visually. If partly narrated and partly acted out, mainly through vocal imitation, the copy would be partly imagined, partly perceived.
18. Goethe accepted the Platonic trichotomy of genres but, like many of his romantic contemporaries, reconceptualized it. Epic was now defined as pure (*klar*) narrative, lyric was marked by enthusiastic excitement, and drama by the acting out of personae. This suggests three stances toward the content of discourse: objective, subjective, and intersubjective. Goethe developed this theory of genre late in his career, appending it in the form of notes to his *West-östlicher Divan*, a book of lyrics influenced by the Persian poet Hafez and published in 1819.
19. *Akraton*, by the way, does not appear in the Platonic passage cited by Genette. Examples by other authors: "In book III of Plato's *Republic*, Socrates speaks of *pure diegesis* when the poet represents the action in his own voice only. In the *mixed mode* of the epic, the poet combines the authorial descriptions and comments with mimetic elements, i.e., direct speech representing the characters' speeches. And when the poet completely effaces his own voice and represents the action in the imitated voices of characters only, this is called *pure mimesis*, to be found in drama" (Berns 2009, 375; emphasis mine). "Plato sees pure diegesis as the only legitimate mode and condemns dramatists and epic poets for their theatrical bent" (Fludernik 2009, 151). See also Chatman (1975, 110).
20. André Gaudreault (1989), better known as a film theorist, pointed out that for Plato *diêgêsis* and *mimêsis* are not dichotomous because all the genres he comments on are varieties of narrative (*diêgêsis*). Moreover, the "mixed" and the "unmixed" (397d) is a second typology clearly distinct from the triadic typology of genres (392d). See also Halliwell (2002, 2009) for illuminating insights into this misinterpretation.
21. See also *Republic* 10.605c–d, 10.607a; and *Theaetetus* 152e, as well as Aristotle's *Poetics* 1448b34–1449a2.
22. In the *Iliad*: Odysseus (11.401–10), Menelaus (17.89–105), Agenor (21.553–70), Hector (22.91–130), and Achilles (22.385). In the *Odyssey*: Odysseus (9.299–302, 20.9–24). See Collins (1996, 75–82) for an extended discussion of these passages. Later Greek and Latin examples of self-address that place the speaker (or author) in the vocative: Theocritus, *Idyll* 11.72–79; Catullus 8, 51, 76, 79; Vergil, *Eclogue* 2.69–73 (an adaptation of Theocritus, *Id.* 11); and Propertius 2.8.17.
23. For a further discussion of "sceptered speech," see Collins (1996, 29–35).

6. Writing for the Voice

1. Cf. Goethe's *Faust*, part 1, act 1. When Wagner, his *famulus* (the medieval equivalent to a graduate research assistant), hears what he thinks is Faust's reading

aloud a passage from a Greek tragedy (actually the doctor has been conversing with cosmic spirits!), he knocks on his chamber door and says: "Excuse me. I hear you're declaiming. You're reading a Greek tragedy, aren't you? I could profit from acquaintance with that art, for nowadays it's quite effective."

2. Hendrickson alludes to *De dialectica*, 5.
3. Pre-Classical Greek and early Latin sometimes separated words by spaces, dots, or a combination of the two, but continuous writing prevailed from the fifth century in Greece and from the second century in Rome and Italy. Why this trend became established if it actually inhibited fluent reading is not yet fully understood. One possibility, proposed by William A. Johnson (2010), is that readers during the late Roman Republic and early Roman Empire regarded books as aesthetic objects and reading aloud well as a sign of high social status. The literate elite, he suggests, were comfortable keeping reading difficult for the *indocti*. Moreover, "reading is not, in my view, exclusively or even mostly a neurophysiological, cognitive act—not in fact an individual phenomenon, but a sociocultural system in which the individual participates" (11). To my mind, however, it is unnecessary—and unhelpful—to exclude in advance either neurophysiological or sociocultural factors when setting out to explore these issues.
4. Gavrilov cites Sokolov (1972), but because he does not mention Hendrickson, I assume he was unaware that the latter had also referred to subvocalization.
5. Quintilian's Latin text, translated above, were it transcribed in its original continuous script, would form a rather long sentence. In the following transcription, I have marked the three ways in which its word boundaries were indicated: (1) by short function words and copulative verbs (crossed out), by prefixes (underlined), and by inflected endings (italicized). The central stems that remain are marked in **bold**.

~~NAM~~<u>PRO</u>**SPIC***ERE***DEXTR***AM*~~QUOD~~**OMN***ES*<u>PRAE</u>**CIP***IUNT*~~ET~~<u>PRO</u>**VID***E*
RE~~NON~~**RATIO***NIS*~~MODOSED~~**US***US*~~QUOQUEESTQUONIAM~~**SEQUENT**
I*A*<u>IN</u>**TU***ENTI***PRI***ORA***DIC***ENDA*~~SUNTETQUOD~~**DIFFICIL***LIMUM*~~EST~~
DIVID*ENDA*<u>IN</u>**TENTIOANIM**~~IUT~~**ALI***UD***VOC***E***ALI***UD***OCUL***IS***AG***ATUR*.

(I recognize that others might annotate this passage slightly differently.)

6. Some of the words we use to mean "answer" imply a process of "talking back." The verb in Old English, *andswerian*, implied swearing against someone (*and-*, *ent-*, and *anti-* are related prefixes denoting "against"; cf. the German *Antwort*). The root of *erwidern*, "to reply," in German is another "against" word, *wider*, as in *widerprechen*, "to contradict." Like the Latin *re-*, the German *wieder/wider* combined the meanings "again" and "against," as English has done. The Spanish word for "answer," *contestar*, is similarly disputatious in origin: the Latin *contestari* meant to call on witnesses, divine or human, to attest to one's veracity. See Collins (1996, 35–39).
7. Besides alternating single lines such as this (*stichomythia* proper), one could compose it in exchanges of two lines each (*distichomythia*). *Antilabe* ("take-over"),

when it takes the form of two voices, each speaking half of a line, is also called *hemistichomythia*. Stichomythia appears frequently in Greek tragedy and may derive from the early function of the actor, the *hupocritês* ("answerer") responding to questions from the chorus or its leader, the *exarchos*. Stichomythia and antilabe, being possible only in metrical dialogues, are features primarily found in verse drama, especially comedy. Aristophanes made extensive use of them, and if the recovered text of Menander's *Dyskolos*, first staged in 317 BCE, is any indicator, both were also staples of Greek New Comedy. There is little doubt that the Roman comedic poets Plautus and Terence borrowed them from playwrights such as Menander.

8. Maecenas was, of course, more than a patron to these poets. He was a friend. It is also worth remembering how young these men were. In 41, the year these two poets first met, both Vergil and Maecenas were twenty-nine years old, Horace twenty-four, and Octavian a mere twenty-two.

9. *Satire* 1.3 is something of an anomaly. It is a dialogue in which Damasippus, a Stoic convert turned proselytizer, accosts Horace and subjects him to a recital of the teachings of his guru, the itinerant preacher Stertinius ("Master Snoreson"), on the nature of insanity. In each of three speeches, Damasippus quotes Stertinius narrating anecdotes and reporting dialogues. Fraenkel (1957) deemed this a tour de force but a signal failure. Sharland (2009, 36) calls it a "monster satire" but finds it exhilarating nonetheless. I myself suspect that a contemporary Roman might have imagined it as a script for a stand-up comic, a Fundanius perhaps (*Satires* 2.10.40–42, 2.8), who, using different voices for Damasippus's rambling cast of characters, might have provoked peals of laughter from a Roman audience.

10. *Rhetoric* 1408b, *Poetics* 1449a23–28. For an excellent study of this meter and rhythm together with its generic "family resemblances," see Rotstein (2010). The meaning of "epode" is somewhat murky and may have changed over time. One standard definition is a couplet of different lengths. Then there is its etymological meaning: an incantation (a "singing upon"). As for "iamb," its origins are even murkier: the word derives either from a word meaning verbal abuse, the distinctive short and long duple meter, or the traditional occasion (cf. "*dithyrambus*") during which abusive, lewd, and "edgy" speech could be indulged in. In the time of the late Republic and early Principate, an "epode" probably meant a declamatory two-line stanza in the style of Archilochus, and "iamb" referred to any vituperative harangue in verse.

11. Note: translating poetry is at best an imperfect art. Translating ancient poetry into a modern language is close to impossible. In attempting the impossible, however, I have tried to find literal equivalences of words and phrases, retain line-for-line regularity, and reproduce some sense of the Latin meter. The latter task, I feel, is important because, as Plato said (*Republic* 10.601), a poem in prose paraphrase is no longer a poem but rather a hollow-sounding piece of rhetoric. I needed, therefore, to find a metrical structure at least vaguely resembling Horace's Second Archilochian Strophe and chose a couplet of iambic hexameter

and iambic pentameter. The Latin text for this, as for succeeding poems, appears in the appendix.

12. Ross Kilpatrick (1970) observed that the setting for this epode may have been the eve of the first battle of Philippi, October 3, 42 BCE. If so, it may also allude to the aftermath of the final battle of Philippi (October 23), occurring at the onset of the cold rainy season, according to Plutarch in his life of Brutus. Directly north of Philippi stretch the Haemus Mountains, now called the Balkan Mountains, where in some cave Boreas was said to live.
13. Epic and drama, traditionally orally performed, are not dedicated in the times of the Republic and early Empire, but the fact that Lucan dedicated his *Pharsalia* to Emperor Nero and Statius dedicated his *Thebaid* to Emperor Vespasian suggests that by the middle of the first century CE newly written epics were considered more as private readings than as performance scripts.
14. Vergil here alludes to his first three *Georgics* but not to this fourth, perhaps because he was at that time in the process of revising it. His reference to "deep Euphrates" apparently concerns Octavian's Middle Eastern campaign following his victory at Actium (31 BCE). Parthenope was the original name of Naples, where Vergil had come to reside early in his career. After declaring his own name as author, he alludes to his first published success, the *Eclogues* (or *Bucolica*), by quoting its first line, addressed to the shepherd, Tityrus, a character that represents Vergil himself in those poems.
15. Heinze's words are: "*Die horazische Ode ist, wie wir sahen, Ansprache an eine als gegenwärtig gedachte Person* [The Horatian ode is, as we saw, an address to a person thought to be present]" (1923, 160). But thought by whom? The speaker (Horace) or the reader? Those who have taken exception to this statement assume it unnecessarily restricts the poet's and the reader's ability to project different kinds and degrees of presentness on the speech event.
16. The meter used in this ode is the Alcmanian strophe (a couplet beginning with a dactylic hexameter line followed by a dactylic tetrameter line). I chose an English accentual-syllabic meter because of its similar flexibility and presented it in a couplet composed of one hexameter and one tetrameter.
17. Familiar Horatian examples would include "*cetera mitte loqui*"—"stop talking about those other things" (*Epode* 13); "*carpe diem*"—"seize the day" (*Odes* 1.11); "*quid sit futurum cras fuge quaerere*"—"beware of asking what tomorrow will be like" (*Odes* 1.9); "*misce stultitiam consiliis brevem*"—"mix a little silliness into your wisdom" (*Odes* 4.12).
18. This principle does not seem to hold for the souls of Agamemnon and Achilles, who were able to converse with the still-articulate souls of the newly arrived Suitors in *Odyssey* 24, the so-called Second *Nekuia*. Perhaps the dead need the taste of blood to marshal their strength only when addressing the living.
19. Confronted with the textual variations, I opted for the more alliterative archaism *dacrumis* rather than the standard *lacrimis*. These lines, which every Roman schoolboy, including young provincials such as Vergil, must have learned, come down to us from only one source, Cicero in his *Tusculan Disputations* 1.15.34.

20. For other Latin instances of this trope see Cicero, *Republic* 1.17.26; Vergil, *Aeneid* 12.234; and Silius Italicus, *Punica* 13.35.
21. Horace did not choose to follow Vergil (who had followed Ennius) in his bird imagery. He probably found that metamorphosis a bit extreme and gave it a send-up in *Odes* 2.20, in which he portrayed himself as a feathered swan in poetical flight.
22. The verb did not mean to "edit" a manuscript but to relinquish control of it. (Cf. the German noun for "edition," *Ausgabe*, i.e., something *given out*.) *Edere* also meant to "bring forth" or "give birth to": in the first line of Horace's first ode of his first book of odes, his dedication to Maecenas, he addresses him as born of or descended from (*edite*) ancient chieftains. In a less-exalted sense of the verb, Horace uses it to mean "tell," "divulge," or, as we say, "spill the beans" (*Satires* 2.4.10, 2.5.61, 2.7.45).
23. Why Ovid at the age of fifty in 9 BCE was exiled is not precisely known. In the *Tristia* and the later *Epistulae ex Ponto* he says that it was somehow occasioned by his erotic writings, the *Ars amatoria* especially, but perhaps his banishment was less a matter of seduction than of sedition.

7. Writing and the Reading Mind

1. A notable "backlooping" oral technique has been called "ring composition," a presentation of topics in a chiastic structure, as: a, b, c, d, c, b, a. This reversal of linear direction may also be viewed as a circular path, a *periodos*, Aristotle's preferred sentence structure (*Rhetoric* 3.9).
2. Ong's final five characteristics of oral language serve to anchor it in the life of the speech community, which provides the only context within which oral discourses have meaning: (5) "close to the human life world" (oral cultures anthropomorphize the nonhuman world), (6) "agonistically toned" (oral discourse is public and competitive), (7) "empathetic and participatory rather than objective" (not coolly distanced), (8) "homeostatic" (language means what it means *now*, and other, older meanings are forgotten, as Goody and Watt [1968] claimed), and (9) "situational rather than abstract" (concepts are concretely realized by being situationally defined, as in *exempla*).
3. Note that, as applied to linguistic processes occurring in time rather than space, parataxis should not be interpreted as derived from the military formation of parallel rows of infantry, its original Greek usage. As parataxis now implies a concatenation of units, hypotaxis implies a dynamic system. This is perhaps the distinction Plato had in mind when in the *Phaedrus* (264c) he enunciated the principle that every discourse ought to be organized like a living body. Likewise, Aristotle (*Rhetoric* 3.9) criticized Herodotus for his "strung-together style" (*lexis eiromenê*), which lacked the unity of the periodic sentence. As James Notopoulos (1949) argued, these two philosophers espoused the new classical Greek aesthetic of organicism and rejected the older oralism, which followed an inorganic model. For an updated discussion of this issue, see Bakker (1997).

4. Ovid in *Metamorphosis* (9.522–525) gives us a vividly animated view of a young woman writing a love letter on a wax tablet: "Her right hand holds the stilus, the other hand holds a blank wax tablet. She begins, then hesitates; writes and condemns the writing; scribbles and deletes it; alters, blames, and approves; in turn lays down what she has lifted, and lifts up what she has laid down [*dextra tenet ferrum, vacuam tenet altera ceram. / incipit et dubitat, scribit damnatque tabellas, / et notat et delet, mutat culpatque probatque / inque vicem sumptas ponit positasque resumit*]."
5. The Greek word *thumos*, also thought to be located in the chest, was regularly associated with strong expressions of emotion such as courage and indignation. The Latin *animus* connoted both the "spiritedness" of the *thumos* and the verbal consciousness of the *phrên*. (Cf. the English words "animus" and "animosity.")
6. As one might expect, the Greeks had preceded the Romans in their use of this scriptorial model (cf. the verbs *strephô, anakukleô,* and *helissô*). The Romans, however, seem to have made it a more idiomatic expression.
7. On the issue of hearing what is read *in animo*, see Seneca (*Letters* 24.15–19): "Keep rolling over in your *animus* these sayings that you have often heard and often said [*Haec in animo voluta, quae saepe audisti, saepe dixisti*]."
8. Here Livy refers to the public viewing in 451 BCE of the first ten of what would be known as the Twelve Tables, the constitution of the Roman Republic. The citizens were asked to "go and read [*legere*] the proposed laws" engraved on large bronze tablets erected in the Forum. When it was asked that they then think over (*versarent*) these laws, this could not be by repeatedly turning a scroll because the laws were on a bronze surface. If my conjecture is correct, the verb *versare* (*in animo* or *animis*) had by the time Livy wrote his history (ca. 20–30 BCE) become a "dead metaphor," a common idiom that few interpreted figuratively.
9. Of these seven Roman elegists, Gallus survives in mere fragments (see Vergil's tenth *Eclogue* for what may be echoes of one of the most celebrated elegists of his generation); Horace's love affairs are not developed into a dominant narrative; and Sulpicia, the only woman of the group, has had her *opera* conflated with those of Tibullus in the textual tradition.
10. Kenneth Quinn ([1963] 2014) noted that Horace seems to have written other imaginary addresses to women, e.g., to Pyrrha (*Odes* 1.5) and to Lyce (*Odes* 4.13), arguing that it is implausible that he would ask his reader to imagine these as face-to-face speeches. Judging by the poet's rhetorical positionings throughout his writing, I would consider the Neobule ode as also dramatic monologue and not soliloquy.
11. In common English usage, there is an overlap between "meter" and "rhythm" that becomes less confusing if we think of meter as a patterned form of rhythm. As a repeated series of pulses, rhythm simply continues. Judging by its Greek name, *metron*, meter is first of all a measurement of rhythmic flow. A secondary useful definition is that the meter of a line is the abstract structure of strong and weak syllables, however prominence is determined in a given language, and

the rhythm of a line is the concrete intonational structure of the line, the "speech rhythm," i.e., how the phrases would be uttered in natural speech. For example, classical dactylic hexameter and English iambic pentameter both allow certain metrical substitutions that preserve the abstract meter while expressing rhythmical variations. The optimal reading of such lines would reflect the tension between the meter and the rhythm and attempt to harmonize the two.

12. In his analysis of poetic structure, Roman Jakobson (1958) noted consistently repetitive elements operating on the levels of semantics, syntax, and phonological expression. Even earlier, Suzanne Langer (1953) had declared rhythmic patterning to be the form of feeling and, as such, an aesthetic universal across all the arts.
13. Late Latin shows a slow but progressive departure from Greek-influenced metrics. When Augustine (*Confessions* 8.12.29) heard a child chanting "*Tolle, lege. Tolle, lege*" ("pick it up and read it"), this could not be classical quantitative meter, but it could be accentual meter: note the lexical stress on the first syllable of "*lege.*" As quoted, this "*cantus*" resembles the trochaic tetrameter of the older contemporary of Augustine, Hilary of Poictiers's "Hymnum dicat turba fratrum" and Thomas Aquinas's "Pange, lingua, gloriosi," two of the more illustrious hymns written in this meter.
14. The broad aesthetic implications of this cognitive polarity of past and present, memory and perception, might lead historians of the arts to consider (1) the relation between visual art (painting and sculpture that imitate memory images) and music (auditory experiences of acoustical events in the here and now) and (2) the relation, within the verbal art genres, between epic (a recollected and retransmitted narrative of a common cultural past) and drama (an imitation of actions, with or without music, dance, and pageantry, in an ongoing present).
15. If metrical patterns could make it easier to remember the words of a composition, it is logical to assume that a more strongly defined pattern would improve verbal recall. Racette and Peretz's 2007 study indicated the opposite: learning to sing a set of lyrics to a given melody was a less-efficient way to memorize the words than simply rehearsing the words themselves. (See chapter 3.)
16. He went on to comment: "If certain sensitive persons listen persistently to the ticking of a watch, or gaze persistently on the monotonous flashing of a light, they fall into the hypnotic trance; and rhythm is but the ticking of a watch made softer, that one must needs listen, and various, that one may not be swept beyond memory or grow weary of listening; while the patterns of the [visual] artist are but the monotonous flash woven to take the eyes in a subtler enchantment" (Yeats 1903, 247–248). To apply this theory of rhythm, we need first to recognize that Yeats is not referring to a strong, dominant beat likely to entrain listeners' bodies to move in time with them. This is obviously a more covert set of movements that operate on a continuum, somewhere between sleep and waking consciousness, between monotony and variety, and it is then and there that occurs the "moment of contemplation," which may also become "the moment of creation." This intermediate state, he says, is equivalent to a hypnotic trance, which, from the Latin *transitus*, is transitional between two states. This in turn brings

to mind the liminal state created for initiands by performers of a rite of passage.

17. Consider, for example, the impact of Iago's account of Cassio's dreaming words (*Othello* 3.3.414–430) and how differently Gertrude's description of the death of Ophelia (*Hamlet* 4.7.167–185) affects Claudius and Laertes. Then there is Enobarbus's speech (*Antony and Cleopatra* 2.2.914–968) describing Cleopatra on the Nile, which must compete with the immediate scene, punctuated by Agrippa's exclamations ("O, rare for Anthony!" . . . "Rare Egyptian!" . . . "Royal wench!"). Finally consider how Edgar's detailed description of the view looking down from the cliffs of Dover (*King Lear* 4.6.11–34) appears secondary in importance to the pathos of this scene: the faithful son's motive for telling his blinded, suicidal father this fabrication at this particular point in the tragedy. For examples of Shakespeare directly inviting his audience to imagine, exceptions that prove the rule, we have the prologues that introduce each act of Henry V.
18. As P1 visuality generally characterizes epic events, P2 characterizes drama, but now from the point of view of the audience member's seat in the "viewing place," the *theatron*. This is the perspective that most approximates hypotaxis. Aristotle's admiration for drama was partly on account of its organic order, its beginning, middle, and end, its progressive movement toward the fulfillment of its periodic structure, its *telos*.
19. *Repetantur*, grammatically parsed, is present tense, third-person plural, passive voice, subjunctive mood, which in this case is "hortatory subjunctive." The word as it appears in this ode has been variously glossed. For example, Orelli (1843) suggests these items should be "frequently sought" (*crebro quaerantur*); Vessey (1985) reads the verb as though it only applied to *susurri* (whispers) and therefore implied a resumption of some earlier daytime exchanges. For its usage as meaning "to assert a legal claim," see Horace, *Epistles* 1.3.18. See also *Epode* 13, lines 4–5: "*dumque virent genua / et decet.*"
20. On the woods and streams images in the first strophe, Kiessling (1884, 37) notes: "*Die Wahrnehmung dieses und des folgenden Zuges beruht nicht sowohl auf der Thätigkeit des Auges, wie auf derjenigen der Phantasie, welche das aufgefangene Bild selbstständig weiter ausmalt.*"

Bibliography

Adornetti, Ines. 2014a. "Dall'azione al discorso: Una prospettiva pragmatica sull'origine del linguaggio." *Reti, saperi, linguaggi,* Società editrice il Mulino, 1:69–84.

——. 2014b. "Making Tools and Planning Discourse: The Role of Executive Functions in the Origin of Language." *Humana: Mente Journal of Philosophical Studies* 27:221–241.

——. 2015. "The Phylogenetic Foundations of Discourse Coherence: A Pragmatic Account of the Evolution of Language". *Biosemiotics* 8 (3): 421–441.

Aiello, L., and R. Dunbar. 1993. "Neocortex Size, Group Size, and the Evolution of Language." *Current Anthropology* 34:184–193.

Aiello, Rita, and John A. Sloboda, eds. 1994. *Musical Perceptions.* New York: Oxford University Press.

Albright, William Foxwell. 1939. "The Israelite Conquest of Canaan in the Light of Archaeology." *BASOR* 74 (4): 11–23.

——. 1957. *From the Stone Age to Christianity: Monotheism and the Historical Process.* Baltimore, Md.: Johns Hopkins University Press.

Allen, Danielle S. 2010. *Why Plato Wrote.* Malden, Mass.: Wiley-Blackwell.

Allen, P. A., K. P. Kaut, and R. R. Lord. 2008. "Emotion and Episodic Memory." *Handbook of Behavioral Neuroscience* 18:115–132.

Altman, Janet Gurkin. 1982. *Epistolarity: Approaches to a Form.* Columbus: Ohio State University Press.

Anderson, Michael L. 2014. *After Phrenology: Neural Reuse and the Interactive Brain.* Cambridge, Mass.: MIT Press.

Anderson, S. E., E. Chui, S. Huette, and M. J. Spivey. 2011. "On the Temporal Dynamics of Language-Mediated Vision and Vision-Mediated Language." *Acta Psychologica* 137:181–189.

Arbib, Michael A. 2012. *How the Brain Got Language: The Mirror System Hypothesis.* New York: Oxford University Press.

——, ed. 2013. *Language, Music, and the Brain: A Mysterious Relationship.* Cambridge, Mass.: MIT Press.

Aristotle. 1998. *The Poetics of Aristotle, Translation and Commentary.* Ed. Stephen Halliwell. Chicago: University of Chicago Press.

Armstrong, D. F., W. C. Stokoe, and S. E. Wilcox. 1995. *Gesture and the Nature of Language.* New York: Cambridge University Press.

Assmann, Jan. 1994. *Egyptian Solar Religion in the New Kingdom: Re, Amun, and the Crisis of Polytheism.* London: Kegan Paul International.

Atran, S., and A. Norezayan. 2004. "Religion's Evolutionary Landscape: Counterintuition, Commitment, Compassion, Communion." *Behavioral and Brain Sciences* 27:713–770.

Baars, B. J. 2013. "A Scientific Approach to Silent Consciousness." *Frontiers in Psychology* 4:1–3.

Baddeley, Alan. 2000. "The Episodic Buffer: A New Component of Working Memory?" *Trends in Cognitive Sciences* 4 (11): 417–423.

Baddeley, Alan, and G. J. Hitch. 1974. "Working Memory." In *Recent Advances in Learning and Motivation,* ed. G. A. Bower, 8:47–89. New York: Academic Press.

Bakhtin, M. 1986. *Speech Genres and Other Essays.* Austin: University of Texas Press.

Bakker, Egbert J. 1997. "Storytelling in the Future: Truth, Time, and Tense in Homeric Epic." In *Written Voices, Spoken Signs: Tradition, Performance, and the Epic Text,* by Egbert J. Bakker and Ahuvia Kahane, 11–36. Cambridge, Mass.: Harvard University Press.

Bakker, Egbert J. 2005. *Pointing at the Past: From Formula to Performance in Homeric Poetics.* Washington, D.C.: Center for Hellenic Studies, Trustees for Harvard University.

Bakker, Egbert J., and Ahuvia Kahane, eds. 1997. *Written Voices, Spoken Signs: Tradition, Performance, and the Epic Text.* Cambridge, Mass.: Harvard University Press.

Bal, Mieke. 1985. *Narratology: Introduction to the Theory of Narrative.* Toronto: University of Toronto Press.

Balogh, Josef. 1927. "Voces Paginarum: Beitrag sur Geschichte des Lauten Lesens und Schreibens." *Philologus* 82:84–109, 202–240.

Barrett, Justin L. 2004. *Why Would Anyone Believe in God?* Walnut Creek, Calif.: AltaMira.

Barrett, Justin L., and Jonathan A. Lanman. 2008. "The Science of Religious Beliefs." *Religion* 38:109–124.

Barsalou, L. W. 2008. "Grounded Cognition." *Annual Review of Psychology* 59:617–645.

——. 2009. "Simulation, Situated Conceptualization, and Prediction." *Philosophical Transactions of the Royal Society B* 364:1281–1289.

Basso, Keith H. 1996. *Wisdom Sits in Places: Landscape and Language Among the Western Apache.* Albuquerque: University of New Mexico Press.

Bates, E., and B. MacWhinney. 1988. "What Is Functionalism?" *Papers and Reports on Child Language Development* 27:137–152.

Bateson, G. 1972. "A Theory of Play and Fantasy." In *Steps to an Ecology of Mind*, 177–193. New York: Chandler.

Batteux, Charles. 1824. *Les beaux arts reduits à un même principe*. Paris: Delalain.

Bax, Marcel, Barend van Heusden, and Wolfgang Wildgen. 2004. *Semiotic Evolution and the Dynamics of Culture*. Bern: Peter Lang.

Bekoff, M., and J. A. Byers. 1998. *Animal Play: Evolutionary, Comparative, and Ecological Perspectives*. Cambridge: Cambridge University Press.

Belfiore, E. 1985. "'Lies unlike the truth': Plato on Hesiod, *Theogony* 27." *Transactions of the American Philosophical Association* 115:47–57.

Bell, Catherine M. 1997. *Ritual Perspectives and Dimensions*. New York: Oxford University Press.

Bellamy, Rufus. 1989. "Bellerophon's Tablet." *The Classical Journal* 84 (4): 289–307.

Ben-Tor, Amnon. 1992. *The Archaeology of Ancient Israel*. New Haven, Conn.: Yale University Press.

Ben-Ze'ev, Aaron. 2000. *The Subtlety of Emotions*. Cambridge, Mass.: MIT Press.

Benveniste, E. (1966) 1973. *Problems in General Linguistics*. Coral Gables, Fla.: University of Miami Press.

Bergson, H. (1910) 1960. *Time and Free Will: An Essay on the Immediate Data of Consciousness*. Trans. F. L. Pogson. New York: Harper. Original work published 1889.

——. 1934. "Le possible et le réel." In *La pensée et le mouvant. Essais et conférences*, 56–65. Les classiques des sciences sociales. http://www.uqac.uquebec.ca/zone30/Classiques_des_sciences_sociales/index.html

Bering, Jesse M., and Todd K. Shackelford. 2004. "Supernatural Agents May Have Provided Adaptive Social Information." Open peer commentary on Atran and Norezayan (2004).

Berns, Ute. 2009. "Performativity." In *Handbook of Narratology*, ed. Peter Hühn et al., 19:370–383. Berlin: de Gruyter.

Besson, M., F. Faita, I. Peretz, et al. 1998. "Singing in the Brain: Independence of Lyrics and Tunes." *Psychological Science* 9 (6): 494–498.

Bever, T. G., and R. J. Chiarello. 1974. "Cerebral Dominance in Musicians and Nonmusicians." *Science* 185:537–539.

Bharucha, J. 1993. "Tonality and Expectation." In *Musical Perceptions*, ed. R. Aiello and J. A. Sloboda, 213–239. New York: Oxford University Press, 1994.

Bickerton, D. 1992. *Language and Species*. Chicago: University of Chicago Press.

Blood, Anne J., and Robert J. Zatorre. 2001. "Intensely Pleasurable Responses to Music Correlate with Activity in Brain Regions Implicated in Reward and Emotion." *Proceedings of the National Academy of Sciences of the United States of America* 98 (20): 11818–11823.

Bowling, D. L. 2013. "A Vocal Basis for the Affective Character of Musical Mode in Melody." *Frontiers in Psychology* 4:464.

Boyarin, Jonathan. 1993. *The Ethnography of Reading*. Berkeley: University of California Press.

Boyer, P., and C. Ramble. 2001. "Cognitive Templates for Religious Concepts: Cross-Cultural Evidence for Recall of Counter-Intuitive Representations." *Cognitive Science* 25:535–564.

Brown, S. W. 1985. "Time Perception and Attention: The Effects of Prospective Versus Retrospective Paradigms and Task Demands on Perceived Duration." *Perception and Psychophysics* 38:115–124.

——. 2001. "The 'Musilanguage' Model of Music Evolution." In *The Origins of Music*, ed. N. L. Wallin et al., 271–300. Cambridge, Mass.: MIT Press.

Brown, S., A. R. Laird, P. Q. Pfordresher, et al. 2009. "The Somatotopy of Speech: Phonation and Articulation in the Human Motor Cortex." *Brain and Cognition* 70 (1): 31–41.

Brown, S., and K. Weishaar. 2010. "Speech Is 'Heterometric:' The Changing Rhythms of Speech." Paper presented at Speech Prosody 2010, Chicago.

Buber, Martin. (1923) 1958. *I and Thou*. 2nd ed. Trans. Ronald Gregor Smith. New York: Scribner.

Bugelski, B. 1971. "The Definition of the Image." In *Imagery: Current Cognitive Approaches*, ed. S. J. Segal, 49–68. New York: Academic Press.

Burghardt, Gordon M. 2005. *The Genesis of Animal Play: Testing the Limits*. Cambridge, Mass.: MIT Press.

Burt, John. 2012. "Hypotaxis and Parataxis." In *The Princeton Encyclopedia of Poetry and Poetics*, 4th ed., 650–651. Princeton, N.J.: Princeton University Press.

Butcher, S. H., and John Gassner. 1951. *Aristotle's Theory of Poetry and Fine Art, with a Critical Text and Translation of the* Poetics. New York: Dover.

Call, J., and M. Tomasello. 2008. "Does the Chimpanzee Have a Theory of Mind? Thirty Years Later." *Trends in Cognitive Sciences* 12 (5): 187–192.

Calvo-Merino B., D. E. Glaser, J. Grèzes, et al. 2005. "Action Observation and Acquired Motor Skills: An FMRI Study with Expert Dancers." *Cerebral Cortex* 15 (8): 1243–1249.

Camps i Calbet, Marta, and Parth R. Chauhan. 2009. *Sourcebook of Paleolithic Transitions: Methods, Theories, and Interpretations*. New York: Springer.

Capuccino, Carlotta. 2011. "Plato's Ion and the Ethics of Praise." In *Plato and the Poets*, ed. Pierre Destrée and Fritz-Gregor Herrmann, 63–92. Leiden: Brill.

Carruthers, Peter. 2002. "The Cognitive Functions of Language." *Behavioral and Brain Sciences* 25 (6): 657–674.

Cartmill, E. A., S. Beilock, and S. Goldin-Meadow. 2012. "A Word in the Hand: Action, Gesture, and Mental Representation in Humans and Nonhuman Primates." *Philosophical Transactions of the Royal Society B: Biological Sciences* 367 (1): 129–143.

Ceccarelli, Paola. 2013. *Ancient Greek Letter Writing: A Cultural History (600 BC–150 BC)*. Oxford: Oxford University Press,

Chanda, Mona Lisa, and Daniel J. Levitin. 2013. "The Neurochemistry of Music." *Trends in Cognitive Sciences* 17 (4): 179–193.

Charalabopoulos, Nikos G. 2012. *Platonic Drama and Its Ancient Reception*. Cambridge: Cambridge University Press.

Chatman, Seymour Benjamin. 1978. *Story and Discourse: Narrative Structure in Fiction and Film*. Ithaca, N.Y.: Cornell University Press.

——. 2004. "The Structure of Narrative Transmission." In *Literary Theory: An Anthology*, ed. J. Rivkin and M. Ryan, 97–124. Malden, Mass.: Blackwell.

Chomsky, Noam. 1965. *Aspects of the Theory of Syntax*. Cambridge, Mass.: MIT Press.

——. 1968. *Language and Mind*. New York: Harcourt, Brace & World.

Clay, Diskin. 2000. *Platonic Questions: Dialogues with the Silent Philosopher*. University Park: Pennsylvania State University Press.

Collins, A. M., and E. F. Loftus. 1975. "A Spreading Activation Theory of Semantic Processing." *Psychological Review* 82:407–428.

Collins, C. 1991a. *Poetics of the Mind's Eye: Literature and the Psychology of Imagination*. Philadelphia: University of Pennsylvania Press.

——. 1991b. *Reading the Written Image: Verbal Play, Interpretation, and the Roots of Iconophobia*. University Park, Pa.: Penn State University Press.

——. 1996. *Authority Figures: Metaphors of Mastery from the Iliad to the Apocalypse*. Latham, Md.: Rowman and Littlefield.

——. 2007. *Homeland Mythology: Biblical Narratives in American Culture*. University Park, Pa.: Penn State University Press.

——. 2013. *Paleopoetics: The Evolution of the Preliterate Imagination*. New York: Columbia University Press.

Combellack, Frederick M. 1950. "Words That Die." *The Classical Journal* 46 (1): 21–26.

Cook, B. F. 1987. *Greek Inscriptions*. Berkeley: University of California Press.

Cook, Nicholas. 2014. *Beyond the Score: Music as Performance*. New York: Oxford University Press.

Cross, E. S., L. Kirsch, L. F. Ticini, et al. 2011. "The Impact of Aesthetic Evaluation and Physical Ability on Dance Perception." *Frontiers in Human Neuroscience* 5 (102): 1–10.

Cross, Ian. 1999. "Is Music the Most Important Thing We Ever Did? Music, Development, and Evolution." In *Music, Mind, and Science*, ed. S. W. Yi, 10–39. Seoul: Seoul National University Press.

——. 2008. "Musicality and the Human Capacity for Culture." *Musicae Scientiae*, special issue: *Narrative in music and interaction*, 147–167.

——. 2010. "Listening as Covert Performance." *Journal of the Royal Musical Association* 135 (1): 67–77.

——. 2012. "Music as a Social and Cognitive Process." In *Language and Music as Cognitive Systems*, ed. P. Rebuschat et al., 313–328. Oxford: Oxford University Press.

Cross, Ian, W. T. Fitch, F. Aboitiz, et al. 2013. "Culture and Evolution." In *Language, Music, and the Brain*, ed. M. A. Arbib, 541–562. Cambridge, Mass: MIT Press.

Cucchiarelli, Andrea. 2001. *La satira e il poeta. Orazio tra Epodi e Sermones*. Pisa: Giardini.

Culler, J. 1985. "Reading Lyric." *Yale French Studies* 69:98–106.

——. 2009. "Lyric, History, and Genre." *New Literary History* 40:879–899.
D'Alessio, Giovan Battista. 2004. "Past Future and Present Past: Temporal Deixis in Greek Archaic Lyric." *Arethusa* 37 (3): 267–294.
D'Ausilio, Alessandro, Ilaria Bufalari, Paola Salmas, et al. 2011. "Vocal Pitch Discrimination in the Motor System." *Brain and Language* 118 (1): 9–14.
David, A. P. 2006. *The Dance of the Muses: Choral Theory and Ancient Greek Poetics.* Oxford: Oxford University Press.
Deacon, T. W. 1997. *The Symbolic Species: The Coevolution of Language and the Brain.* New York: Norton.
Deane, P. D. 1992. *Grammar in Mind and Brain: Explorations in Cognitive Syntax.* Berlin: Mouton de Gruyter.
Dehaene, S. 2009. *Reading in the Brain.* New York: Viking.
Dehaene, S., and L. Cohen. 2011. "The Unique Role of the Visual Word Form Area in Reading." *Trends in Cognitive Sciences* 15 (6): 254–262.
De Man, Paul. 1984. *The Rhetoric of Romanticism.* New York: Columbia University Press.
Dentico, D., B. L. Cheung, J. Y. Chang, et al. 2014. "Reversal of Cortical Information Flow During Visual Imagery as Compared to Visual Perception." *NeuroImage* 100:237–243.
De Puma, Richard Daniel, Sinclair Bell, and Helen Nagy. 2009. *New Perspectives on Etruria and Early Rome.* Madison: University of Wisconsin Press.
Derderian, Katharine. 2001. *Leaving Words to Remember: Greek Mourning and the Advent of Literacy.* Leiden: Brill.
Derrida, Jacques. 1980. *La carte postale: de Socrates à Freud et au-delà.* Paris: Aubier-Flammarion.
Destrée, Pierre, and Fritz-Gregor Herrmann. 2011. *Plato and the Poets.* Leiden: Brill.
Detienne, Marcel, and Elizabeth Jones. 2001. "The Art of Founding Autochthony: Thèbes, Athens, and Old-Stock French." *Arion* (Boston) 9 (1): 4555.
Deutsch, Diana. 2013. *The Psychology of Music.* 3rd ed. Amsterdam: Academic Press.
De Vos, Mark. 2014. "The Evolutionary Origins of Syntax: Optimization of the Mental Lexicon Yields Syntax for Free." *Lingua* 150 (1): 25–44.
Dewey, John. 1979. *Art as Experience.* New York: Paragon.
Donald, Merlin. 1991. *Origins of the Modern Mind: Three Stages in the Evolution of Culture and Cognition.* Cambridge, Mass.: Harvard University Press.
——. 2001. *A Mind So Rare: The Evolution of Human Consciousness.* New York: Norton.
——. 2007. "The Slow Process: A Hypothetical Cognitive Adaptation for Distributed Cognitive Networks." *Journal of Physiology–Paris* 101:214–222.
Dowden, Ken, and Niall Livingstone, eds. 2011. *A Companion to Greek Mythology.* Chichester: Wiley-Blackwell.
Draaisma, Douwe. 2000. *Metaphors of Memory: A History of Ideas About the Mind.* Cambridge: Cambridge University Press.
Dunbar, R. 1996. *Grooming, Gossip, and the Evolution of Language.* London: Faber and Faber.

——. 2004. "Gossip in Evolutionary Perspective." *Review of General Psychology* 8: 100–110.

Edmunds, Lowell. 1997. "The Seal of Theognis." In *Poet, Public, and Performance in Ancient Greece*, ed. Lowell Edmunds and Robert W. Wallace, 29–48. Baltimore, Md.: Johns Hopkins University Press.

Egger, V. 1881. *La parole intérieure. Essai de psychologie descriptive.* Paris: Librairie Germer Baillière et Cie.

Ericsson, K. A., and W. Kintsch. 1995. "Long-Term Working Memory." *Psychological Review* 102: 211–245.

Eskin, Michael. 2000. *Ethics and Dialogue in the Works of Levinas, Bakhtin, Mandelśhtam, and Celan.* Oxford: Oxford University Press.

Evans, Vyvyan. 2014. *The Language Myth: Why Language Is Not an Instinct.* Cambridge: Cambridge University Press.

Ezzyat, Youssef, and Lila Davachi. 2011. "What Constitutes an Episode in Episodic Memory?" *Psychological Science* 22 (2): 243–252.

Fabb, Nigel. 2015. *What Is Poetry? Language and Memory in the Poems of the World.* Cambridge: Cambridge University Press.

Fadiga, Luciano, Laila Craighero, Giovanni Buccino, et al. 2002. "Speech Listening Specifically Modulates the Excitability of Tongue Muscles: A TMS Study." *European Journal of Neuroscience* 15 (2): 399–402.

Fantuzzi, Marco, and R. L. Hunter. 2004. *Tradition and Innovation in Hellenistic Poetry.* Cambridge: Cambridge University Press.

Fernyhough, Charles. "Life in the Chatter Box." *New Scientist* (June 1, 2013): 32–35.

Ferretti, F., I. Adornetti, E. Cosentino, et al. 2013. "Keeping the Route and Speaking Coherently: The Hidden Link Between Spatial Navigation and Discourse Processing." *Journal of Neurolinguistics* 26 (2): 327–334.

Ferretti, Francesco, and Erica Cosentino. 2013. "Time, Language, and Flexibility of the Mind: The Role of Mental Time Travel in Linguistic Comprehension and Production." *Philosophical Psychology* 26 (1): 24–46.

Feyerabend, Paul. 1975. *Against Method: Outline of an Anarchistic Theory of Knowledge.* London: NLB.

Finkelberg, Margalit. 2005. *Greeks and Pre-Greeks: Aegean Prehistory and Greek Heroic Tradition.* Cambridge: Cambridge University Press.

Finley, M. I. (1954) 2002. *The World of Odysseus.* New York: NYRB.

Finnegan, Ruth H. 1977. *Oral Poetry: Its Nature, Significance, and Social Context.* Cambridge: Cambridge University Press.

Fischer, Steven R. 2003. *A History of Reading.* London: Reaktion.

Fitch, W. Tecumseh. 2006. "On the Biology and Evolution of Music." *Music Perception* 24 (1): 85–88.

——. 2013. "Rhythmic Cognition in Humans and Animals: Distinguishing Meter and Pulse Perception." *Frontiers Media S.A.*, article 68. http://www.ncbi.nlm.nih.gov/pmc/articles/PMC3813894.

Flaherty, M. G. 1999. *A Watched Pot: How We Experience Time.* New York: New York University Press.

Fludernik, Monika. 2009. *An Introduction to Narratology*. London: Routledge.

Ford, Andrew. 1992. *Homer: The Poetry of the Past*. Ithaca, N.Y.: Cornell University Press.

——. 2003. "From Letters to Literature: Reading the 'Song Culture' of Classical Greece." In *Written Texts and the Rise of Literate Culture in Ancient Greece*, ed. H. Yunis, 15–37. Cambridge: Cambridge University Press.

——. 2006. "The Genre of Genres: Paeans and Paian in Early Greek Poetry." *Poetica* 38 (3–4): 277–296.

——. 2009. "Performance, Text, and the History of Criticism." In *The Oxford Handbook of Hellenic Studies*, ed. George Boys-Stones, 628–638. Oxford: Oxford University Press.

——. 2013. "The Poetics of Dithyramb." In *Dithyramb in Context*, ed. Barbara Kowalzig and Peter Wilson, 313–331. Oxford: Oxford University Press.

Fowler, D. P. 1991. "Narrate and Describe: The Problem of Ekphrasis." *Journal of Roman Studies* 81:25–35.

Fowler, R. L. 2004. *The Cambridge Companion to Homer*. Cambridge: Cambridge University Press.

Fraenkel, Eduard. 1957. *Horace*. New York: Oxford University Press.

Freud, S. (1904) 1917. *Zur Psychopathologie des Alltagslebens*. Berlin: Karger.

Friedrich, R. 1983. "ΕΠΕΙΣΟΔΙΟΝ in Drama and Epic: A Neglected and Misunderstood Term of Aristotle's 'Poetics.'" *Hermes* 111 (H. 1): 34–52.

Galantucci, B., C. A. Fowler, and M. T. Turvey. 2006. "The Motor Theory of Speech Perception Reviewed." *Psychonomic Bulletin and Review* 13 (3): 361–377.

Garner, R. Scott. 2012. "Oral Tradition and Sappho." Center for Studies in Oral Tradition. http://muse.jhu.edu/journals/oral_tradition/v026/26.2.garner03.html.

Gavrilov, Alexander K. 1997. "Techniques of Reading in Classical Antiquity." *The Classical Quarterly* 47 (1): 56–73.

Gaudreault, André. 1989. "Mimésis et Diègèsis chez Platon." *Revue de Métaphysique et de Morale* 94 (1): 79–92.

Geertz, Armin W. 2011. *Origins of Religion, Cognition, and Culture*. London: Equinox.

Gelb, I. J. 1952. *A Study of Writing: The Foundations of Grammatology*. Chicago: University of Chicago Press.

Genette, Gérard. 1979. "Frontières de récit." *Figures* 2: 49–69.

——. 1980. *Narrative Discourse: An Essay in Method*. Ithaca, N.Y.: Cornell University Press.

——. 1992. *The Architext: An Introduction*. Berkeley: University of California Press.

Gentili, Bruno, and Carmine Catenacci. 2007. "Saffo 'Politicamente corretta.'" *Quaderni Urbinati di Cultura Classica*, n.s., 86 (2): 79–87.

Giardini, Francesca. 2012. "Deterrence and Transmission as Mechanisms Ensuring Reliability of Gossip." *Cognitive Process* 13 (suppl. 2): 465–475.

Gibson, James J. 1950. *The Perception of the Visual World*. Boston: Houghton Mifflin.

Givón, T. 2002. *Biolinguistics: The Santa Barbara Lectures*. Amsterdam: John Benjamins.

——. 2009. *The Genesis of Syntactic Complexity: Diachrony, Ontogeny, Neurocognition, Evolution*. Philadelphia: John Benjamins.

Goody, Jack, ed. 1968. *Literacy in Traditional Societies*. Cambridge: Cambridge University Press.

Goody, Jack, and Ian P. Watt. 1968. "The Consequences of Literacy." In *Literacy in Traditional Societies*, ed. J. Goody, 27–68. Cambridge: Cambridge University Press.

Gottfried, Jay A., ed. 2011. *Neurobiology of Sensation and Reward*. Boca Raton, Fla.: CRC.

Green, André. 1986. *On Private Madness*. London: Hogarth.

Greenberg, Joseph H. 1990. *On Language: Selected Writings of Joseph H. Greenberg*. Ed. Keith M. Denning and Suzanne Kemmer. Stanford, Calif.: Stanford University Press.

Greenberg, Nathan A. 1961. "The Use of Poiema and Poiesis." *Harvard Studies in Classical Philology* 65:263–289.

Greene, Ellen. 1996. *Reading Sappho: Contemporary Approaches*. Berkeley: University of California Press.

Grethlein, Jonas. 2008. "Memory and Material Objects in the Iliad and the Odyssey." *Journal of Hellenic Studies* 125:S.27–51.

——. 2010. *The Greeks and Their Past: Poetry, Oratory, and History in the Fifth Century BCE*. Cambridge: Cambridge University Press.

Grethlein, Jonas, and A. Rengakos, eds. 2009. *Narratology and Interpretation: The Content of the Form in Ancient Texts*. Berlin: de Gruyter.

Grice, H. P. 1989. *Studies in the Way of Words*. Cambridge, Mass.: Harvard University Press.

Griffin, Jasper. 2004. "The Speeches." In *The Cambridge Companion to Homer*, ed. R. L. Fowler, 156–168. Cambridge: Cambridge University Press.

Guthrie, Stewart. 1980. "A Cognitive Theory of Religion." *Current Anthropology* 21 (2): 181–194.

Hall, Jonathan M. 2013. *A History of the Archaic Greek World, ca. 1200–479 BCE*. Hoboken, N.J.: Wiley-Blackwell.

Halliwell, Stephen. 2002. *The Aesthetics of Mimesis: Ancient Texts and Modern Problems*. Princeton, N.J.: Princeton University Press.

——. 2009. "The Theory and Practice of Narrative in Plato." In *Narratology and Interpretation: The Content of the Form in Ancient Texts*, ed. J. Grethlein and A. Rengakos, 15–41. Berlin: de Gruyter.

——. 2014. "Diegesis—Mimesis." In *The Living Handbook of Narratology*, ed. Peter Hühn et al. Hamburg: Hamburg University.

Hardie, Alex. 2005. "Sappho, the Muses, and Life After Death." *Zeitschrift für Papyrologie und Epigraphik* 154:13–32.

Harmand, S., J. E. Lewis, C. S. Feibel, et al. 2015. "3.3-Million-Year-Old Stone Tools from Lomekwi 3, West Turkana, Kenya." *Nature* 521:310–315.

Harrison, S. J. 2007. *The Cambridge Companion to Horace*. Cambridge: Cambridge University Press.

Hausen M., R. Torppa, T. Sarkamo, et al. 2013. "Music and Speech Prosody: A Common Rhythm." *Frontiers in Psychology* 4:1–16.

Hauskeller, Michael. 2012. "My Brain, My Mind, and I: Some Philosophical Assumptions of Mind-Uploading." *International Journal of Machine Consciousness* 4 (1): 187–200.

Havelock, Eric Alfred. 1963. *Preface to Plato*. Cambridge, Mass.: Harvard University Press.

——. 1986. *The Muse Learns to Write: Reflections on Orality and Literacy from Antiquity to the Present*. New Haven, Conn.: Yale University Press.

Haverkate, H. 1990. "A Speech Act Analysis of Irony." *Journal of Pragmatics* 14 (1): 71–109.

Heiden, Bruce. 2007. "The Muses' Uncanny Lies: Hesiod, 'Theogony' 27, and Its Translators." *American Journal of Philology* 128 (2): 153–175.

Heinze, Richard. 1923. *Die horazische Ode*. Leipzig: Teubner.

Hendrickson, G. L. 1929. "Ancient Reading." *Classical Journal* 25:182–196.

Henige, David P. 1974. *The Chronology of Oral Tradition: Quest for a Chimera*. Oxford: Clarendon.

Henkel, John. 2014. "Metrical Feet on the Road of Poetry: Foot Puns and Literary Polemic in Tibullus." *Classical World* 107 (4): 451–475.

Henrich, J., and R. McElreath. 2003. "The Evolution of Cultural Evolution." *Evolutionary Anthropology* 12:123–135.

——. 2007. "Dual Inheritance Theory: The Evolution of Human Cultural Capacities and Cultural Evolution." In *Oxford Handbook of Evolutionary Psychology*, ed. Robin Dunbar and Louise Barrett. Oxford: Oxford University Press.

Hermida, João, and Mariana Ferreo. 2010. *Music Education*. New York: Nova Science.

Hertel, D. 2003. *Die Mauern von Troia: Mythos und Geschichte im antiken Ilion*. Munich: C. H. Beck Verlag.

Heusden, Barend van. 2004. "A Bandwidth Model of Semiotic Evolution." In *Semiotic Evolution and the Dynamics of Culture*, ed. M. Bax, B. van Heusden, and W. Wildgen, 3–33. Berlin: Peter Lang.

Hewes, G. W. 1992. "Primate Communication and the Gestural Origin of Language." *Current Anthropology* 33 (1): 65–84.

Hickok G. 2012. "The Cortical Organization of Speech Processing: Feedback Control and Predictive Coding the Context of a Dual-Stream Model." *Journal of Communication Disorders* 45 (6).

Hickok, G., and D. Poeppel. 2004. "Dorsal and Ventral Streams: A Framework for Understanding Aspects of the Functional Anatomy of Language." *Cognition* 92:67–99.

Hockett, C. F. (1960) 1982. "The Origin of Speech." *Scientific American* 293 (88). Reprinted in *Human Communication: Language and Its Psychobiological Bases. Readings from Scientific American*, ed. Shih Yi Wang, 5–12. San Francisco: W. H. Freeman.

Holland, Margaret G. 1996. "What's Wrong with Telling the Truth? An Analysis of Gossip." *American Philosophical Quarterly* 33 (2): 197–209.

Holmes, J. B., H. S. Waters, and S. Rajaram. 1998. "The Phenomenology of False Memories: Episodic Content and Confidence." *Journal of Experimental Psychology: Learning, Memory, and Cognition* 24 (4): 1026–1040.

Höschele, Regina. 2007. "The Traveling Reader: Journeys Through Ancient Epigram Books." *Transactions of the American Philological Association (1974–)* 137 (2): 333–369.

Howard, Rosaleen. 2002. "Spinning a Yarn: Landscape, Memory, and Discourse Structures in Quechua Narratives." In *Narrative Threads: Accounting and Recounting in Andean Khipu*, ed. J. Quilter and G. Urton, 26–49. Austin: University of Texas Press.

Howe, Nicholas. 2003. "The Cultural Construction of Reading in Anglo-Saxon England." In *The Ethnography of Reading*, ed. J. Boyarin, 58–79. Berkeley: University of California Press.

Howe, Nicholas. 2008. *Writing the Map of Anglo-Saxon England: Essays in Cultural Geography*. New Haven, Conn.: Yale University Press.

Hudson, R. 2001. "Review of Terrence Deacon, *The Symbolic Species: The Coevolution of Language and the Human Brain*." *Journal of Pragmatics* 33:129–135.

Hühn, Peter. 2009. *Handbook of Narratology*. Berlin: de Gruyter.

Hunter, R. L., and Ian Rutherford. 2009. *Wandering Poets in Ancient Greek Culture: Travel, Locality, and Pan-Hellenism*. Cambridge: Cambridge University Press.

Hurford, J. R. 2007. *Language in the Light of Evolution*. Vol. 1: *Foundations of Meaning*. Oxford: Oxford University Press.

Huron, David Brian. 2006. *Sweet Anticipation: Music and the Psychology of Expectation*. Cambridge, Mass.: MIT Press.

Husserl, Edmund. 1991. *On the Phenomenology of the Consciousness of Internal Time (1893–1917)*. Trans. John B. Brough. Dordrecht: Kluwer Academic.

Hutto, Daniel D. 2008. *Folk-Psychological Narratives: The Sociocultural Basis of Understanding Reasons*. Cambridge, Mass.: MIT Press.

Hutto, Daniel D., and Michael D. Kirchhoff. 2015. "Looking Beyond the Brain: Social Neuroscience Meets Narrative Practice." *Cognitive Systems Research* 34–35:45–55.

Hyde, Krista L., Isabelle Peretz, and Robert J. Zatorre. 2008. "Evidence for the Role of the Right Auditory Cortex in Fine Pitch Resolution." *Neuropsychologia* 46 (2): 632–639.

Ihde, Don. (2007) 1976. *Listening and Voice: Phenomenologies of Sound*. Albany: State University of New York Press.

Ingold, Tim. 2007. *Lines: A Brief History*. London: Routledge.

Jackendoff, R. 2009. "Parallels and Nonparallels Between Language and Music." *Music Perception: An Interdisciplinary Journal* 26 (3): 195–204.

Jager, Eric. 2000. *The Book of the Heart*. Chicago: University of Chicago Press.

Jakobson, R. 1960. "Closing Statement: Linguistics and Poetics." In *Style in Language*, ed. T. A. Sebeok, 350–377. New York: Wiley.

James, C. E., J. Britz, P. Vuilleumier, et al. 2008. "Early Neuronal Responses in Right Limbic Structures Mediate Harmony Incongruity Processing in Musical Experts." *NeuroImage* 42:1597–1608.

James, W. (1890) 1950. *The Principles of Psychology*. Vol. 1. New York: Dover.

Jaworski, Adam, and Justine Coupland. 2005. "Othering in Gossip: 'you go out you have a laugh and you can pull yeah okay but like . . .'" *Language in Society* 34:667–694.

Jaynes, Julian. 1976. *The Origin of Consciousness in the Breakdown of the Bicameral Mind*. Boston: Houghton Mifflin.

Jeannerod, M. 2006. *Motor Cognition: What Actions Tell the Self*. Oxford: Oxford University Press.

Jeannerod, M., M. A. Arbib, G. Rizzolatti, et al. 1995. "Grasping Objects: The Cortical Mechanisms of Visuomotor Transformation." *Trends in Neuroscience* 18:314–320.

Jeffreys, Mark. 1998. *New Definitions of Lyric: Theory, Technology, and Culture*. New York: Garland.

Johnson, Mark. 1987. *The Body in the Mind: The Bodily Basis of Meaning, Imagination, and Reason*. Chicago: University of Chicago Press.

Johnson, W. R. 1982. *The Idea of Lyric: Lyric Modes in Ancient and Modern Poetry*. Berkeley: University of California Press.

Johnson, William A. 2010. *Readers and Reading Culture in the High Roman Empire: A Study of Elite Communities*. New York: Oxford University Press.

Johnson, William A., and Holt N. Parker. 2009. *Ancient Literacies: The Culture of Reading in Greece and Rome*. Oxford: Oxford University Press.

Jordania, J. M. 2006. *Who Asked the First Question: The Origins of Human Choral Singing, Intelligence, Language, and Speech*. [Tbilisi, Georgia]: Logos.

Juslin, P. N. 2013a. "From Everyday Emotions to Aesthetic Emotions: Towards a Unified Theory of Musical Emotions." *Physics of Life Reviews* 10:235–266.

——. 2013b. "What Does Music Express? Basic Emotions and Beyond." *Frontiers in Psychology* 4.

Juslin, P. N., and P. Laukka. 2003. "Communication of Emotions in Vocal Expression and Music Performance: Different Channels, Same Code?" *Psychological Bulletin* 129:770–814.

Juslin, P. N., and Sloboda, J. A. 2010. "The Past, Present, and Future of Music and Emotion Research." In *Handbook of Music and Emotion: Theory, Research, Applications*, ed. P. N. Juslin and J. A. Sloboda, 933–955. New York: Oxford University Press.

——. 2013. "Music and Emotion." In *The Psychology of Music*, 3rd ed., ed. Diana Deutsch, 583–645. Amsterdam: Academic Press.

Keats, John. 1899. *The Complete Poetical Works and Letters of John Keats*. Cambridge ed. Houghton, Mifflin and Co.

Kemmerer, David. 2014. "Does the motor system contribute to the perception and understanding of actions? Reflections on Gregory Hickok's *The Myth of Mirror Neurons: The Real Neuroscience of Communication and Cognition*." *Language and Cognition* 1–26.

Kennedy, George A. 1980. *Classical Rhetoric and Its Christian and Secular Tradition from Ancient to Modern Times*. Chapel Hill: University of North Carolina Press.

——. 1992. "A Hoot in the Dark: The Evolution of General Rhetoric." *Philosophy and Rhetoric* 25 (1): 1–21.

Kidd, Stephen E. 2014. *Nonsense and Meaning in Ancient Greek Comedy*. New York: Cambridge University Press.

Kiessling, Adolf. 1884. *Q. Horatius Flaccus: Oden und Epoden*. Berlin: Weidmannsche Buchhandlung.

Kilner, J. M., and R. N. Lemon. 2013. "What We Know Currently About Mirror Neurons." *Current Biology* 23:R1057–R1062.

Kilpatrick, Ross S. 1970. "An Interpretation of Horace, Epodes 13." *Classical Quarterly* 20 (1): 135–141.

Kirby-Smith, H. T. 1999. *The Celestial Twins: Poetry and Music Through the Ages*. Amherst: University of Massachusetts Press.

Kita, Sotaro, and Hedda Lausberg. 2008. "Generation of Co-Speech Gestures Based on Spatial Imagery from the Right Hemisphere: Evidence from Split-Brain Patients." *Cortex* 44 (2): 131–139.

Klapp, Stuart T. 1971. "Implicit Speech Inferred from Response Latencies in Same-Different Decisions." *Journal of Experimental Psychology* 91 (2): 262–267.

Klapproth, Danièle M. 2004. *Narrative as Social Practice: Anglo-Western and Australian Aboriginal Oral Traditions*. Berlin: de Gruyter.

Knight, C. 1998. "Ritual/Speech Coevolution: A Solution to the Problem of Deception." In *Approaches to the Evolution of Language: Social and Cognitive Bases*, ed. J. R. Hurford, M. Studdert-Kennedy, and C. Knight, 68–91. Cambridge: Cambridge University Press.

Knox, Bernard M. W. 1968. "Silent Reading in Antiquity." *Greek, Roman, and Byzantine Studies* 9:421–435.

Koelsch S. 2005. "Investigating Emotion with Music: Neuroscientific Approaches." *Annals of the New York Academy of Sciences* 1060: 412–418.

——. 2010. "Towards a Neural Basis of Music-Evoked Emotions." *Trends in Cognitive Sciences* 14 (3): 131–137.

Koelsch, S., S. Skouras, T. Fritz, et al. 2013. "The Roles of Superficial Amygdala and Auditory Cortex in Music-Evoked Fear and Joy." *NeuroImage* 81 (9/10): 49–60.

Koelsch, S., T. Fritz, D. Yves von Cramon, et al. 2006. "Investigating Emotion with Music: An fMRI Study." *Human Brain Mapping* 27 (3): 239.

Koelsch, S., T. Fritz, and G. Schlaug G. 2008. "Amygdala Activity Can Be Modulated by Unexpected Chord Functions During Music Listening." *Neuro-report* 19:1815–1819.

Kosslyn, S. M. 1980. *Image and Mind*. Cambridge, Mass.: Harvard University Press.

Kowalzig, Barbara, and Peter Wilson, eds. 2013. *Dithyramb in Context*. Oxford: Oxford University Press.

Krašovec, Jože. 2010. *The Transformation of Biblical Proper Names*. New York: T&T Clark.

Laird, Andrew. 2005. *Ancient Literary Criticism*. Oxford: Oxford University Press.

Lakoff, George. 1987. *Women, Fire, and Dangerous Things: What Categories Reveal About the Mind*. Chicago: University of Chicago Press.

Lakoff, G., and M. Johnson. 1980. *Metaphors We Live By*. Chicago: University of Chicago Press.

Langer, S. 1953. *Feeling and Form*. New York: Scribner.

Lardinois, A. 2008. "'Someone, I say, will remember us': Oral Memory in Sappho's Poetry." In *Orality, Literacy, Memory in the Ancient Greek and Roman World*, ed. E. Anne Mackay, 79–96. Leiden: Brill.

Lavigne, Donald Edward. 2005. *Iambic Configurations: Iambos from Archilochus to Horace*. Ph.D diss. Stanford University.

Leavitt, Jonathan D., and Nicholas J. S. Christenfeld. 2011. "Story Spoilers Don't Spoil Stories." *Psychological Science* 22 (9): 1152–1154.

Lende, Daniel H., and Greg Downey. 2012. *The Encultured Brain: An Introduction to Neuroanthropology*. Cambridge, Mass.: MIT Press.

Leslie, A. M. 1987. "Pretense and Representation: The Origins of 'Theory of Mind.'" *Psychological Review* 94 (4): 412–426.

Levitin, Daniel J., and Anna K. Tirovolas. 2009. "Current Advances in the Cognitive Neuroscience of Music." *Annals of the New York Academy of Sciences* 1156: 211–231.

Liberman, A. M., and D. H. Whalen. 2000. "On the Relation of Speech to Language." *Trends in Cognitive Sciences* 4:187–196.

Liberman, A. M., and I. G. Mattingly. 1985. "The Motor Theory of Speech Perception Revised." *Cognition* 21:1–36.

Liew S.-L., T. Sheng, L. Aziz-Zadeh, et al. 2013. "Both Novelty and Expertise Increase Action Observation Network Activity." *Frontiers in Human Neuroscience* (SEP) 1–15.

Livingstone, Niall, and Gideon Nisbet. 2010. *Epigram*. Cambridge: Cambridge University Press.

Longus. 1972. *Daphnis and Chloé*. Trans. Christopher Collins. Barre, Mass.: Imprint Society.

Louis, Renee Pualani. 2004. "Indigenous Hawaiian Cartographer: In Search of Common Ground." *Cartographic Perspectives* 48:7–23.

Lowrie, Michèle. 1997. *Horace's Narrative Odes*. Oxford: Clarendon.

——. 2009. *Writing, Performance, and Authority in Augustan Rome*. Oxford: Oxford University Press.

MacEachren, Alan M. 1986. "A Linear View of the World: Strip Maps as a Unique Form of Cartographic Representation." *American Cartographer* 13:7–25.

Mackay, E. Anne. 1998. *Signs of Orality: The Oral Tradition and Its Influence in the Greek and Roman World*. Leiden: Brill.

——. 2008. *Orality, Literacy, Memory in the Ancient Greek and Roman World*. Leiden: Brill.

MacWhinney, B. 1999. "The Emergence of Language from Embodiment." In *The Emergence of Language*, ed. B. MacWhinney, 213–256. Mahwah, N.J.: Lawrence Erlbaum.

——. 2001. "Emergentist Approaches to Language." In *Frequency and the Emergence of Linguistic Structure*, ed. J. Bybee and P. Hopper. New York: Benjamins.

——. 2005. "The Emergence of Grammar from Perspective." In *The Grounding of Cognition: The Role of Perception and Action in Memory, Language, and Thinking*, ed. D. Pecher and R. A. Zwaan, 198–223. Mahwah, N.J.: Lawrence Erlbaum.

Mallarmé, Stéphane. 1967. *Poésies. Poésies, choix de vers de circonstance, poèmes d'enfance et de jeunesse*. Paris: Gallimard.

Mandler, Jean Matter. 2004. *The Foundations of Mind: Origins of Conceptual Thought*. Oxford: Oxford University Press.

Marin, Manuela M., and Joydeep Bhattacharya. 2010. "Music-Induced Emotions: Some Current Issues and Cross-Modal Comparisons." In *Music Education*, ed. João Hermida and Mariana Ferreo, 1–38. New York: Nova Science.

Marks, H. 1995. "Biblical Names and Poetic Etymology." *Journal of Biblical Literature* 114 (1): 21–42.

Mayor, Adrienne. 2011. *The First Fossil Hunters: Dinosaurs, Mammoths, and Myth in Greek and Roman Times: With a New Introduction by the Author*. Princeton, N.J.: Princeton University Press.

McCune-Nicolich, L. 1981. "Toward Symbolic Functioning: Structure of Early Pretend Games and Potential Parallels with Language." *Child Development* 52 (3): 785–797.

McNeill, D. 1992. *Hand and Mind: What Gestures Reveal About Thought*. Chicago: University of Chicago Press.

Melzack, R., and P. D. Wall. 1965. "Pain Mechanisms: A New Theory." *Science* 150 (3699): 971–979.

Meschiari, Matteo. 2009a. "Terra Sapiens: How Landscape Invented Man." In *Landscape in Mind: Dialogue on Space Between Anthropology and Archaeology*, ed. G. Dimitriadis. Oxford: Archaeopress.

——. 2009b. "Roots of the Savage Mind. Apophenia and Imagination as Cognitive Process." *Quaderni di Semantica* 60 (2): 183–222.

Mill, John Stuart. 1859. *Dissertations and Discussions Political, Philosophical, and Historical: Reprinted Chiefly from the Edinburgh and Westminster Reviews*. Vol. 1. London: John Parker.

Mill, John Stuart, and J. B. Schneewind. 1965. *Mill's Essays on Literature and Society*. New York: Collier.

Miller, Paul Allen. 1994. *Lyric Texts and Lyric Consciousness: The Birth of a Genre from Archaic Greece to Augustan Rome*. London: Routledge.

Milner, A. D., and M. A. Goodale. 1995. *The Visual Brain in Action*. New York: Oxford University Press.

Minchin, E. 2001. *Homer and the Resources of Memory: Some Applications of Cognitive Theory to the* Iliad *and the* Odyssey. Oxford: Oxford University Press.

——. 2016. "Heritage in the Landscape: The 'Heroic Tumuli' in the Troad Region." In *Valuing Landscape in Classical Antiquity*, ed. J. McInerney and I. Sluiter. Leiden: Brill.

Mithen, S. J. 2006. *The Singing Neanderthals: The Origins of Music, Language, Mind, and Body*. Cambridge, Mass.: Harvard University Press.

Mourelatos, A. P. D. 1965. "φραζω and Its Derivatives in Parmenides." *Classical Philology* 60 (4): 261–262.

Muir, J. V. 2008. *Life and Letters in the Ancient Greek World.* Milton Park: Routledge.

Müller, F. M. 1868. *Lectures on the Science of Language Delivered at the Royal Institution of Great Britain . . . 1861 [and 1863].* New York: Scribner.

——. 1889. *Three Lectures on the Science of Language and Its Place in General Education* [delivered at the Oxford University Extension Meeting 1889]. London: Longmans, Green.

Müller-Zettelmann, Eva. 2000. *Lyrik und Metalyrik: Theorie einer Gattung und ihrer Selbstbespiegelung anhand von Beispielen aus der englisch- und deutschsprachigen Dichtkunst.* Heidelberg: Winter.

Müller-Zettelmann, Eva, and Margarete Rubik. 2005. *Theory Into Poetry: New Approaches to the Lyric.* Amsterdam: Rodopi.

Mullins, D. A., H. Whitehouse, and Q. D. Atkinson. 2013. "The Role of Writing and Recordkeeping in the Cultural Evolution of Human Cooperation." *Journal of Economic Behavior and Organization* 90:141–151.

Murray, Penelope. 2005. "Poetic Inspiration in Early Greece." In *Ancient Literary Criticism,* by Andrew Laird, 37–61. Oxford: Oxford University Press.

Murray, Penelope, and Peter Wilson. 2004. *Music and the Muses: The Culture of "Mousikē" in the Classical Athenian City.* Oxford: Oxford University Press.

Nagy, Gregory. 1979. *The Best of the Achaeans: Concepts of the Hero in Archaic Greek Poetry.* Baltimore, Md.: Johns Hopkins University Press.

——. 1990. *Greek Mythology and Poetics.* Ithaca, N.Y.: Cornell University Press.

——. 1996. *Poetry as Performance: Homer and Beyond.* Cambridge: Cambridge University Press.

——. 2007. "Lyric and Greek Myth." In *The Cambridge Companion to Greek Mythology,* ed. R. D. Woodard, 19–51. Cambridge: Cambridge University Press.

Napier, J. R. 1980. *Hands.* New York: Pantheon.

Narayanan, S. 1997. "Talking the Talk Is Like Walking the Walk." *Proceedings of the 19th Meeting of the Cognitive Science Society*: 55–59.

Neisser, U. 1976. *Cognition and Reality: Principles and Implications of Cognitive Psychology.* San Francisco: W. H. Freeman.

Nietzsche, F. W. (1872) 1999. *The Birth of Tragedy and Other Writings,* ed. and trans. R. Geuss and R. Speirs. Cambridge: Cambridge University Press.

Nigro, G., and U. Neisser. 1983. "Point of View in Personal Memories." *Cognitive Psychology* 15:467–482.

Nikulin, D. V. 2006. *On Dialogue.* Lanham, Md.: Lexington.

Nimis, S. A. 1998. "Ring-Composition and Linearity in Homer." In *Signs of Orality: The Oral Tradition and Its Influence in the Greek and Roman World,* ed. E. Anne Mackay, 65–78. Leiden: Brill.

Norden, Eduard. 1898. *Die antike Kunstprosa vom VI. Jahrhundert v. Chr. bis in die Zeit der Renaissance.* Leipzig: Teubner.

Norenzayan, A., S. Atran, J. Faulkner, et al. 2006. "Memory and Mystery: The Cultural Selection of Minimally Counterintuitive Narratives." *Cognitive Science* 30:531–553.

Notopoulos, James A. 1949. "Parataxis in Homer: A New Approach to Homeric Literary Criticism." *Transactions and Proceedings of the American Philological Association* 8:1–23.

Oesterreicher, Wulf. 1997. "Types of Orality in Text." In *Written Voices, Spoken Signs: Tradition, Performance, and the Epic Text,* ed. Egbert J. Bakker and Ahuvia Kahane. Cambridge, Mass.: Harvard University Press.

Ogden, Daniel. 2009. *Magic, Witchcraft, and Ghosts in the Greek and Roman Worlds: A Sourcebook.* Oxford: Oxford University Press.

Olson, David. 2009. "Why Literacy Matters, Then and Now." In *Ancient Literacies: The Culture of Reading in Greece and Rome,* ed. William A. Johnson and Holt N. Parker, 385–403. Oxford: Oxford University Press.

Ong, Walter J. (1982) 2013. *Orality and Literacy: The Technologizing of the Word.* New York: Routledge.

Orelli, J. K. v. 1843. *Q. Horatius Flaccus.* 2nd ed. Vol. 1. Zürich: Orell Füssli.

Ornstein, R. E. 1969. *On the Experience of Time.* Baltimore, Md.: Penguin.

Paivio, A. 1971. *Imagery and Verbal Processes.* New York: Holt, Rinehart and Winston.

——. 1990. *Mental Representations: A Dual-Coding Approach.* New York: Oxford University Press.

——. 2007. *Mind and Its Evolution: A Dual-Coding Theoretical Approach.* Mahwah, N.J.: Lawrence Erlbaum.

Parker, Holt N. 2009. "Books and Reading Latin Poetry." In *Ancient Literacies: The Culture of Reading in Greece and Rome,* ed. William A. Johnson and Holt N. Parker, 385–403. Oxford: Oxford University Press.

Parmentier, R. J. 2006. "Semiotic Anthropology." *Encyclopedia of Language and Linguistics* (Second Edition), 199–212.

Parry, Milman. 1937. "About Winged Words." *Classical Philology* 32:59–63.

Pasquali, Giorgio. 1920. *Orazio lirico; studio.* Firenze: Le Monnier.

Patel, A. D. 2012. "Language, Music, and the Brain: A Resource-Sharing Framework." In *Language and Music as Cognitive Systems,* ed. P. Rebuschat et al., 204–223. Oxford: Oxford University Press.

Patel, A. D., I. Peretz, M. Tramo, et al. 1998. "Processing Prosodic and Musical Patterns: A Neuropsychological Investigation." *Brain and Language* 61:123–144.

Patterson, Lee E. 2010. *Kinship Myth in Ancient Greece.* Austin: University of Texas Press.

Pausanias. 1918. *Pausanias' Description of Greece.* Ed. and trans. W. H. S. Jones, Henry Arderne Ormerod, and R. E. Wycherley. London: W. Heinemann.

Peretz, I. 2009. "Music, Language, and Modularity Framed in Action." *Psychologica Belgica* 49 (2/3): 157–175.

Peretz, I., and M. Coltheart. 2003. "Modularity of Music Processing." *Nature Neuroscience* 6 (7): 688–691.

Pinker, Steven. 1994. *The Language Instinct.* New York: Morrow.

——. (1997) 2009. *How the Mind Works.* New York: Norton.

Pokorny, J. (1959) 1989. *Indogermanisches etymologisches Wörterbuch.* Bern: Francke.

Poster, Carol, and Linda C. Mitchell. 2007. *Letter-Writing Manuals and Instruction from Antiquity to the Present: Historical and Bibliographic Studies.* Columbia, S.C.: University of South Carolina Press.

Poulet, Georges. 1969. "The Phenomenology of Reading." *New Literary History* 1 (1): 53–68.

Pouzoulet, Christine. 2012. "A propos de l'essai de Victor Egger (1881): « parole intérieure » et formes littéraires du monologue intérieur." *Fabula / Les colloques.* http://www.fabula.org/colloques/document1641.php.

Powell, Barry B. 2002. *Writing and the Origins of Greek Literature.* Cambridge: Cambridge University Press.

Pratt, Louise H. 1993. *Lying and Poetry from Homer to Pindar: Falsehood and Deception in Archaic Greek Poetics.* Ann Arbor: University of Michigan Press.

——. 1995. "The Seal of Theognis, Writing, and Oral Poetry." *American Journal of Philology* 116 (2): 171–184.

Premack, D., and G. Woodruff. 1978. "Does the Chimpanzee Have a Theory of Mind?" *Behavioral and Brain Sciences* 1:515–526.

Probert, Philomen. 2015. *Early Greek Relative Clauses.* Oxford: Oxford University Press.

Quilter, Jeffrey, and Gary Urton, eds. 2002. *Narrative Threads: Accounting and Recounting in Andean Khipu.* Austin: University of Texas Press.

Quinn, Kenneth. (1963) 2014. *Latin Explorations: Critical Studies in Roman Literature.* Abingdon: Routledge.

Racette, Amélie, and Isabelle Peretz. 2007. "Learning Lyrics: To Sing or Not to Sing?" *Memory and Cognition* 35 (2): 242–253.

Rajaram, S. 1993. "Remembering and Knowing: Two Means of Access to the Personal Past." *Memory and Cognition* 21:89–102.

Rebuschat, P., M. Rorhrmeier, J. A. Hawkins, et al., eds. 2012. *Language and Music as Cognitive Systems.* Oxford: Oxford University Press.

Regier, T., and P. Kay. 2009. "Language, Thought, and Color: Whorf Was Half Right." *Trends in Cognitive Sciences* 13 (10): 439–446.

Reich, L., M. Szwed, L. Cohen, et al. 2011. "A Ventral Visual Stream Reading Center Independent of Visual Experience." *Current Biology* 21:363–368.

Reich, Wendelin. 2011. "The Cooperative Nature of Communicative Acts." *Journal of Pragmatics* 43:1349–1365.

Reik, T. 1953. *The Haunting Melody: Psychoanalytic Experiences in Life and Music.* New York: Farrar, Straus and Young.

Reisberg, Daniel. 2014. *Auditory Imagery.* Hoboken: Taylor and Francis.

Repp, B. H., and A. Penel. 2004. "Rhythmic Movement Is Attracted More to Auditory Than to Visual Rhythms." *Psychological Research* 68:252–270.

Ribeiro, S., A. Loula, I. de Araújo, et al. 2007. "Symbols Are Not Uniquely Human." *BioSystems* 90:263–272.

Rice, H. J., and D.C. Rubin. 2009. "I Can See It Both Ways: First- and Third-Person Visual Perspectives at Retrieval." *Consciousness and Cognition* 18 (4): 877–890.

Richardson, Alan. 2002. "Apostrophe in Life and in Romantic Art: Everyday Discourse, Overhearing, and Poetic Address." *Style* 36 (3): 363–385.

Richerson, Peter J., and Robert Boyd. 2010. "Why Possibly Language Evolved." *Biolinguistics* 4 (2/3): 289–306.

Ricoeur, P. 1980. "Narrative Time." *Critical Inquiry* 7 (1): 169–190.

Rijksbaron, Albert. 2007. *Plato, Ion; or: On the Iliad.* Leiden: Brill.

Rimell, Victoria. 2007. *Seeing Tongues, Hearing Scripts: Orality and Representation in the Ancient Novel.* Groningen: Groningen University Library.

Rosati, A. G., L. R. Santos, and B. Hare. 2010. "Primate Social Cognition: Thirty Years After Premack and Woodruff." In *Primate Neuroethology*, ed. A. Ghazanfar and M. Platt, 117–143. Oxford: Oxford University Press.

Rosenmeyer, Patricia A. 2001. *Ancient Epistolary Fictions.* Cambridge: Cambridge University Press.

——. 2006. *Ancient Greek Literary Letters: Selections in Translation.* London: Routledge.

Rotstein, Andrea. 2010. *The Idea of Iambos.* Oxford: Oxford University Press.

Rozik, Eli. 2002. *The Roots of Theatre: Rethinking Ritual and Other Theories of Origin.* Iowa City: University of Iowa Press.

Rubin, D. C. 1995. *Memory in Oral Traditions: The Cognitive Psychology of Epic, Ballads, and Counting-Out Rhymes.* New York: Oxford University Press.

Russo, Joseph Anthony, and Bennett Simon. 1968. *Homeric Psychology and the Oral Epic Tradition.* New York.

Russo, Joseph Anthony. 2012. "Rethinking Homeric Psychology: Snell, Dodds, and Their Critics." In *Quaderni Urbinati di cultura classica*, ed. Bruno Gentili, 130:11–28.

Saenger, Paul Henry. 1996. *Space Between Words: The Origins of Silent Reading.* Stanford, Calif.: Stanford University Press.

Salimpoor, Valorie N., David H. Zald, Robert J. Zatorre, et al. 2015. "Predictions and the Brain: How Musical Sounds Become Rewarding." *Trends in Cognitive Sciences* 19 (2): 86–91.

Salimpoor, Valorie N., and Robert J. Zatorre. 2013. "Neural Interactions That Give Rise to Musical Pleasure." *Psychology of Aesthetics, Creativity, and the Arts* 7 (1): 62–75.

Sammler, D., S. Koelsch, T. Ball, et al. 2013. "Co-localizing Linguistic and Musical Syntax with Intracranial EEG." *NeuroImage* 64:134–146.

Sandberg, A., and N. Bostrom. 2008. *Whole Brain Emulation: A Roadmap.* Technical report 2008-3. Oxford: Future of Humanity Institute, Oxford University.

Sansone, David. 1996. "Plato and Euripides." *Illinois Classical Studies* 21:35–67.

Santos-Granero, Fernando. 1998. "Writing History Into the Landscape: Space, Myth, and Ritual in Contemporary Amazonia." *American Ethnologist* 25 (2): 128–148.

Sapir, Edward. 1921. *Language: An Introduction to the Study of Speech.* Harcourt, Brace.

Schacter, D. L. 1996. *Searching for Memory: The Brain, the Mind, and the Past.* New York: Basic Books.

Scheid, John, and Jesper Svenbro. 1996. *The Craft of Zeus: Myths of Weaving and Fabric.* Cambridge, Mass.: Harvard University Press.

Scherer, K. R. 2001. "Appraisal Considered as a Process of Multilevel Sequential Checking." In *Appraisal Processes in Emotion Theory, Methods, and Research*, ed. K. R. Scherer, A. Schorr, and T. Johnstone, 92–120. New York: Oxford University Press.

Scherer, K. R., A. Schorr, and T. Johnstone, eds. 2001. *Appraisal Processes in Emotion Theory, Methods, and Research*. New York: Oxford University Press.

Scherer, K. R., and K. R. Zentner. 2001. "Emotional Effects of Music: Production Rules." In *Music and Emotion: Theory, Research, and Applications*, ed. P. N. Juslin and J. A. Sloboda, 361–392. New York: Oxford University Press.

Schlegel, Catherine. 2005. *Satire and the Threat of Speech: Horace's Satires, Book 1*. Madison: University of Wisconsin Press.

Schmitz, T. A. 2010. "Epigrammatic Communication in Callimachus' Epigrams." *Greek, Roman, and Byzantine Studies* 50 (3): 370–390.

Schubart, W. 1921. *Das Buch bei den Griechen und Römern*. 2nd ed. Berlin: de Gruyter.

Schultz, W. 1998. "Predictive Reward Signal of Dopamine Neurons." *Journal of Neurophysiology* 80:1–27.

Schweik, Susan. 2008. "On the River-Merchant's Wife: A Letter." *Modern American Poetry*. Urbana: University of Illinois Board of Trustees.

Searle, John R. 1983. *Intentionality: An Essay in the Philosophy of Mind*. Cambridge: Cambridge University Press.

Sebeok, T. A., ed. 1960. *Style in Language*. New York: Wiley.

Segal, Charles. 1996. "Eros and Incantation: Sappho and Oral Poetry." In *Reading Sappho: Contemporary Approaches*, ed. E. Greene, 58–76. Berkeley: University of California Press.

Senghas, Ann. 2010. "The Emergence of Two Functions for Spatial Devices in Nicaraguan Sign Language." *Human Development* 53 (5): 287–302.

Sharland, Suzanne. 2009. *Horace in Dialogue: Bakhtinian Readings in the Satires*. Oxford: Peter Lang.

Skulsky, Harold. 1958. "Aristotle's Poetics Revisited." *Journal of the History of Ideas* 19 (2): 147–160.

Small, Jocelyn Penny. 1999. "Time *in* Space: Narrative in Classical Art." *Art Bulletin* 81 (4): 562–575.

——. 2009. "Is Linear Perspective Necessary?" In *New Perspectives on Etruria and Early Rome: In Honor of Richard Daniel De Puma*, ed. Sinclair Bell and Helen Nagy, 149–157. Madison: University of Wisconsin Press.

Smith, J. David, Margaret Wilson, and Daniel Reisberg. 1995. "The Role of Subvocalization in Auditory Imagery." *Neuropsychologia* 33 (11): 1433–1454.

Snell, Bruno. 1953. *The Discovery of the Mind: The Greek Origins of European Thought*. Trans. T. G. Rosenmeyer. Cambridge, Mass.: Harvard University Press.

Snow, David. 2000. "The Emotional Basis of Linguistic and Nonlinguistic Intonation: Implications for Hemispheric Specialization." *Developmental Neuropsychology* 17 (1): 1–28.

Sokolov, A. N. 1972. *Inner Speech and Thought*. Trans. G. T. Onischenko. New York: Plenum.

Solari, Amara. 2013. *Maya Ideologies of the Sacred: The Transfiguration of Space in Colonial Yucatan*. Austin: University of Texas Press.

Sonesson, Göran. 2006. "The Meaning of Meaning in Biology and Cognitive Science: A Semiotic Reconstruction." *Sign Systems Studies* 34 (1): 135–213.

——. 2010. "From Mimicry to Mime by Way of Mimesis." *Sign Systems Studies* 38 (1): 18–65.

Spariosu, Mihai. 1991. *God of Many Names: Play, Poetry, and Power in Hellenic Thought from Homer to Aristotle*. Durham, N.C.: Duke University Press.

Sperber, Dan, and Deirdre Wilson. 1986. *Relevance: Communication and Cognition*. Cambridge, Mass.: Harvard University Press.

Stehle, Eva. 1997. *Performance and Gender in Ancient Greece: Nondramatic Poetry in Its Setting*. Princeton, N.J.: Princeton University Press.

Stetson, C., M. P. Fiesta, and D. M. Eagleman. 2007. "Does Time Really Slow Down During a Frightening Event?" *PLoS ONE* 2 (12): e1295.

Stock, Brian. 2010. *Augustine's Inner Dialogue: The Philosophical Soliloquy in Late Antiquity*. Cambridge: Cambridge University Press.

Stoddard, Kathryn B. 2003. "The Programmatic Message of the 'Kings and Singers' Passage: Hesiod, *Theogony* 80–103." *Transactions of the American Philological Association* 133 (1): 1–16.

——. 2004. *The Narrative Voice in the* Theogony *of Hesiod*. Leiden: Brill.

Svenbro, Jesper. 1976. *La parole et le marbre. Aux origines de la poétique grecque*. Lund: Klassiska Institutionen.

——. 1993. *Phrasikleia: An Anthropology of Reading in Ancient Greece*. Ithaca, N.Y.: Cornell University Press.

Sucala, M., B. Scheckner, and D. David. 2010. "Psychological Time: Interval-Length Judgments and Subjective Passage-of-Time Judgments." *Current Psychology Letters* 26 (2): 2–9.

Suddendorf, T., and M. C. Corballis. 2007. "The Evolution of Foresight: What Is Mental Time Travel, and Is It Unique to Humans?" *Behavioral and Brain Sciences* 30:299–313.

Sugiyama, M. Scalise. 2001. "Food, Foragers, and Folklore: The Role of Narrative in Human Subsistence." *Evolution and Human Behavior* 22:221–240.

Sweetser, Eve. 1991. *From Etymology to Pragmatics: Metaphorical and Cultural Aspects of Semantic Structure*. Cambridge: Cambridge University Press.

Taylor, Insup, and Martin M. Taylor. 1983. *The Psychology of Reading*. New York: Academic Press.

Thomas, Rosalind. 1989. *Oral Tradition and Written Record in Classical Athens*. Cambridge: Cambridge University Press.

——. 2009. "Writing, Reading, Public and Private 'Literacies': Functional Literacy and Democratic Literacy in Greece." In *Ancient Literacies*, ed. W. A. Johnson and H. N. Parker, 13–45. Oxford: Oxford University Press.

Thoreau, Henry David, and J. Lyndon Shanley. (1854) 2004. *Walden*. Princeton, N.J.: Princeton University Press.

Tierney, A., F. Dick, D. Deutsch, et al. 2012. "Speech Versus Song: Multiple Pitch-Sensitive Areas Revealed by a Naturally Occurring Musical Illusion." *Cerebral Cortex* 23.

Tomasello, Michael. 1999. *The Cultural Origins of Human Cognition*. Cambridge, Mass.: Harvard University Press.

——. 2008. *Origins of Human Communication*. Cambridge, Mass.: MIT Press.

Traxler, Matthew J. 2012. *Introduction to Psycholinguistics: Understanding Language Science*. Chichester: Wiley-Blackwell.

Trost, W., T. Ethofer, M. Zentner, et al. 2012. "Mapping Aesthetic Musical Emotions in the Brain." *Cerebral Cortex* 22 (12): 2769–2783.

Trubshaw, Bob. 2012. "How Anglo-Saxons Found Their Way." PDF file. Marlborough: Heart of Albion. http://www.hoap.co.uk.

Tsagalis, Christos. 2008. *Inscribing Sorrow: Fourth-Century Attic Funerary Epigrams*. Berlin: de Gruyter.

Tse, Peter Ulric. 2006. "Symbolic Thought and the Evolution of Human Morality." In *Moral Psychology*, vol. 1: *The Evolution of Morality*, ed. W. Sinnott-Armstrong, 269–317. Cambridge: MIT Press.

Tseng, Wen-Shing. 2001. *Handbook of Cultural Psychiatry*. San Diego, Calif.: Academic Press.

Tulving, E. 1972. "Episodic and Semantic Memory." In *Organization of Memory*, ed. E. Tulving, W. Donaldson, and G. H. Bower, 386–402. New York: Academic Press.

——. 1983. *Elements of Episodic Memory*. New York: Oxford University Press.

Tulving, E., W. Donaldson, and G. H. Bower, eds. 1972. *Organization of Memory*. New York: Academic Press.

Turner, Frederick, and Ernst Pöppel. 1983. "The Neural Lyre: Poetic Meter, the Brain, and Time." *Poetry* 142 (5): 277–309.

Ungerleider, L. G., and M. Mishkin. 1982. "Two Cortical Visual Systems: Separation of Appearance and Location of Objects." In *Analysis in Visual Behavior*, edited by D. J. Ingle, M. A. Goodale, and R. J. W. Mansfield, 549–586. Cambridge, Mass.: MIT Press.

Urban, R. T. 1966. "All or Nothing at All: Another Look at the Unity of Time in Aristotle." *Classical Journal* 61 (6): 262–264.

Vansina, Jan. 1985. *Oral Tradition as History*. Madison: University of Wisconsin Press.

Vendler, Z. 1957. "Verbs and Times." *Philosophical Review* 66 (2): 143–160.

Vermeule, Emily. 1964. *Greece in the Bronze Age*. Chicago: University of Chicago Press.

——. 1979. *Aspects of Death in Early Greek Art and Poetry*. Berkeley: University of California Press.

Vessey, D. W. T. 1985. "From Mountain to Lovers' Tryst: Horace's Soracte Ode." *Journal of Roman Studies* 75:26–38.

Vigneau, M., V. Beaucousin, P. Y. Hervé, et al. 2011. "What Is the Right-Hemisphere Contribution to Phonological, Lexico-Semantic, and Sentence Processing? Insights from a Meta-Analysis." *NeuroImage* 54 (1): 577–593.

Vygotsky, L. S. (1934) 1986. *Thought and Language*. Ed. A. Kozulin. Cambridge, Mass.: MIT Press.

Walker, Jeffrey. 1998. "The View from Halicarnassus: Aristotelianism and the Rhetoric of Epideictic Song." In *New Definitions of Lyric: Theory, Technology, and Culture*, ed. Mark Jeffreys, 17–48. New York: Garland.

——. 2000. *Rhetoric and Poetics in Antiquity*. Oxford: Oxford University Press.

Wallin, Nils Lennart, Björn Merker, and Steven Brown. 2000. *The Origins of Music*. Cambridge, Mass.: MIT Press.

Walton, J. Michael. 1980. *Greek Theatre Practice*. Westport, Conn.: Greenwood.

Warren, J. E., R. J. Wise, and J. D. Warren. 2005. "Sounds Do-able: Auditory-Motor Transformations and the Posterior Temporal Plane." *Trends in Neurosciences* 28 (12): 636–643.

Webster, T. B. L. 1958. *From Mycenae to Homer*. London: Methuen.

West, M. L. 1974. *Studies in Greek Elegy and Iambus*. Berlin: de Gruyter.

——. 1992. *Ancient Greek Music*. Oxford: Clarendon.

Whiten, Andrew, Robert Hinde, Chris Stringer, and Kevin N. Laland, ed. 2012. *Culture Evolves*. Oxford: Oxford University Press.

Whorf, Benjamin L. 1956. *Language, Thought, and Reality: Selected Writings of Benjamin Lee Whorf*. Ed. John B. Carroll. Cambridge, Mass.: MIT Press.

Wilson, Deirdre, and Dan Sperber. 1990. "Rhetoric and Relevance." In *The Ends of Rhetoric: History, Theory, Practice*, ed. David Wellbery and John Bender, 140–155. Stanford, Calif.: Stanford University Press.

Wimsatt, W. K., and Monroe Beardsley. 1954. *The Verbal Icon: Studies in the Meaning of Poetry*. Lexington: University of Kentucky Press.

Wolff, K. J., and P. Holmes. 2011. "Linguistic Relativity." *Wiley Interdisciplinary Reviews: Cognitive Science* 2:253–265.

Woodard, R. D., ed. 2007. *The Cambridge Companion to Greek Mythology*. Cambridge: Cambridge University Press.

Wray, A., ed. 2002. *Transition to Language: Studies in the Evolution of Language*. New York: Oxford University Press.

Wray, A. 1998. "Protolanguage as a Holistic System for Social Interaction." *Language & Communication* 18:47–67.

Wray, A., and G. W. Grace. 2007. "The Consequences of Talking to Strangers: Evolutionary Corollaries of Sociocultural Influences on Linguistic Form." *Lingua* 117:543–578.

Yates, Frances A. (1966) 2010. *The Art of Memory*. London: Routledge.

Yeats, William Butler. B. 1903. "Symbolism in Poetry." In *Ideas of Good and Evil*, 237–256. London: A. H. Bullen.

——. 1999. *Autobiographies*. In *The Collected Works of W. B. Yeats*, ed. William H. O'Donnell and Douglas N. Archibald. New York: Scribner.

Yi, S. W., ed. *Music, Mind, and Science*. Seoul: Seoul National University Press.

Yunis, Harvey. 2003. *Written Texts and the Rise of Literate Culture in Ancient Greece*. Cambridge: Cambridge University Press.

Zald, David H., and Robert J. Zatorre. 2011. "Music." In *Neurobiology of Sensation and Reward*, ed. Jay A. Gottfried, 405–428. Boca Raton, Fla.: CRC.

Zatorre, R. J., A. C. Evans, E. Meyer, et al. 1992. "Lateralisation of Phonetic and Pitch Discrimination in Speech Processing." *Science* 256:846–849.

Zatorre, R. J., J. L. Chen, and V. B. Penhune. 2007. "When the Brain Plays Music: Auditory-Motor Interactions in Music Perception and Production." *Nature Reviews Neuroscience* 8:547–558.

Zentner, M. R., D. Grandjean, and K. R. Scherer. 2008. "Emotions Evoked by the Sound of Music: Characterization, Classification, and Measurement." *Emotion* 8:494–521.

Zentner, M. R., and T. Eerola. 2011. "Self-Report Measures and Models of Musical Emotions." In *Handbook of Music and Emotion: Theory, Research, Applications*, ed. P. Juslin and J. Sloboda, 187–223. Oxford: Oxford University Press.

Zlatev, J., and J. Blomberg. 2015. "Language May Indeed Influence Thought." *Frontiers in Psychology* 6.

Zwaan, R. A. 2004. "The Immersed Experiencer: Toward an Embodied Theory of Language Comprehension." *Psychology of Learning and Motivation* 44:35–62.

Zwaan, R. A., C. A. Madden, R. H. Yaxley, et al. 2004. "Moving Words: Dynamic Representations in Language Comprehension." *Cognitive Science* 28:611–619.

Zwaan, R. A., and M. Kaschak. 2006. "Language, Visual Cognition, and Motor Action." In *Encyclopedia of Language and Linguistics*, 2nd ed., ed. E. K. Brown and Anne Anderson, 648–651. Boston, Mass: Elsevier.

Index

Note: for a brief definition of certain technical terms, see page numbers in *italic* type.